Methods in Molecular Biology

For further volumes:
http://www.springer.com/series/7651

For over 35 years, biological scientists have come to rely on the research protocols and methodologies in the critically acclaimed *Methods in Molecular Biology* series. The series was the first to introduce the step-by-step protocols approach that has become the standard in all biomedical protocol publishing. Each protocol is provided in readily-reproducible step-by step fashion, opening with an introductory overview, a list of the materials and reagents needed to complete the experiment, and followed by a detailed procedure that is supported with a helpful notes section offering tips and tricks of the trade as well as troubleshooting advice. These hallmark features were introduced by series editor Dr. John Walker and constitute the key ingredient in each and every volume of the *Methods in Molecular Biology* series. Tested and trusted, comprehensive and reliable, all protocols from the series are indexed in PubMed.

Calcium Signaling

Methods and Protocols

Edited by

Caroline M. Gorvin

Metabolism and Systems Science, University of Birmingham, Birmingham, UK

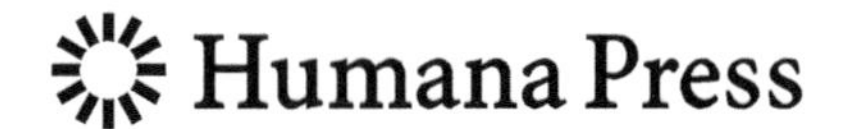

Editor
Caroline M. Gorvin
Metabolism and Systems Science
University of Birmingham
Birmingham, UK

ISSN 1064-3745 ISSN 1940-6029 (electronic)
Methods in Molecular Biology
ISBN 978-1-0716-4163-7 ISBN 978-1-0716-4164-4 (eBook)
https://doi.org/10.1007/978-1-0716-4164-4

This Humana imprint is published by the registered company Springer Science+Business Media, LLC, part of Springer Nature.
The registered company address is: 1 New York Plaza, New York, NY 10004, U.S.A.

Preface

Calcium signaling is essential for diverse cellular processes including cell motility, gene transcription, and mitochondrial activity, and it drives a range of physiological responses including neuronal excitation, heart conductance, and skeletal muscle contraction. A diverse array of techniques has been developed to improve the assessment of calcium signaling, including fluorescent probes and luminescent reporters that have facilitated the monitoring of calcium in live cells, tissues, and whole organisms. This volume contains protocols with step-by-step instructions for performing calcium signaling in diverse cell types, alongside detailed methods for data analysis, and highlights the potential challenges of each technique, with tips on how to avoid these pitfalls.

The book is divided into five parts, each containing a collection of chapters linked by a common theme written by researchers who use these techniques regularly. Part I is focused on cellular assays to assess calcium signaling. These assays can be performed in multiple cell types to help researchers answer a wide variety of questions. The section begins with protocols to measure calcium signaling using high-throughput assays (Chap. 1), homogeneous time resolved fluorescence (HTRF) (Chap. 2), NanoBiT luminescence biosensors (Chap. 3), bioluminescence resonance energy transfer (BRET) (Chap. 4), flow cytometry (Chap. 5), and dual luciferase reporter assays (Chap. 6). Part II focuses on imaging methods to assess calcium signaling in brain tissue slices (Chap. 7), using two photon microscopy to measure calcium signaling in the living brain (Chap. 8), and using total internal reflection fluorescence (TIRF) microscopy to assess G protein-coupled receptor trafficking and signaling (Chap. 9). Part III contains methods to assess calcium within three organelles: primary cilia (Chap. 10), the endoplasmic reticulum (Chap. 11), and the mitochondria (Chap. 12). Part IV focuses on tissue-specific calcium signaling, including the establishment of primary cell or tissue models and measuring calcium in whole organisms such as *Drosophila*. Part IV contains chapters detailing calcium measurement techniques in bone cells (Chap. 13), kidney (Chap. 14), mammary tissue (Chap. 15), intestinal organoids (Chap. 16), pancreas (Chap. 17), sperm (Chap. 18), and *Drosophila* (Chap. 19). The final part contains a single chapter describing computational analysis of calcium flux using R software (Chap. 20).

Where possible, the chapters describe analysis using Open Source software, and the variety of methods will enable laboratories of all types to measure calcium robustly, ensuring that this is the essential reference textbook for calcium signaling protocols.

Birmingham, UK *Caroline M. Gorvin*

Contents

Contributors

MARIA TERESA ALONSO • *Instituto de Biomedicina y Genética Molecular de Valladolid (IBGM), Unidad de Excelencia, Universidad de Valladolid y Consejo Superior de Investigaciones Científicas (CSIC), Valladolid, Spain*
AYAN K. BARUI • *Department of Biomedical and Pharmaceutical Sciences, Chapman University, Irvine, CA, USA; Department of Pharmaceutical Sciences, School of Pharmacy, University of California Irvine, Irvine, CA, USA*
YOSSI BUSKILA • *School of Medicine, Western Sydney University, Campbelltown, NSW, Australia; The MARCS Institute, Western Sydney University, Penrith, NSW, Australia*
BELEN CALVO • *Instituto de Biomedicina y Genética Molecular de Valladolid (IBGM), Unidad de Excelencia, Universidad de Valladolid y Consejo Superior de Investigaciones Científicas (CSIC), Valladolid, Spain*
SILKE B. CHALMERS • *Department of Biomedicine, Aarhus University, Aarhus, Denmark*
CLAIRE E. J. CHEETHAM • *Department of Neurobiology, University of Pittsburgh, Pittsburgh, PA, USA*
ROY COHEN • *Baker Institute for Animal Health, Cornell University College of Veterinary Medicine, Ithaca, NY, USA; Department of Public & Ecosystem Health, Cornell University College of Veterinary Medicine, Ithaca, NY, USA*
KENDALL A. CURTIS • *Department of Neurobiology, University of Pittsburgh, Pittsburgh, PA, USA*
FELICITY M. DAVIS • *Department of Biomedicine, Aarhus University, Aarhus, Denmark; School of Biomedical Sciences, University of New South Wales, Sydney, NSW, Australia; Aarhus Institute of Advanced Studies, Aarhus University, Aarhus, Denmark; Danish Research Institute of Translational Neuroscience, Aarhus University, Aarhus, Denmark; School of Pharmacy, University of Queensland, Brisbane, QLD, Australia*
JURIJ DOLENŠEK • *Faculty of Medicine, University of Maribor, Maribor, Slovenia; Faculty of Natural Sciences and Mathematics, University of Maribor, Maribor, Slovenia*
MAJA DUH • *Faculty of Medicine, University of Maribor, Maribor, Slovenia; Faculty of Natural Sciences and Mathematics, University of Maribor, Maribor, Slovenia*
ANDREJA FIGUREK • *Institute of Anatomy, University of Zurich, Zürich, Switzerland*
MATHILDE FOLACCI • *Department of Biomedicine, Aarhus University, Aarhus, Denmark*
JAVIER GARCIA-SANCHO • *Instituto de Biomedicina y Genética Molecular de Valladolid (IBGM), Unidad de Excelencia, Universidad de Valladolid y Consejo Superior de Investigaciones Científicas (CSIC), Valladolid, Spain*
DAVID V. GAZZO • *Department of Chemical and Biomolecular Engineering, University of Notre Dame, Notre Dame, IN, USA; Bioengineering Graduate Program, University of Notre Dame, Notre Dame, IN, USA*
NASAB GHAZAL • *Graduate Program in Biochemistry, Cell and Developmental Biology, Graduate Division of Biological and Biomedical Sciences, Emory University, Atlanta, GA, USA; Division of Pediatric Cardiology, Department of Pediatrics, Emory University School of Medicine, and Children's Healthcare of Atlanta, Atlanta, GA, USA*
VANSH K. GOEL • *Department of Biomedical and Pharmaceutical Sciences, Chapman University, Irvine, CA, USA*

Caroline M. Gorvin • *Metabolism and Systems Science and Centre for Diabetes, Endocrinology and Metabolism (CEDAM), University of Birmingham, Birmingham, UK; Centre for Membrane Proteins and Receptors (COMPARE), Universities of Birmingham and Nottingham, Birmingham, UK*

Marko Gosak • *Faculty of Medicine, University of Maribor, Maribor, Slovenia; Faculty of Natural Sciences and Mathematics, University of Maribor, Maribor, Slovenia; Alma Mater Europaea University, Maribor, Slovenia*

Andrew M. Hall • *Institute of Anatomy, University of Zurich, Zürich, Switzerland; Department of Nephrology, University Hospital Zurich, Zürich, Switzerland; Zurich Kidney Center, University of Zurich, Zürich, Switzerland*

Morten S. Hansen • *Molecular Endocrinology Department (KMEB), Department of Endocrinology, Odense University Hospital, Odense, Denmark*

Sarah A. Howles • *Nuffield Department of Surgical Sciences, University of Oxford, Oxford, UK; National Institute for Health Research Oxford Biomedical Research Centre, Oxford, UK*

Aqfan Jamaluddin • *Metabolism and Systems Science, University of Birmingham, Birmingham, UK; Centre of Membrane Proteins and Receptors (COMPARE), Universities of Birmingham and Nottingham, Birmingham, UK*

Nevena Jankovic • *Institute of Anatomy, University of Zurich, Zürich, Switzerland*

Nisal Keembiyage • *School of Medicine, Western Sydney University, Campbelltown, NSW, Australia*

Orsolya Kékesi • *School of Biomedical Engineering, The University of Sydney, Camperdown, NSW, Australia*

Lidija Križančić Bombek • *Faculty of Medicine, University of Maribor, Maribor, Slovenia*

Shree S. Kumar • *Faculty of Medical and Health Sciences, Department of Molecular Medicine and Pathology, School of Medical Sciences, The University of Auckland, Auckland, New Zealand*

Tenzin Kunkhyen • *Department of Neurobiology, University of Pittsburgh, Pittsburgh, PA, USA*

Jennifer Q. Kwong • *Division of Pediatric Cardiology, Department of Pediatrics, Emory University School of Medicine, and Children's Healthcare of Atlanta, Atlanta, GA, USA; Department of Cell Biology, Emory University School of Medicine, Atlanta, GA, USA*

Kathleen G. Mountjoy • *Faculty of Medical and Health Sciences, Department of Molecular Medicine and Pathology, School of Medical Sciences, The University of Auckland, Auckland, New Zealand*

Surya M. Nauli • *Department of Biomedical and Pharmaceutical Sciences, Chapman University, Irvine, CA, USA; Department of Medicine, University of California Irvine, Irvine, CA, USA*

Eva Paradiž Leitgeb • *Faculty of Medicine, University of Maribor, Maribor, Slovenia*

Viljem Pohorec • *Faculty of Medicine, University of Maribor, Maribor, Slovenia*

Macarena Pozo-Morales • *Institut de Recherche Interdisciplinaire en Biologie Humaine et Moléculaire (IRIBHM), Université Libre de Bruxelles (ULB), Brussels, Belgium*

Eva Rath • *Chair of Molecular Nutritional Medicine, School of Life Sciences, Technische Universität München, Freising-Weihenstephan, Germany*

Macarena Rodriguez-Prados • *Instituto de Biomedicina y Genética Molecular de Valladolid (IBGM), Unidad de Excelencia, Universidad de Valladolid y Consejo Superior de Investigaciones Científicas (CSIC), Valladolid, Spain; Universidad Alfonso X el Sabio,, Madrid, Spain*

JONATHAN ROJO-RUIZ • *Instituto de Biomedicina y Genética Molecular de Valladolid (IBGM), Unidad de Excelencia, Universidad de Valladolid y Consejo Superior de Investigaciones Científicas (CSIC), Valladolid, Spain*
SUMEET PAL SINGH • *Institut de Recherche Interdisciplinaire en Biologie Humaine et Molé culaire (IRIBHM), Université Libre de Bruxelles (ULB), Brussels, Belgium*
MAŠA SKELIN KLEMEN • *Faculty of Medicine, University of Maribor, Maribor, Slovenia*
FRANZISKA LENA SÖRGEL • *Pediatric Nutritional Medicine & Else Kröner-Fresenius-Centre for Nutritional Medicine (EKFZ), Technical University of Munich (TUM), Freising, Germany*
ANDRAŽ STOŽER • *Faculty of Medicine, University of Maribor, Maribor, Slovenia*
ALEXANDER J. TRAVIS • *Baker Institute for Animal Health, Cornell University College of Veterinary Medicine, Ithaca, NY, USA; Department of Public & Ecosystem Health, Cornell University College of Veterinary Medicine, Ithaca, NY, USA*
MELISSA A. WHITE • *Baker Institute for Animal Health, Cornell University College of Veterinary Medicine, Ithaca, NY, USA*
JEREMIAH J. ZARTMAN • *Department of Chemical and Biomolecular Engineering, University of Notre Dame, Notre Dame, IN, USA; Department of Biological Sciences, University of Notre Dame, Notre Dame, IN, USA; Bioengineering Graduate Program, University of Notre Dame, Notre Dame, IN, USA*
TAMARA ZIETEK • *Doctors Against Animal Experiments, Köln, Germany*

Part I

Cellular Assays to Assess Calcium Signaling

Chapter 1

Measuring GPCR-Induced Intracellular Calcium Signaling Using a Quantitative High-Throughput Assay

Shree S. Kumar and Kathleen G. Mountjoy

Abstract

Alterations in intracellular calcium are integral to signal transduction pathways for many G-protein-coupled receptors, but this signaling is not well studied. This is mostly due to a lack of reliable, robust, high-throughput, quantitative methods to monitor intracellular calcium concentrations in live cells. Recently, we developed a reliable, robust, quantitative method to measure intracellular calcium levels in which HEK293 cell suspensions loaded with Fura-2/AM are placed in 96-well plates. Minimum and maximum intracellular calcium levels, which are required for converting fluorescence into calcium concentrations, are calibrated using EGTA to chelate calcium and ionomycin to load calcium into cells, respectively. Fluorescence is monitored with a PHERAstar FS plate reader. We provide a detailed method for this high-throughput assay that can be used to quantitate intracellular calcium in endogenous and exogenously (stable or transient) expressed GPCRs in HEK293 cells.

Key words Calcium mobilization, Fluorescence, Fura-2 dye, G-protein-coupled receptors, Intracellular calcium quantitation, PHERAstar FS plate reader, Ratiometric

1 Introduction

G-protein-coupled receptors (GPCRs) commonly signal through activation of two critical intracellular second messengers, cyclic adenosine monophosphate (cAMP) and calcium (Ca^{2+}). In contrast to a vast number of GPCR/cAMP signaling studies, GPCR/Ca^{2+} signaling studies are relatively rare. This is due to a lack of reliable, robust, high-throughput, and quantitative methods to monitor intracellular Ca^{2+} concentrations ([Ca^{2+}]i) in living cells [1, 2]. Ca^{2+} plays crucial regulatory roles in normal cellular functions, and alterations in [Ca^{2+}]i are integral to signal transduction pathways for many GPCRs [3, 4]. Therefore, there is a need for a reliable, high-throughput, quantitative method for measuring [Ca^{2+}]i in living cells.

Quantitation of [Ca^{2+}] can only accurately be obtained using ratiometric dye. Fluorescent indicator dyes are used for monitoring

Caroline M. Gorvin (ed.), *Calcium Signaling: Methods and Protocols*, Methods in Molecular Biology, vol. 2861, https://doi.org/10.1007/978-1-0716-4164-4_1,

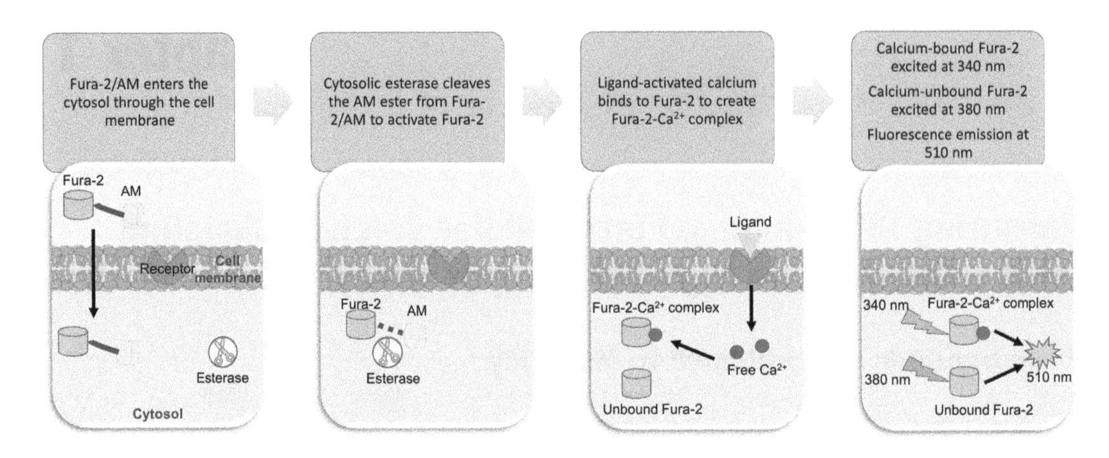

Fig. 1 Schematic of the principle for Fura-2/AM quantitation of intracellular Ca^{2+} in vitro

Ca^{2+} signals, and these indicators are either dual-wavelength ratiometric dyes or single-wavelength non-ratiometric dyes [5, 6]. Non-ratiometric dyes, such as Fluo-3 and Fluo-4, are most commonly used today to monitor GPCRs regulating $[Ca^{2+}]i$ in living cells [7]. These dyes become brighter when they bind Ca^{2+}; this fluorescence increase implies an increase in $[Ca^{2+}]i$. However, the fluorescence signal is susceptible to interference from numerous factors [8, 9] and is not quantitative. Data are reported as fluorescence changes instead of reporting $[Ca^{2+}]$. Any comparisons between ligands, concentrations, or receptors regulating $[Ca^{2+}]i$ must be performed within a single experiment. In contrast to non-ratiometric dyes, the ratiometric dye indicator, Fura-2, exhibits fluorescence changes with changing $[Ca^{2+}]$, and the Ca^{2+}-free and Ca^{2+}-bound Fura-2 have distinct spectra at different wavelengths. These are not affected by interference observed with single-wavelength calcium dyes [10]. Significant shifts are observed in the activation but not in the emission spectra. The peak-emitted fluorescence of Fura-2 shifts from 340 nm in the Ca^{2+}-bound state to 380 nm in the Ca^{2+}-free state. This allows for quantitation of Fura-2-Ca^{2+} by computing the fluorescence ratio at these two wavelengths, independent of dye concentration [11, 12] (Fig. 1). The ratio between the emission at each excitation wavelength is quantitatively related to $[Ca^{2+}]$ when minimum and maximum Ca^{2+} levels are calibrated for the cells [11]. For each experiment, the cells minimum (Rmin) and maximum (Rmax) Ca^{2+} levels are determined using EGTA to chelate all intracellular calcium and ionomycin to fully load extracellular Ca^{2+} into cells, respectively.

More than 20 years ago, Fura-2 was used to quantitatively measure GPCR-mediated $[Ca^{2+}]i$ using cell suspensions pipetted into a cuvette and with fluorescence monitored using a dual-wavelength spectrofluorometer equipped with a thermostated cuvette holder [13, 14]. More recently, multimode plate readers have been developed to monitor fluorescence from cells in 96-well plates with simultaneous injection of ligands into wells through in-built injectors. High-throughput assays have subsequently been developed for monitoring calcium produced in cell monolayers growing in 96-well plates. However, these Ca^{2+} assays are not quantitative as they use non-ratiometric dyes and produce highly variable data much of which results from activation of mechanoreceptors [8, 15]. Cell lines used for measuring GPCR signaling responses possess endogenous mechanosensitive receptors at the cell membrane which mobilize $[Ca^{2+}]i$. Stretch-receptor responses are triggered by turbulence resulting from ligand injection into the media covering the cell monolayer and also sudden plate movements [2, 16–22]. This interferes with ligand-induced Ca^{2+} mobilization and contributes to assay variability. Furthermore, there is potential for dye accumulation in multiple discrete subcellular compartments in cell monolayers compared to cell suspensions [12]. Plate-reader-based methods also suffer from reduced sensitivity due to low cell numbers in monolayers in 96-well plates.

To circumvent these problems, we recently developed a reliable, robust, quantitative method to measure $[Ca^{2+}]$ in which HEK293 cell suspensions are placed in 96-well plates. Fluorescence is monitored with a PHERAstar FS plate reader [23, 24]. We load cells in suspension with Fura-2/AM prior to placing the cells in 96-well plates. After this, the cells settle evenly as single cells on the bottom of each well where they are stimulated with ligand anytime between 15 min and 4 h after being placed in the well. We eliminate mechanoreceptor activation by optimizing fluid volume in the well, injection volume, and injector speed. We enhance sensitivity for measuring Ca^{2+} by having more cells/well compared to a monolayer of cells/well.

In 2001, we reported quantitatively measuring $[Ca^{2+}]i$ in suspensions of HEK293 cells using a dual-wavelength spectrophotometer. We tested vehicle and two different ligands in duplicates and five different concentrations of each ligand on a single receptor in one assay. This was technically demanding and would take at least 5 h, not including the time it took to prepare cells and load them with Fura-2/AM dye. In comparison, our novel method in which HEK293 cell suspensions are placed in 96-well plates and fluorescence is monitored with a PHERAstar FS plate reader [23, 24], is relatively high-throughput. A greater number of Ca^{2+} measurements can be made using the 96-well plate assay in approximately the same time it took to measure Ca^{2+} responses using the dual-wavelength spectrophotometer method. The 96-well plate enables measuring in triplicate, three different ligands, each at ten different concentrations activating a single receptor. Furthermore, our 96-well plate assay has enhanced reliability and accuracy of ligand-mediated GPCR-induced $[Ca^{2+}]i$ changes. The built-in injector in the plate reader enables simultaneous injection of ligand and

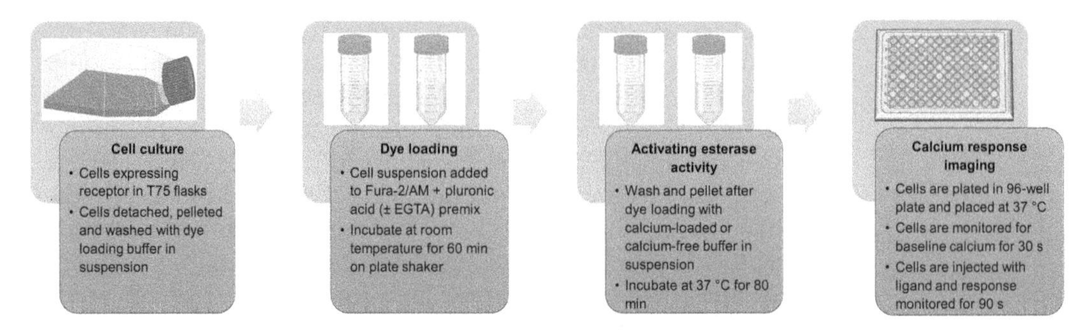

Fig. 2 Flow chart summarizing method to measure Ca^{2+} response using Fura-2/AM dye

monitoring of the well fluorescence output. We observe a change in $[Ca^{2+}]i$ in < 3 s. Importantly, this 96-well plate assay has enhanced accuracy and efficiency for quantitation of ligand-mediated GPCR-induced $[Ca^{2+}]i$ changes. The software operating the plate reader can be used to measure baseline and peak responses accurately. Previously, these measurements obtained from dual-wavelength spectrophotometers were subjective since they were made by estimating baseline and peak responses by viewing the output on a computer screen or data printout.

This high-throughput assay can be used to quantitate $[Ca^{2+}]i$ in endogenous and exogenously (stable or transient) expressed GPCRs in HEK293 cells (Fig. 2). However, the assay is not automated and is time-consuming for the operator. It takes ~4 h to perform injections and monitor fluorescence responses for a full 96-well plate using 2 injectors which are washed and primed back-to-back. If only one injector is used, the time to complete a full 96-well plate experiment monitoring the Ca^{2+} response in each well for 90 s is > 4 h. A further limiting factor is the size of each experiment. The Fura-2 dye will slowly leak out of the cells as they sit at 37 °C. Therefore, each experiment should be limited to one 96-well plate so that cells loaded with Fura-2 dye are used within 4 h of dye loading.

2 Materials

2.1 Equipment

1. Black polystyrene, flat, clear bottom, TC-treated, sterile microplate (96-well).
2. 15 mL Conical sterile polypropylene centrifuge tubes.
3. 50 mL Conical sterile polypropylene centrifuge tubes.
4. 2 mL LoBind DNA tubes.
5. 2 mL LoBind protein tubes.
6. 2 mL serological pipette.
7. 5 mL serological pipette.

8. 10 mL serological pipette.
9. 25 mL serological pipette.
10. T75 flasks.
11. 100 mm tissue culture dish.
12. Plate reader for reading fluorescent signal with 340/380 excitation and 510 emission wavelength module and equipped with 2 injectors. We use a PHERAstar FS, BMG LABTECH.

2.2 Chemicals

1. Fugene 6 transfection reagent.
2. Glacial acetic acid.
3. Trypsin/Versene.
4. Versene.

2.3 Buffers and Stock Solutions

Prepare all solutions using MilliQ water (to attain a sensitivity of 18.2 MΩ.cm at 25 °C) and analytical grade reagents. Prepare and store all reagents at room temperature (unless indicated otherwise). Follow waste disposal regulations when disposing of waste materials.

1. Cell culture media: Dulbecco's modified eagle medium (DMEM) (high glucose), 10% newborn calf serum (NCS), 1% penicillin/streptomycin (P/S). Add geneticin (also known as G418) or another suitable antibiotic (dependent on the expression plasmid expressing the receptor) for stably transfected cells.
2. Phosphate-buffered solution (PBS): 137 mM NaCl, 2.7 mM KCl, 1.44 mM Na_2HPO_4, 1.8 mM KH_2PO_4. Each chemical is weighed and then added to 400 mL of MilliQ water in a beaker with a magnetic stirrer. The chemicals are thoroughly mixed at room temperature. The pH 7.4 is achieved by adding 1 M HCl to the dissolved solution. MilliQ water is then added to get a final volume of 500 mL. The PBS is then sterilized by autoclaving and stored at 4 °C.
3. Dye loading buffer: DMEM with high glucose and 25 mM HEPES (*see* **Note 1**). Add 48.75 mL of DMEM to a 50 mL centrifuge tube. Add 1.25 mL of 1 M HEPES to the DMEM and vortex gently to mix the solution.
4. Calcium-loaded buffer (CLB): 132 mM NaCl, 5 mM KCl, 5 mM Na_2HPO_4, 1.2 mM NaH_2PO_4, 1 mM $CaCl_2$, 0.8 mM $MgCl_2$, 10 mM Glucose (*see* **Note 2**). Each chemical is individually weighed and added to 45 mL of MilliQ water in a beaker with a magnetic stirrer and is made into a 50 mL stock solution. Stock solutions comprise 5 M NaCl, 1 M KCl, 1 M Na_2HPO_4, 0.1 M NaH_2PO_4, 1 M $CaCl_2.2H_2O$, 1 M $MgCl_2.6H_2O$, and 1 M D-Glucose. Stock solutions are

added to 900 mL MilliQ water as follows: 26.4 mL NaCl, 5 mL KCl, 5 mL Na_2HPO_4, 12 mL $NaH_2PO_4.H_2O$, 1 mL $CaCl_2.2H_2O$, 0.8 mL $MgCl_2.6H2O$. 10 mL D-Glucose. The pH of the mixed solution is adjusted to pH 7.4 and the final volume adjusted to 1 L with MilliQ water. The solution is filter sterilized and stored at room temperature. Approximately 120 mL CLB is required for each independent assay.

5. Calcium-free buffer (CFB): 132 mM NaCl, 5 mM KCl, 5 mM Na_2HPO_4, 1.2 mM NaH_2PO_4, 0.8 mM $MgCl_2$, 10 mM glucose. The stock solutions used to make this buffer are the same as used for making CLB. Approximately 3 mL CFB solution is required for each independent assay.
6. Fura-2/AM stock solution (1 mM): Add 50 μL of DMSO to 50 μg of Fura-2/AM ester. Stock stored at −20 °C. Each independent assay requires 48 μL of the stock.
7. Probenecid Stock Solution (250 mM): Add 1 mL of Hank's balanced salt solution (HBSS) buffer to the 77 mg of Probenecid inside the vial. Stock stored at −20 °C. Each independent assay requires 240 μL of the stock.
8. EGTA (ethylene glycol-bis(β-aminoethyl ether)-N,N,N′,N′--tetraacetic acid) stock solution (165 mM): Add 1.26 g of EGTA to 10 mL of MilliQ water and stir in a 25 mL beaker with a magnetic stirrer. Bring the solution to pH 11.0 with 1 M NaOH to dissolve the EGTA. Then, adjust the pH to 8.0 with 1 M HCl and add MilliQ water to obtain a final volume of 20 mL (*see* **Note 3**).
9. Ionomycin stock solution (0.242 mM): Weigh and add 1.8 g of ionomycin salt to 10 mL DMSO (*see* **Note 4**).
10. Poly-L-Lysine (PLL) stock solution (10 mg/mL): Add 2.5 mL of MilliQ water to dissolve 5 mg of PLL. PLL stock (100 μL of 10 mg/mL) is diluted in 4.9 mL of PBS, and the solution is mixed by vortexing (final PLL concentration; 0.2 mg/mL) to make the working solution.
11. Ligand stock solution. The concentration, media for dissolving, and storage conditions are specific to the ligand. Just prior to assay performance, a "working stock solution" is made from a stored ligand aliquot. The "working stock solution" (2 mL) needs to be made at 8 x the highest concentration of ligand that will be tested for stimulating a calcium response (*see* **Note 5**).

3 Methods

3.1 PLL-Coated 96-Well Plates

1. Treat each 96-well plate with 75% glacial acetic acid (50 μL/well) in a fume hood and incubate for 3 minutes.
2. Add 150 μL/well MilliQ water to dilute the acetic acid.
3. Immediately wash the plates three times with 300 μL/well MilliQ water and blot by tipping upside down onto a paper towel and tapping the plate in between each wash.
4. Add 50 μL/well of 0.2 mg/mL PLL to the 96-well plate, then wrap the plate in cling film/glad wrap (to prevent wells from drying), and place inside an incubator at 37 °C for 3 h.
5. Wash the plate twice with PBS (200 μL/well) to remove PLL suspension (PLL in suspension is toxic to cells).
6. UV-sterilize the plate for 3 h in a Class II hood (*see* **Note 6**).

3.2 Cell Culture

3.2.1 Generation of Stably Transfected Cells

1. HEK293 cells are seeded at a density of 2.4×10^6 cells/10 mL in cell culture media in a 100 mm tissue culture dish maintained at 37 °C with 5% CO_2 (*see* **Notes 7** and **8**).
2. At ~50% confluency, the cells are transfected with plasmid DNA transfection mix (680 μL) containing the receptor of interest in a dropwise manner. The plate is mixed gently by holding and rocking it back and forth.
3. Plasmid DNA transfection mix (690 μL) is prepared in a LoBind DNA Eppendorf tube by adding 570 μL of DMEM (serum- and antibiotic-free) to 45 μL of Fugene 6 and 37.5 μL of plasmid DNA (0.2 μg) (final concentration in plasmid DNA mix is 7.5 μg/mL). The mixture is gently mixed by vortexing and incubated at room temperature for 15 min (*see* **Notes 9** and **10**).
4. The cells are left to grow for 48 h at 37 °C with 5% CO_2.
5. The media is then aspirated using a Pasteur pipette and gentle vacuum suction, and 10 mL of cell culture media with 1 mg/mL G418 is added to select the growth of cells expressing the plasmid DNA.
6. The 1 mg/mL G418 media is replaced every 3 days for 2–3 weeks until only foci are visible on the plates.
7. The media is then aspirated, and 2 mL versene is added to the plate and gently swirled before being removed using a Pasteur pipette and vacuum suction. Trypsin/versene (2 mL) is added to detach the cells while the plate is incubated at 37 °C with 5% CO_2 for 45 s. The cells are then gently suspended by repeated pipetting before being added to 5 mL of fresh cell culture media with 0.5 mg/mL G418 in a 15 mL centrifuge tube.

The cells are then pelleted with centrifugation at 950 *g* for 5 min, and the supernatant is removed.

8. Fresh cell culture media (with 0.5 mg/mL G418) (10 mL) is added to the pelleted cells and repeatedly pipetted gently before being plated in a 100 mm tissue culture dish. When the cells are grown to ~99% confluency, they are plated in T75 flasks for the calcium assay.

3.2.2 Generation of Transiently Transfected Cells

1. HEK293 cells are seeded at a density of 3.2×10^6 cells/10 mL in T75 flasks with cell culture media and incubated at 37 °C with 5% CO_2 (*see* **Notes** 7 and **8**).
2. When the cells reach ~50% confluency, they are transfected with plasmid DNA transfection mix (853 μL) containing the receptor of interest in a dropwise manner. The plate is mixed gently by holding and rocking it back and forth.
3. The plasmid DNA transfection mix (863 μL) is prepared in a LoBind DNA Eppendorf tube by adding 754 μL of DMEM (serum- and antibiotic-free) with 59.4 μL of Fugene 6 and 49.6 μL of plasmid DNA (0.2 μg) (final concentration of DNA is 9.93 μg/mL). The mixture is gently mixed by vortexing and incubated at room temperature for 15 min (*see* **Note 9**).
4. The cells are then left to grow in an incubator at 37 °C with 5% CO_2 for 48 h, after which they are used for calcium assays.

3.3 Fura-2/AM Calcium Assay

All procedures are performed at room temperature unless otherwise specified.

All volumes stated below are to perform a single assay on a full 96-well plate for testing multiple ligands on a single receptor.

3.3.1 Loading Fura-2/AM into Cells

1. Mix 48 μL of the Fura-2/AM (1 mM) with 48 μL of 20% pluronic acid (1:1) in a 50 mL centrifuge tube immediately prior to use. Gently mix by vortexing.
2. Remove 4 μL of the premixed Fura-2/AM and pluronic acid F-127 mixture into a 15 mL centrifuge tube and add 2 μL of 5 mM EGTA stock solution (*see* **Note 11**). Gently mix by vortexing. Both 50 mL and 15 mL centrifuge tubes from the above steps are wrapped individually in aluminum foil to protect the reagents from light and then kept at room temperature while performing the following steps.
3. Aspirate cell culture media using a Pasteur pipette from confluent cells grown in T75 tissue culture flasks. Add 4 mL versene and place the flask at 37 °C for 2–3 min to detach the cells from the T75 flask.

4. Dilute the detached cells in 6 mL fresh cell culture media in a new 15 mL centrifuge tube, and then pellet the cells by centrifugation at 950 *g* for 5 min at room temperature.
5. Remove the cell culture media from the cell pellet using vacuum suction with a Pasteur pipette. Wash the cell pellet with 10 mL loading buffer and pellet the cells again by centrifugation as above.
6. The cells are then counted using a hemocytometer to obtain the number of cells/mL (*see* **Note 12**).
7. Resuspend the cells in the loading buffer (minimum volume of 26 mL) in a 50 mL centrifuge tube to give sufficient cells (8.5×10^5 cells/mL for HEK293 cells) to plate an entire 96-well plate.
8. The resuspended cells (23 mL) are added to the 50 mL centrifuge tube containing the Fura-2/AM and pluronic acid premix (refer to Subheading 3.3.1, **step 1** and Subheading 3.3.3) to give a 2 μM Fura-2/AM final concentration. These cells are loaded in the absence of EGTA. The remaining cells (1 mL) are added to the 15 mL centrifuge tube containing Fura-2/AM, pluronic acid, and EGTA pre-mix (refer to Subheading 3.3.1, **step 2**), and these cells are loaded in the presence of EGTA (*see* **Note 13**).
9. Ensure the caps on the centrifuge tubes are screwed on tightly before each tube is completely wrapped in aluminum foil. The tubes are then placed in lengthwise position on a shaking platform and incubated at room temperature for 60 min with gentle mixing.
10. Remove the aluminum foil and pellet the cells with and without EGTA by centrifugation at 950 *g* for 5 min at room temperature. Remove supernatant with vacuum suction and a Pasteur pipette, and wash the cells without EGTA by adding CLB containing 1 mM probenecid (use the volume in mL equal to the volume of the resuspended cells). Use a 25 mL pipette to resuspend cells gently. Remove supernatant with vacuum suction and a Pasteur pipette, and wash the cells with EGTA by adding CFB containing 1 mM probenecid (use the volume in mL equal to the volume of the resuspended cells) and then gently pipette the cell pellet repeatedly using a 2 mL pipette to resuspend cells.
11. Ensure the caps on the centrifuge tubes are screwed on tightly and then completely wrap each tube again in aluminum foil. The tubes are then placed lengthwise and incubated at 37 °C for 80 min with gentle mixing (*see* **Note 14**).

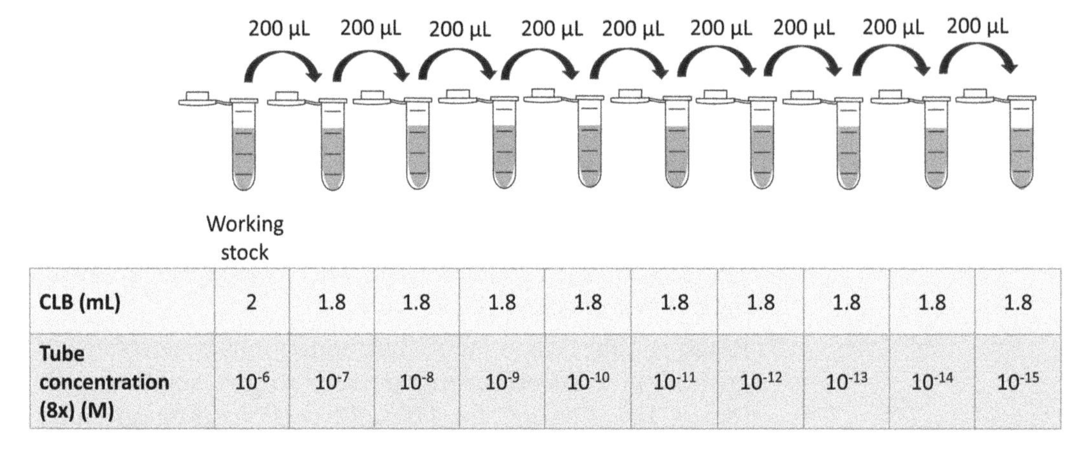

CLB (mL)	2	1.8	1.8	1.8	1.8	1.8	1.8	1.8	1.8	1.8
Tube concentration (8x) (M)	10^{-6}	10^{-7}	10^{-8}	10^{-9}	10^{-10}	10^{-11}	10^{-12}	10^{-13}	10^{-14}	10^{-15}

Fig. 3 Ligand dilution series for a single ligand. CLB is added to all Eppendorf tubes using a P1000 pipette. The minimum volume of the ligand (X) used to make a working stock solution should be 5 µL to maintain accuracy between experiments. This is an example of a tenfold dilution series

3.3.2 Preparation of Ligand Concentrations Ready to be Injected into 96-Well Plate

A working range of ligand concentrations is made from the "working stock ligand" solution during the 80 min incubation period (refer Subheading 3.3.1, **step 11**). Each working concentration is made in a minimum of 1.8 mL CLB (Fig. 3). The final concentration in the well represents the ligand concentration used to stimulate the cells (*see* **Note 15**).

3.3.3 Plating Cells into 96-Well Plate

Remove the tubes of cells from their 37 °C incubation. Gently agitate each tube to keep a homogenous cell suspension while pipetting 0.21 mL cells into specific wells of a PLL-coated 96-well plate (~1.8×10^5 cells/0.21 mL/well) using a 1000 µL pipettor (e.g., P1000). Cells without EGTA would be added to 93 wells in a 96-well plate. The cells with EGTA would be added to the three remaining wells in the 96-well plate (*see* **Note 16**, Fig. 4).

3.4 Setting Up the PHERAstar FS Plate Reader

1. Reading direction: To avoid activation of the stretch-receptor calcium channels, the movement of plate inside the plate reader needs to be minimized. Therefore, we use a bidirectional horizontal reading direction (*see* **Note 17**).
2. Number of flashes: To improve the accuracy of the response, ten flashes per well are used to set to average one intensity value per well.
3. A first kinetic window is set from 0 s to 30 s prior to injections. At 30 s, either injector 1 or 2 will be set up to inject the treatment into the appropriate well. A second kinetic window is set from 31 s to 90 s to resume the measurement after treatment.
4. The bottom optic is set to measure fluorescence.

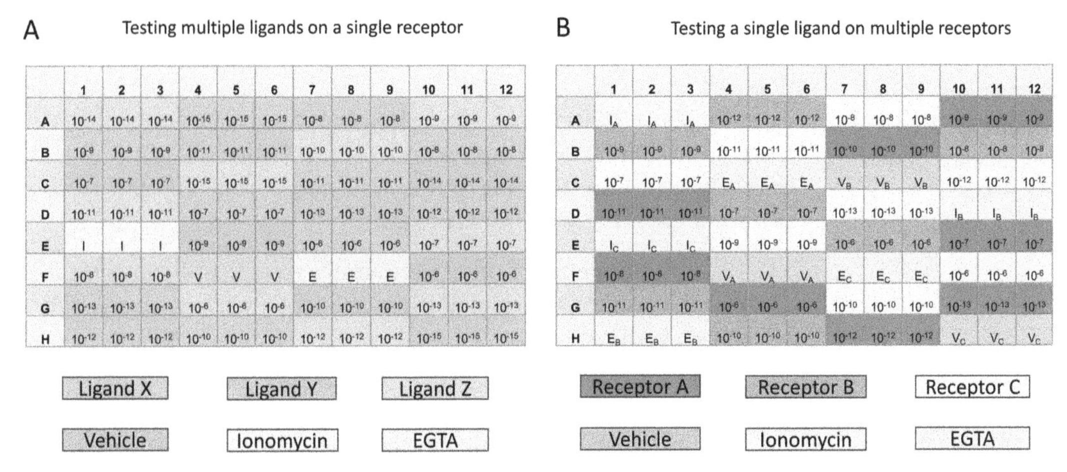

Fig. 4 Assay design on a 96-well plate with a single receptor subtype or multiple receptor subtypes. (**a**) Single receptor subtype. Clusters of triplicate ligand (X or Y or Z) dilutions are randomized across the plate to stimulate a single receptor (A). Triplicates of vehicle (V) alone are injected first, followed by triplicates of ligand starting with the lowest concentration (e.g., 10^{-15} M, 10^{-14} M, ..., 10^{-7} M, and 10^{-6} M, as shown in Fig. 3). Triplicate injections of EGTA (E) and ionomycin (I) and then follow the ligand injections. (**b**) Multiple receptor subtypes. Clusters of triplicate ligand (X) dilutions are randomized across the plate to stimulate multiple different receptors (A, B, C). Triplicates of vehicle (V_A or V_B or V_C) alone are injected first for each receptor, followed by triplicates of ligand starting with the lowest concentration (e.g., 10^{-12} M, 10^{-11} M, ..., 10^{-7} M, and 10^{-6} M, as shown in Fig. 3). Triplicate injections of EGTA (E_A or E_B or E_C) and ionomycin (I_A or I_B or I_C) and then follow the ligand injections for each receptor. Randomization differs between independent experiments

5. The focal height is 4.1 for 0.21 mL assay buffer in a 0.32 mL depth well (*see* **Note 18**). The focal height adjustment of the optical system ensures the best signal-to-noise ratio for the volume of solution in every well.
6. The gain is adjusted prior to running each assay (*see* **Note 19**). The gain adjustment optimizes signal amplification so that maximum sensitivity is achieved.
7. The plate setup is randomized for each experiment, and the injections are randomly distributed across the plate. The vehicle is injected first, followed by the lowest concentration of the ligand (Fig. 3: 10^{-15} M) and then progressively increasing to higher concentrations (Fig. 3: 10^{-14} M, 10^{-13} M, ..., 10^{-7} M, and 10^{-6} M). The EGTA and ionomycin for the standard calcium measurements are performed after all ligand concentrations are tested. Randomization is used because a ligand is injected and the calcium response measured in only one well at a time. There is ~4 h difference between the first and final well being injected. Randomization allows for assessing whether the calcium response is reproduced across independent experiments regardless of the time cells are left in the buffer, the dye leakage, and variability in dye loading (Fig. 4).

3.5 Simultaneous Vehicle/Ligand Injection and Monitoring Fluorescence Output

The plate reader requires either injectors or a compound plate (*see* **Note 20**) so that the transient Ca^{2+} response is measured simultaneously with injections and monitored for up to 90 s post-injections (*see* **Note 21**). The cells plated into the 96-well plate from homogenous cell suspension (refer to Subheading 3.3.3) will have settled to the bottom of the wells and cover ~90–95% of the surface area of the well (*see* **Note 22**).

1. Place the Eppendorf tubes containing vehicle or ligands on ice to help prevent the degradation of ligands while they are waiting to be injected.
2. Warm up in your hands two LoBind protein bind tubes containing the vehicle or ligand just before they are used for priming the two injectors. Next, place the warmed-up Eppendorf tubes in an Eppendorf tube rack, and place the tubing connected to the specific injector to be used into each specific tube (*see* **Note 23**).
3. Prime the two injectors one at a time, immediately before injecting vehicle or the ligand solution (500 μL/injector). The prime button in the software enables setting the volume for priming. This will pre-coat the injectors and tubing with the vehicle or ligand. This enables consistent ligand concentration to be used for injection for multiple wells. Although two injectors are primed consecutively, only one injector injects at any one time. Priming two injectors consecutively for this assay saves assay time since six wells, not three wells, can be injected and read in between washing and priming injectors and tubing.
4. Set up the software to measure baseline fluorescence intensity in each well for 30 s using 340/380 nm excitation and 510 emission wavelengths (*see* **Note 24**).
5. Set the plate reader to inject the wells with either vehicle or ligand concentrations at a speed of 150 μL/s (Fig. 4), and monitor the 340 and 380 fluorescence wavelengths post-injection for 90 s (*see* **Note 25**).
6. Start the software to inject and read.
7. After the injection of vehicle or ligand from the two injectors into the wells, remove the tubing out of each Eppendorf tube and place these into a 50 mL conical tube holding 50 mL MilliQ water ensuring that the opening of the tubing is fully submerged in the water. The injectors and tubing are each washed 3 × with 2 mL MilliQ water to flush out the remaining ligand in the injector and tubing and prevent cross-contamination of ligands. Follow previous steps (refer to Subheading 3.5, **steps 2–3**) to prepare for injecting further ligand concentrations (*see* **Note 26**).

8. Inject ionomycin (10 μL of 0.242 mM stock in DMSO) 30 s after the baseline measurement into the three wells that contain cells without EGTA (refer to Subheading 3.3.3). Ionomycin will increase the extracellular Ca^{2+} flux across the plasma membrane and saturate intracellular Fura-2 binding. This is the maximum calcium value (Rmax and Fmax) (*see* **Notes 27** and **28**).
9. Inject EGTA (10 μL of 165 mM stock in water) diluted in Ca^{2+}-free buffer into the three wells that contain cells with EGTA (refer to Subheading 3.3.3) to chelate all intracellular Ca^{2+}. This will be considered as the minimum Ca^{2+} value (Rmin and Fmin) (*see* **Note 28**).
10. At the end of every experiment, the injectors and the tubing are flushed 5 x with 3 mL of MilliQ water. Then, the injectors and tubing are loaded with 10% bleach and left loaded for 15 min prior to flushing and washing them 5 × with 3 mL MilliQ water.

3.6 Data Analysis

3.6.1 Calibration of Maximum and Minimum Ca^{2+} Concentration

The Grynkiewicz et al. equation [11] is used to obtain a quantitative Ca^{2+} value from the f340 and f380 measurements. For HEK293 cells, the $K_d = 224$ (cell-type-dependent value). The equation requires measurement of the maximum and minimum Ca^{2+} values detected by loading Fura-2/AM into the cell.

3.6.2 Calculation of Ligand-Induced Ca^{2+} Response

1. The intracellular Ca^{2+} is calculated using the Grynkiewicz et al. equation [11] for individual experiments. The plate reader measures f340 and f380 measurements.
2. Using the plate reader-associated data analysis software (e.g., MARS analysis for PHERAstar FS plate reader), f340/f380 is used to calculate the ratio for each well (Fig. 5).
3. To calculate the Rmin, the measurement range is set at 31–90 s in the cells injected with EGTA. The fluorescence f340/f380 ratio for this range is calculated. Similarly, to calculate Rmax, the measurement range is set at 31–90 s in the wells injected with ionomycin. The fluorescence ratio in this range is calculated.
4. To calculate background fluorescence (Fmin and Fmax), the measurement range is set at 0–30 s in the cells injected with EGTA and ionomycin, respectively. The background Ca^{2+} is obtained by calculating the fluorescence f380 for this range.
5. The Grynkiewicz et al. equation [11] is entered in the data calculation field. Using the Rmax, Rmin, Fmax, and Fmin values generated above and the Kd value, a quantitative Ca^{2+} value is obtained.

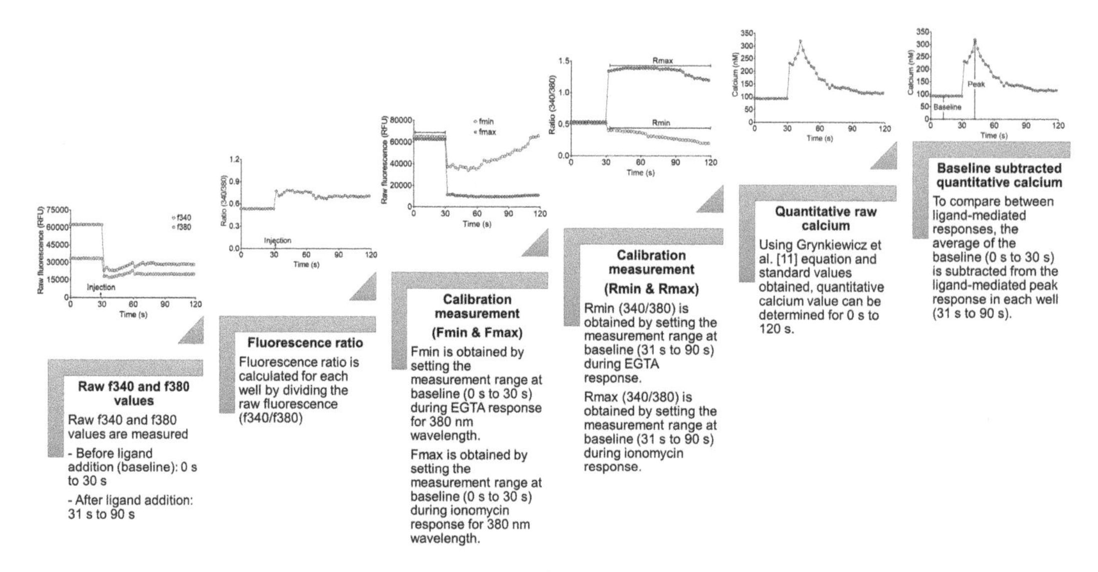

Fig. 5 Flow chart for quantitating ligand-mediated Ca^{2+} responses from raw fluorescence measurements

6. The baseline intracellular Ca^{2+} concentration should be around 100 nM at the beginning of an experiment. As the experiment progresses and the Fura-2 slowly leaks out of cells, the baseline can increase from 100 nM to ~140 nM–150 nM.
7. The data is exported into Microsoft Excel format. To compare ligand-mediated responses, an average of the baseline measurement is subtracted from the ligand-mediated calcium response (*see* **Note 29**).
8. The quantitative Ca^{2+} data can be pooled from three or more independent experiments for statistical analyses (*see* **Note 30**).

4 Notes

1. The solution is prepared fresh prior to performing the experiment. Do not add probenecid to this solution, as we have identified that it could prevent dye uptake.
2. Prepare 50 mL sterile stock solution of individual buffer components since $CaCl_2$ does not mix well with NaH_2PO_4 and precipitates from the solution when kept long term. The sterile stock solutions are used to prepare quantities of Ca^{2+}-loaded and Ca^{2+}-free buffers as required.
3. Stock solution for the loading buffer (5 mM) is prepared in a 15 mL centrifuge tube by adding 9.697 mL MilliQ water and 0.303 mL of stock EGTA (165 mM). This diluted EGTA is used in the Fura-2/AM dye premix for cells that contain EGTA (refer to Subheading 3.3.1, **step 2**).

4. Stock solution of the ionomycin should be prepared and left at −20 °C. After use, the leftover dissolved ionomycin salt in DMSO can be stored at 4 °C and used for further assays.
5. In our assay design, 30 μL of the ligand is injected into 0.21 mL of buffer in each well of a 96-well plate. Therefore, the stock solution of the ligand needs to be at least 8 × more concentrated than the concentration required for testing since the injected ligand will be diluted in the well before it reaches the cells.
6. The PLL-coated plates are prepared in bulk before the experiments and stored at 4 °C. The plates are brought to room temperature for at least 3 h before an assay commences.
7. The cell numbers for plating other cell types need to be determined to produce at least 8.5×10^5 cells/mL at ~99% confluency.
8. We have described a method for cell culture using a single receptor subtype transfected in a cell line. The assay design can be modified to accommodate multiple receptor subtypes, each separately transfected to produce a cell line (see Fig. 4b). Each cell line is then grown in an individual T75 flask, and the same procedure is followed for each flask.
9. We use a transfection ratio of 1: 6 (plasmid DNA: Fugene 6) that we optimized for our cells. The appropriate transfection ratio needs to be optimized to achieve maximal transfection efficiency for specific cell types. Of note, do not incubate the transfection mixture for more than 30 min since longer incubations decrease transfection efficiency.
10. We have used Fugene 6 as the transfection reagent. Other transfection reagents can be used, and the method for efficient transfection optimized accordingly.
11. Adding EGTA to the dye premix for the cells that are used to measure the minimum standard calculation is critical. Without the EGTA during dye loading, EGTA injections are unable to chelate all Ca^{2+} in the cells, therefore affecting the final quantitation of the $[Ca^{2+}]$.
12. For HEK293 cells, a minimum of 24 mL cells at 8.5×10^5 cells/mL is required to perform an assay with an entire 96-well plate. A single confluent T75 flask of HEK293 cells produces sufficient cells for a whole 96-well plate. If a cell line is found to produce fewer cells, then use multiple confluent T75 flasks and follow the steps above for each flask.
13. For HEK293 cells, the cells are suspended in 24 mL of loading buffer. From this, 22.90 mL of the resuspended cells is added to the 50 mL centrifuge tube containing Fura-2/AM and pluronic acid (refer to Subheading 3.3.1, **step 1**) and

0.99 mL of resuspended cells is added to the 15 mL centrifuge tube containing Fura-2/AM, pluronic acid, and EGTA (refer to Subheading 3.3.1, **step 2**).

14. Ideally, use a shaking platform at 37 °C. Alternatively, the cells are incubated at 37 °C in an incubator and removed at 10 min intervals for gentle mixing by manually rotating the tubes.
15. We recommend diluting peptide ligands in LoBind protein Eppendorf tubes. Do not include bovine serum albumin (BSA) to reduce peptide/protein sticking to tubes since BSA will bind Fura-2/AM and reduce the ligand response. A minimum of 1.8 mL of ligand solution is required to prime the injectors and inject the solution into wells.
16. Our assay design has three replicates of each treatment (vehicle and ligand). When a single receptor is tested, the vehicle and ligand treatments must cover no more than 90 wells in a 96-well plate (see Fig. 4a). Three wells with the cells not containing EGTA will be used for a maximum Ca^{2+} (Rmax and Fmax) calibration calculation (refer to Subheading 3.6.1, **step 2**). The remaining three wells containing cells with EGTA will be used for a minimum Ca^{2+} (Rmin and Fmin) calibration calculation (refer to Subheading 3.6.1, **step 3**). When multiple receptors (e.g., 3 receptors) are tested, the vehicle and ligand treatments must cover no more than 78 wells in a 96-well plate (see Fig. 4b). Nine wells with cells not containing EGTA will be used for maximum Ca^{2+} calibration for individual receptors. The remaining nine wells containing cells with EGTA will be used for a minimum Ca^{2+} calibration calculation for individual receptors. For example, Receptor A is treated with three different ligands (Ligands X, Y, and Z) at nine or ten different concentrations for each ligand. The remaining wells are used for vehicle, EGTA, and ionomycin calibration calculation (Figs. 4 and 5). When multiple receptors (A, B, and C) are treated with a single ligand at seven or eight different concentrations, the vehicle, ionomycin, and EGTA calibration calculations need to be performed for individual receptors, as the dye loading is performed in independent tubes for each receptor. This reduces variability to the quantitation calculation as the dye loading is kept consistent among the cells treated with vehicle/ligands and the calibration measurements (Fig. 4b). The assay design can be modified depending on the number of receptor subtypes and ligand treatments in a single 96-well plate.
17. In the plate layout, the addition of vehicle, ligands, EGTA, and ionomycin are clustered in three wells next to each other (Fig. 4). This is done to reduce the movement of the plate

and the injectors between wells and avoid activation of the mechanosensitive Ca^{2+} channels.

18. The focal height must be adjusted if a final volume different from 0.21 mL in the wells is used. This is done by employing the automatic focal height adjustment option in the plate reader software.
19. The Ca^{2+} response for some GPCRs can be huge; therefore, we set the target value at 10%. This allows for capturing maximal calcium response from higher ligand concentrations. Furthermore, adjusting the gain for each assay removes some of the variability associated with dye loading and the esterase activation of the Fura-2/AM. The gain changes very little if the assay conditions for loading and imaging are maintained similarly between independent experiments.
20. We have used injectors for this assay since the PHERAstar FS is equipped with two injectors. However, we expect a compound plate [25, 26] would also be suitable for this assay. The compound plate layout would resemble the layout for the plate with the correct ligands placed in each well. The compound plate is inserted independently into the plate reader, and the injectors automatically remove the ligand dilutions from the compound plate and add them to the experimental 96-well plate containing the cells.
21. Run smaller experiments if the imaging time for a well is longer than 120 s since the dye will slowly leak out of the cells. This is noticeable after 4 h, with an increase in the baseline response resulting from the leaked dye binding to extracellular Ca^{2+} present in the buffer. Dye leakage out of cells needs to be avoided as it will increase response variability between replicate wells and reduce the ligand-activated Ca^{2+} response.
22. View the cells in the 96-well plate using an inverted microscope to check that the cells are evenly distributed in the well and not clumped or piled on top of each other. Clumping of cells can produce variable Ca^{2+} measurements if the cells move positions following injection of vehicle or ligand.
23. Ensure the opening of the tubing connected to each injector is fully submerged in the solution in the tube to avoid getting bubbles in the injectors.
24. Measuring the baseline Ca^{2+} level for 30 s is sufficient to obtain a stable reading. However, if there is a lot of fluorescence variability, the baseline Ca^{2+} measurements can be extended to 40–50 s or until the baseline reading is stable.
25. The volume of the imaging buffer on the cells, the volume of vehicle or ligand being injected, and the injection speed need to be optimized [23] for the cell line being examined to avoid

triggering a Ca^{2+} response driven by activation of endogenous mechanosensitive channels present in the cell membrane. In our experience, there is a minimum mechanosensitive-triggered Ca^{2+} response when 30 μL fluid is injected at a speed of 150 μL/s into a well holding 0.21 mL assay buffer [23]. This injection rate is a sufficient speed to mix the 30 μL into the 0.21 mL in each well.

26. Washing injectors and tubing in between the injections is not required when going from a lower concentration to a higher concentration of the same ligand. When switching between ligands, it is necessary to thoroughly wash the injectors and tubing.
27. Ionomycin is expensive, and therefore the ionomycin within the injectors can be retrieved (i.e., backflushed) into the ionomycin-containing Eppendorf tube, and then ionomycin can be reused in subsequent experiments. As the ionomycin is dissolved in highly viscous DMSO, it is essential to flush the injectors twice with DMSO after the injections are completed. After this, the standard protocol for washing injectors and tubing with water is performed (refer to Subheading 3.5, **step** 7).
28. The minimum and maximum Ca^{2+} measurements are dependent on the cell type. If there is more than one cell type included in the 96-well plate, it will be necessary to measure maximum and minimum values in all the different cell types. The minimum and maximum Ca^{2+} values of a specific cell type must then be used in the quantitation calculation of the same cell type treated with vehicle or ligands.
29. Both baseline and ligand-mediated peak Ca^{2+} responses are of interest for GPCR signaling. The average baseline fluorescence f340/f380 ratio and Ca^{2+} response are obtained (refer to Subheading 3.6.2). The ligand-mediated peak fluorescence f340/f380 ratio is identified in each well from 31 to 90 s postinjection. The next step involves subtracting the averaged baseline Ca^{2+} concentration from the peak Ca^{2+} concentration, which gives the ligand-mediated Ca^{2+} level increase up to 90 s post-injection. For temporal kinetic data, the average baseline Ca^{2+} response is subtracted from the ligand-mediated Ca^{2+} concentration obtained at every 3 s interval up to 90 s postinjection.
30. Three or four independent experiments can produce robust quantitative ligand-receptor mediated Ca^{2+} levels. However, a ligand–receptor interaction that only weakly increases intracellular Ca^{2+} levels may require more than four independent experiments due to increased variability between replicate wells and between experiments.

References

1. Gillyard T, Fowler K, Williams SY, Cone RD (2019) Obesity-associated mutant melanocortin-4 receptors with normal Galphas coupling frequently exhibit other discoverable pharmacological and biochemical defects. J Neuroendocrinol 31(10):e12795. https://doi.org/10.1111/jne.12795
2. Heusinkveld HJ, Westerink RH (2011) Caveats and limitations of plate reader-based high-throughput kinetic measurements of intracellular calcium levels. Toxicol Appl Pharmacol 255(1):1–8. https://doi.org/10.1016/j.taap.2011.05.020
3. Dhyani V, Gare S, Gupta RK, Swain S, Venkatesh KV, Giri L (2020) GPCR mediated control of calcium dynamics: a systems perspective. Cell Signal 74:109717. https://doi.org/10.1016/j.cellsig.2020.109717
4. Werry TD, Wilkinson GF, Willars GB (2003) Mechanisms of cross-talk between G-protein-coupled receptors resulting in enhanced release of intracellular Ca2+. Biochem J 374(Pt 2): 281–296. https://doi.org/10.1042/BJ20030312
5. Kao JP (1994) Practical aspects of measuring [Ca2+] with fluorescent indicators. Methods Cell Biol 40:155–181
6. Kao JP, Li G, Auston DA (2010) Practical aspects of measuring intracellular calcium signals with fluorescent indicators. Methods Cell Biol 99:113–152. https://doi.org/10.1016/B978-0-12-374841-6.00005-0
7. Woszczek G, Fuerst E, Maguire TJA (2021) FLIPR calcium mobilization assays in GPCR drug discovery. Methods Mol Biol 2268:193–205. https://doi.org/10.1007/978-1-0716-1221-7_13
8. Kurko D, Bekes Z, Gere A, Baki A, Boros A, Kolok S, Bugovics G, Nagy J, Szombathelyi Z, Ignacz-Szendrei G (2009) Comparative pharmacology of adrenergic alpha(2C) receptors coupled to Ca(2+) signaling through different Galpha proteins. Neurochem Int 55(7): 467–475. https://doi.org/10.1016/j.neuint.2009.04.015
9. Galaz-Montoya M, Wright SJ, Rodriguez GJ, Lichtarge O, Wensel TG (2017) beta2-adrenergic receptor activation mobilizes intracellular calcium via a non-canonical cAMP-independent signaling pathway. J Biol Chem 292(24):9967–9974. https://doi.org/10.1074/jbc.M117.787119
10. Paredes RM, Etzler JC, Watts LT, Zheng W, Lechleiter JD (2008) Chemical calcium indicators. Methods 46(3):143–151. https://doi.org/10.1016/j.ymeth.2008.09.025
11. Grynkiewicz G, Poenie M, Tsien RY (1985) A new generation of Ca2+ indicators with greatly improved fluorescent properties. J Biol Chem 260:3440–3450
12. Malgaroli A, Milani D, Meldolesi J, Pozzan T (1987) Fura-2 measurement of cytosolic free Ca2+ in monolayers and suspensions of various types of animal cells. J Cell Biol 105(5): 2145–2155. https://doi.org/10.1083/jcb.105.5.2145
13. Roe MW, Lemasters JJ, Herman B (1990) Assessment of Fura-2 for measurements of cytosolic free calcium. Cell Calcium 11(2–3): 63–73. https://doi.org/10.1016/0143-4160(90)90060-8
14. Mountjoy KG, Kong PL, Taylor JA, Willard DH, Wilkison WO (2001) Melanocortin receptor-mediated mobilization of intracellular free calcium in HEK293 cells. Physiol Genomics 5(1):11–19
15. Valentine WJ, Tigyi G (2012) High-throughput assays to measure intracellular Ca (2)(+) mobilization in cells that express recombinant S1P receptor subtypes. Methods Mol Biol 874:77–87. https://doi.org/10.1007/978-1-61779-800-9_7
16. Demer LL, Wortham CM, Dirksen ER, Sanderson MJ (1993) Mechanical stimulation induces intercellular calcium signaling in bovine aortic endothelial cells. Am J Phys 264(6 Pt 2):H2094–H2102. https://doi.org/10.1152/ajpheart.1993.264.6.H2094
17. Naruse K, Sokabe M (1993) Involvement of stretch-activated ion channels in Ca2+ mobilization to mechanical stretch in endothelial cells. Am J Phys 264(4 Pt 1):C1037–C1044. https://doi.org/10.1152/ajpcell.1993.264.4.C1037
18. Sigurdson WJ, Sachs F, Diamond SL (1993) Mechanical perturbation of cultured human endothelial cells causes rapid increases of intracellular calcium. Am J Phys 264(6 Pt 2): H1745–H1752. https://doi.org/10.1152/ajpheart.1993.264.6.H1745
19. Rosales OR, Isales CM, Barrett PQ, Brophy C, Sumpio BE (1997) Exposure of endothelial cells to cyclic strain induces elevations of cytosolic Ca2+ concentration through mobilization of intracellular and extracellular pools. Biochem J 326(Pt 2):385–392. https://doi.org/10.1042/bj3260385
20. Tong J, Du GG, Chen SR, MacLennan DH (1999) HEK-293 cells possess a carbachol-

and thapsigargin-sensitive intracellular Ca2+ store that is responsive to stop-flow medium changes and insensitive to caffeine and ryanodine. Biochem J 343(Pt 1):39–44

21. Hayakawa K, Tatsumi H, Sokabe M (2008) Actin stress fibers transmit and focus force to activate mechanosensitive channels. J Cell Sci 121(Pt 4):496–503. https://doi.org/10.1242/jcs.022053
22. Meijer M, Hendriks HS, Heusinkveld HJ, Langeveld WT, Westerink RH (2014) Comparison of plate reader-based methods with fluorescence microscopy for measurements of intracellular calcium levels for the assessment of in vitro neurotoxicity. Neurotoxicology 45: 31–37. https://doi.org/10.1016/j.neuro.2014.09.001
23. Kumar SS, Ward ML, Mountjoy KG (2021) Quantitative high-throughput assay to measure MC4R-induced intracellular calcium. J Mol Endocrinol 66(4):285–297. https://doi.org/10.1530/JME-20-0285
24. Botha R, Kumar SS, Grimsey NL, Mountjoy KG (2023) A unique MC4R signaling profile for obesity-associated constitutively active variants. J Mol Endocrinol 71. https://doi.org/10.1530/JME-23-0008
25. Hunt BD, Lambert DG (2013) Ratiometric [Ca(2)(+)]i measurements in adherent cell-lines using the NOVOstar microplate reader. Methods Mol Biol 937:111–120. https://doi.org/10.1007/978-1-62703-086-1_6
26. Sum CS, Murphy BJ, Li Z, Wang T, Zhang L, Cvijic ME (2004) Pharmacological characterization of GPCR agonists, antagonists, allosteric modulators and biased ligands from HTS hits to lead optimization. In: Markossian S, Grossman A, Brimacombe K et al (eds) Assay guidance manual. Bethesda, Rockville

Chapter 2

Quantifying Gq Signaling Using the IP_1 Homogenous Time-Resolved Fluorescence (HTRF) Assay

Aqfan Jamaluddin

Abstract

G protein-coupled receptors (GPCRs) play pivotal roles in cellular signaling and can regulate several cellular functions such as proliferation, secretion, protein expression, and cellular metabolism. Coupling of GPCRs to members of the Gq/11 protein family results in activation of inositol trisphosphate (IP3) and accumulation of calcium intracellularly. This protocol chapter outlines a step-by-step guide for utilizing the inositol phosphate-1 (IP_1) accumulation assay, a time-resolved fluorescence resonance energy transfer (TR-FRET) method, to investigate Gq-IP3 signaling. The assay serves as a valuable tool for those conducting pharmacological investigations and compound screening targeting this critical cellular pathway. This protocol chapter covers experimental setup, sample preparation, and data analysis, providing researchers with an in-depth guide to explore the pharmacology of Gq-coupled receptors.

Key words Calcium signaling, G protein recruitment, Inositol phosphate, IP_3, Receptor activity, Secondary messenger, TR-FRET assay

1 Introduction

The study of G protein-coupled receptor (GPCR) signaling is central to many intricate cellular responses and holds significance for drug development. Signaling by the Gq/11 family has an important role in cell differentiation, metabolism, and proliferation [1, 2]. Examples of Gq-coupled receptors include the α_1-adrenergic receptor and the angiotensin AT_1 receptor that regulate cardiovascular functions, the endothelin ET_A receptor that mediates vasoconstriction and the ghrelin receptor that has metabolic roles [1]. Among the various methodologies employed to investigate the Gq pathway, the inositol phosphate-1 (IP_1) homogenous time-resolved fluorescence (HTRF) accumulation assay is a quick and robust technique that can be used in primary or transfected cells [3].

Caroline M. Gorvin (ed.), *Calcium Signaling: Methods and Protocols*, Methods in Molecular Biology, vol. 2861,
https://doi.org/10.1007/978-1-0716-4164-4_2,

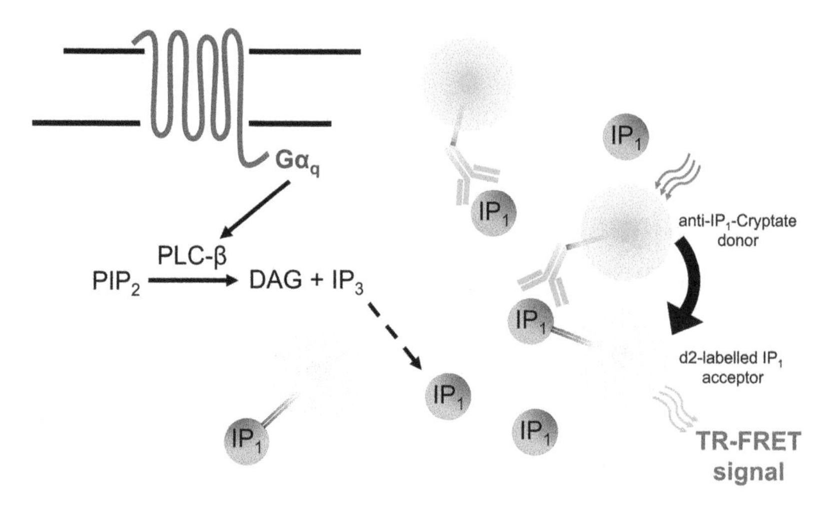

Fig. 1 Activation of receptor and Gq signaling results in the production of the secondary messenger IP_3. IP_1 is a byproduct of IP_3. The IP-One assay is a competitive TR-FRET assay, where native IP_1 produced by cells or unlabeled IP_1 standards compete with d2 fluorophore-labeled IP_1 (acceptor) for binding to the anti-IP_1 antibody tagged with Terbium (Tb) cryptate fluorophore (donor). The output TR-FRET signal is therefore inversely proportional to the amount of native IP_1 produced by the cells

The IP_1 accumulation assay measures IP_1, a byproduct of a crucial second messenger in the Gq signaling pathway (Fig. 1). Upon receptor activation, the Gq protein stimulates the enzyme beta-phospholipase C (PLC-β), leading to the hydrolysis of phosphatidylinositol 4,5-bisphosphate (PIP_2) into 2 s messengers: inositol 1,4,5-trisphosphate (IP_3) and diacylglycerol (DAG) [4, 5]. DAG remains attached to the membrane, while IP_3 binds to IP_3 receptors on the ER membrane [5, 6]. This triggers the release of calcium ions from intracellular stores which induces intracellular calcium signaling [7]. IP_3 is then hydrolyzed to the downstream metabolite IP_1 [8].

Sold as IP-One and manufactured by Cisbio (now part of Revvity's portfolio), the assay is essentially a competitive immunoassay, where native IP_1 produced by cells or unlabeled IP_1 standards compete with d2 fluorophore-labeled IP_1 for binding to the anti-IP_1 antibody tagged with Terbium (Tb) cryptate fluorophore [9, 10]. The binding of the antibody to the d2-labeled IP_1 results in transfer of the donor fluorophore emission energy (Tb cryptate) to the acceptor fluorophore (d2) when in close proximity, which in turn emits fluorescence at a specific wavelength. Acceptor and donor fluorescence emissions have distinct wavelengths, enabling quantification of the fluorophore interactions [9, 10].

There are several advantages to using the IP-One assay. Calcium flux assays, conventionally used to measure Gq signaling, do not have the ability to measure either inverse agonism or constitutive receptor activity [11]. Additionally, IP_1 is a more stable downstream metabolite of IP_3 that accumulates in cells in the presence of lithium chloride [9, 10]. Although IP_1 is not directly involved in the Gq signaling pathway, very good correlation can be obtained between intracellular calcium assays and the IP-One assay, signifying it as a good measure of Gq signaling [9]. TR-FRET detection also offers reduced background noise and an enhanced signal-to-noise ratio when compared to conventional fluorescence methods due to the time delay between excitation and emission, thereby achieving heightened sensitivity [3]. Moreover, the assay's mix-and-read format eliminates the need for washing steps when compared to other antibody-dependent methods, such as Western blot and ELISA. This, coupled with the assay's inherent stability and adaptability for miniaturization, facilitates automation and high-throughput applications.

This chapter will include an in-depth protocol that provides a step-by-step guide for the experimental setup, sample preparation, and data analysis required to conduct the IP_1 accumulation assay. This protocol uses HEK293 cells transiently transfected with a GPCR that signals via Gq recruitment. The protocol is divided into four sequential procedures. The first is cell plating, followed by transfection of cells with the appropriate constructs. This is then followed by GPCR stimulation or conditional tests, sample preparation, and data collection (Fig. 2). The assay can accommodate cells in suspension with very minor modifications needed. Additionally, this chapter will delve into common issues encountered

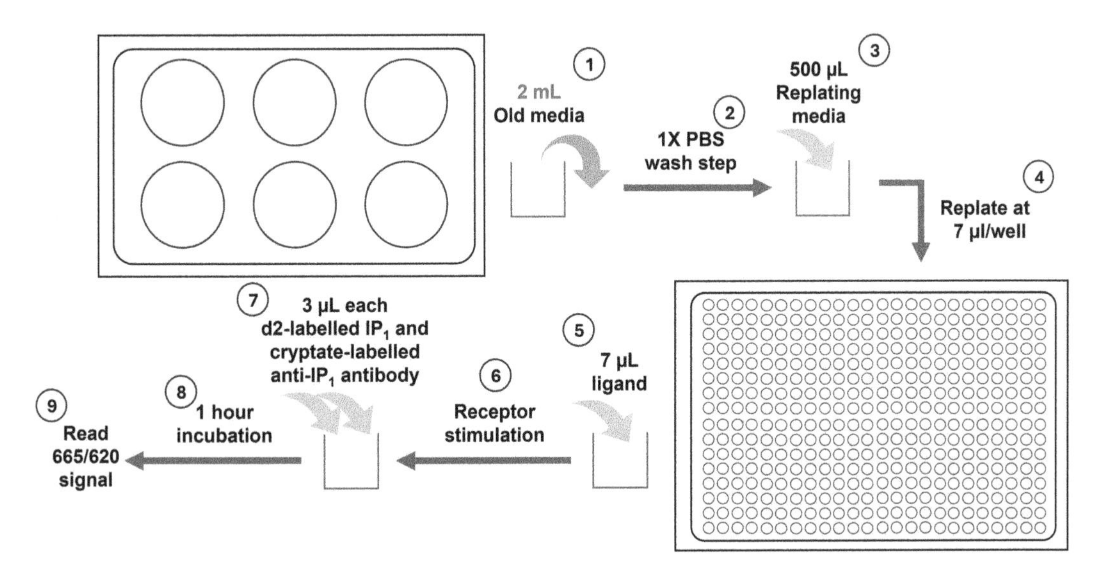

Fig. 2 IP-One assay steps following transfection of cells. In steps 1–4, cells are replated from 6-well plates to a 384-well plate. Ligands are added to the cells in step 5 and incubated at 37 °C in a humidified chamber under 5% CO_2 for a specific time to produce IP_1 in step 6. In steps 7–8, the IP-One assay components are added to the cells and incubated at room temperature for 1 h (shaking optional), before reading the TR-FRET signal in step 9 in a TR-FRET-compatible plate reader

when performing these assays, provide troubleshooting solutions and will also provide recommended optimizations associated with the protocol.

2 Materials

1. Adherent HEK293 cells.
2. Plasmid construct. Prepare plasmid construct containing your GPCR of choice. Obtain the concentration of plasmid in the solution after plasmid purification and store at 4 °C for short–/medium-term storage, or −20 °C for longer-term storage.
3. 6-well clear cell culture plates: TC-treated, flat bottom with lid.
4. 384-well white OptiPlate assay plate: flat non-clear bottom (*see* **Note 1**).
5. Growth media: High glucose Dulbecco's modified Eagle medium (DMEM) media, 10% bovine serum (*see* **Note 2**).
6. Transfection media: DMEM media.
7. 1× TrypLE™ Express Enzyme.
8. 1× sterile phosphate-buffered saline (PBS).
9. Hemocytometer for cell counting and trypan blue (*optional*).
10. Lipofectamine™ 2000.
11. Replating media: FluoroBrite™ DMEM, 10% bovine serum, 100 mM lithium chloride (*see* **Note 3**).
12. IP-One Gq kit (Revvity). Follow kit manufacturer's instructions for preparation of stock reagents and storage (*see* **Note 4**). The stock IP_1 standard, d2-labeled IP_1, and cryptate-labeled anti-IP_1 antibody should be stored in single-use aliquots at −20 °C. The 5× stimulation buffer and the lysis and detection buffer can be stored at 4 °C.
13. HTRF®-certified plate reader: we have used both BMG Labtech PHERAstar and PerkinElmer EnVision (*see* **Note 5**).

3 Methods

3.1 Cell Culture

1. Gently wash the flask containing HEK293 cells once with 1× PBS.
2. Detach cells from flasks with 2–10 mL of 1× TrypLE (sufficient volume to cover flask surface), and observe cells for detachment from surface. Cells should not take longer than 5–10 min to detach with gentle tapping.

3. Neutralize cells with an equal volume of growth media. Pipette cells up and down to remove clumps and to detach any remaining cells on the flask surface.
4. Obtain a small sample of cells from the cell suspension and count cells with a hemocytometer to determine total cell amount in the suspension.
5. Plate cells into 6-well plates at approximately 2–3 × 10^5 cells per well and 2 mL/well (*see* **Note 6**).
6. Incubate cells overnight at 37 °C in a humidified chamber under 5% CO_2.

3.2 Transfection

1. Dilute the appropriate amount of DNA construct for your experiment in 100 μL of the transfection media. For a 6-well plate format, 1–2 μg of your GPCR construct is a good starting point (*see* **Note** 7).
2. Add 2 μL of Lipofectamine™ 2000 into the transfection mix (*see* **Note 8**).
3. Incubate transfection mix for a minimum of 5 min at room temperature.
4. Add 100 μL of the transfection mix dropwise into the well.
5. Incubate cells for an additional 1–2 days at 37 °C and 5% CO_2 before GPCR stimulation or further conditional tests to ensure robust GPCR expression.

3.3 GPCR Stimulation

1. Prepare the GPCR ligand/treatment in the recommended kit's stimulation buffer (*see* **Notes 3** and **9**).
2. Wash HEK293 cells with 1× PBS, and then replace with 500 μL of replating media (*see* **Note 10**).
3. Detach and de-clump cells by pipetting up and down.
4. Pipette cells into 384-well plate at 7 μL/well.
5. Add ligand/treatment at 7 μL/well, making a total volume of 14 μL in the well consisting of cells and ligand/treatment.
6. Incubate for 1 h at 37 °C and 5% CO_2 to allow for Gq signaling and IP_1 accumulation (*see* **Note 11**).

3.4 Sample Preparation and Data Collection

1. During incubation, prepare the IP_1 standards in the recommended kit's stimulation buffer (*see* **Notes 3** and **12**). Also prepare the d2-labeled IP_1 and cryptate-labeled anti-IP_1 antibody dilutions in the recommended kit's lysis and detection buffer, to be added to both samples and standards.
2. After incubation, add the d2-labeled IP_1 and cryptate-labeled anti-IP_1 antibody to the cells in the 384-well plate at 3 μL volume each (total volume in the well is now 20 μL).

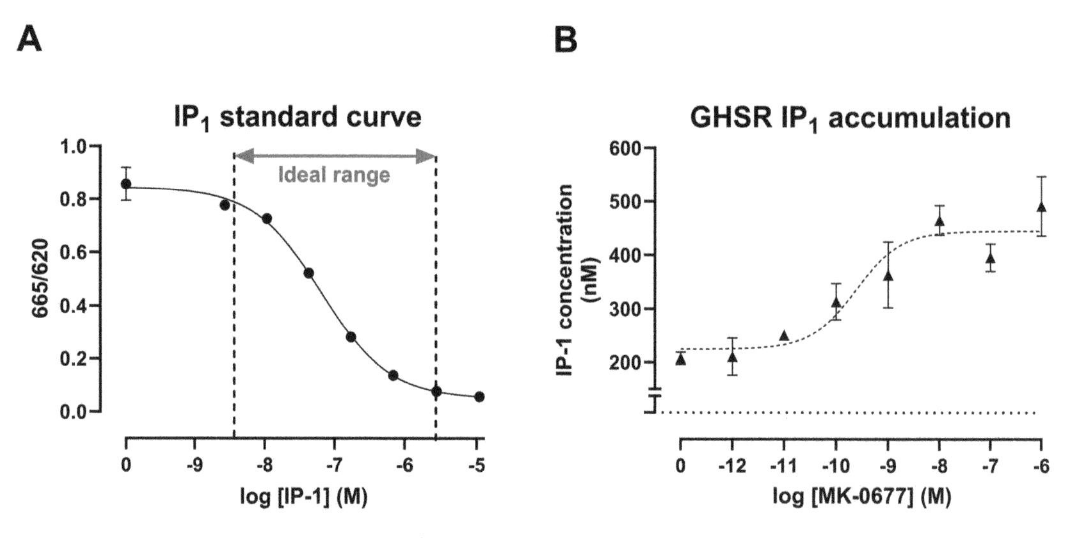

Fig. 3 Sample analysis of IP-One assay. Graph (**a**) shows an IP_1 standard curve plot, with the IC_{10} and IC_{90} range where sample data should ideally fall for extrapolation to determine their IP_1 concentration. Once IP_1 concentrations have been determined, IP_1 concentration can be plotted against the concentrations of ligand, as shown in graph (**b**)

3. In addition, add 14 μL of IP_1 standards to additional empty wells, along with the d2-labeled IP_1 and cryptate-labeled anti-IP_1 antibody at 3 μL volume each.
4. Incubate at room temperature while shaking at a low relative centrifugal force (~100–200 × *g*) for 1 h at room temperature (*see* **Note 13**).
5. After incubation, pulse spin down the 384-well plate (2–3 s; ~400–600 × *g*) to ensure all volumes of lysates/standards remain at the bottom of the wells (*see* **Note 14**).
6. Measure IP_1 levels in both cell lysates and standards with the HTRF®-certified plate reader. Excitation is at 320 nm. Emissions are measured at 620 nm and 665 nm. The raw data is calculated as the emission at 665 nm divided by the emission at 620 nm. Increased accumulation of IP-1 will result in a decrease in signal.
7. Plot a standard curve comprised of 665/620 signal against IP-1 standard concentration using nonlinear least squares fit (sigmoidal dose response variable slope, four-parameter fit) (Fig. 3). Measured signals from samples are interpolated from the standard curve to determine the IP_1 concentration produced (*see* **Note 15**). The amount of IP_1 concentration can then be plotted against the concentration of ligand (Fig. 3).

4 Notes

1. The IP-One assay can be adjusted proportionally to run the assay in 96-well to 1536-well microplates. The volume of each assay component must be proportionally adjusted to approximately maintain the reagent concentrations as in the 20 μL final assay volume in a 384-well plate.
2. Antibiotics can be supplemented into the media, but it is not recommended.
3. The manufacturer recommended using the stimulation buffer provided within the kit for replating, but this can be substituted to save kit reagent. Other alternatives to the stimulation buffer can be used, such as FluoroBrite and DMEM with no phenol. However, substitutions will require the addition of LiCl to ensure IP_1 does not degrade during GPCR stimulation.
4. IP-One Gq kit requires stock preparation of its reagents. Note that preparations can differ according to kit sizes, which will lead to differences in stock concentrations. The d2-labeled IP_1 and cryptate-labeled anti-IP_1 antibody are prepared to 6× stock solution in the 1000 tests kit and 20× stock solution in the 20,000 or 100,000 tests kit. Follow the manufacturer's instructions for reconstitution volumes.

   Reconstitute the IP_1 standard with MilliQ water following instructions on the vial to make the stock standard. The stimulation buffer is diluted from the 5× stock to 1× with distilled water prior to using the buffer to dilute ligands and the IP_1 standard (can be done prior to assay).

   According to the manufacturer, the d2-labeled IP_1, cryptate-labeled anti-IP_1 antibody and the IP_1 standard stored at −20 °C are stable for 3 months, but we have found our single-thaw stock aliquots have a similar performance when stored at −20 °C for longer than 1 year.
5. We find there is little difference in sensitivity and limit of detection between plate readers. Additionally, it is highly recommended to utilize the IP_1 standard provided for normalization of raw data, thereby allowing comparable results across different plate readers. Please make sure to set up the plate reader to read for Terbium cryptate.
6. If other plate formats are used for plating cells, the cell density must be optimized depending on the well's volume and surface. For example, 96-well plates may only require 15,000–20,000 cells per well in a 100 μL volume.
7. Transfected DNA amount will require optimization to ensure sufficient receptor expression. Additionally, different plate formats will require different DNA amounts (e.g., cells in 96-well

plate will require less DNA per well). We have found that lower amounts of receptor DNA per well (0.5–1 μg) in a 6-well plate format can provide sufficient receptor expression to elicit measurable Gq signaling responses.

8. Alternative transfection methods can be used. For example, PEI can yield better transfection efficiency and is more affordable than Lipofectamine™ 2000 but can cause more cell death [12]. It is highly recommended that transfection efficiency is validated and optimized.
9. Prepare ligand/treatment to 2× the final concentration in the well, as this concentration factors in volume of ligands added and the total volume in the wells after addition.
10. Volume needed to resuspend cells should be optimized. The ideal resuspended cell density should obtain robust IP_1 accumulation levels during the IP-One assay. A lesser volume of replating media used to resuspend the cells would concentrate the cell density further, increasing IP_1 levels, and vice versa. We would recommend resuspending cells to a consistent cell density for replating across independent experiments, ideally with cell counting and further dilutions as needed prior to replating to maintain consistent results.
11. Optimization of the incubation period should be performed to determine the ideal time it takes to generate robust and detectable IP_1 levels. Too short of an incubation time may result in low IP_1 levels generated by the cells, while too long of an incubation time may lead to IP_1 degradation, causing variations in the detectable IP_1 levels. The manufacturer recommends anywhere between 30 min and 2 h.
12. The manufacturer recommends a specific concentration range and serial dilutions of IP_1 standards (a range between 11 μM and 2.7 μM).
13. The kit's manufacturer did not call for shaking in their protocol. However, we have found that shaking, specifically orbital shaking, helps with cell lysis, the release of IP_1 from the cells as well as the mixing of IP_1 and IP_1 standards with the d2-labeled IP_1 and cryptate-labeled anti-IP_1 antibody. This can result in less assay variability between independent repeats. The 1-h incubation time should be adhered to as a minimum to ensure sufficient lysing of cells and for the TR-FRET components to come to an equilibrium in the solution for competitive binding. Inversely, we have found that longer incubation time (1.5–8 h) did not significantly alter the assay's detection window nor the standard curve's IC_{50} value. In rare events where longer storage is needed prior to read, plates can be frozen at −20 °C and stored for up to 2 weeks, resulting in slightly reduced (due to freeze-thawing) but measurable responses.

Note that the IP_1 standards should also be prepared and stored in the same way to ensure similar comparisons between standards and samples.

14. A quick pulse spin is necessary to ensure all of the solution is at the bottom of the well in the 384-well plate after 1 h of shaking. If shaking was omitted from the previous step, this step may not be necessary. Additionally, the quick pulse spin can eradicate any air bubbles that may have formed, which will affect assay signal captured by the plate reader. Careful to not overspin as it can affect the TR-FRET.
15. Sample's 665/620 signal should ideally fall on the linear range of the standard curve (between IC_{10} and IC_{90}) as this is the range that the standard curve (and by extension the assay itself) can accurately determine the concentration of IP_1 produced (Fig. 3). Optimization of the protocol will be needed if most signals fall on the nonlinear or horizontal segments of the standard curve. This may require reducing the concentration of IP_1 produced (e.g., by reducing the amount of cells/cell density during replating), or vice versa, increasing the concentration of IP_1 produced (e.g., by increasing the amount of cells/cell density during replating) (*see* **Note 10**).

Acknowledgments

This work was performed in the laboratory of Dr. Caroline Gorvin, University of Birmingham. Assay establishment was supported by a Sir Henry Dale Fellowship jointly funded by the Wellcome Trust and the Royal Society (Grant Number 224155/Z/21/Z) and an Academy of Medical Sciences Springboard Award supported by the British Heart Foundation, Diabetes UK, the Global Challenges Research Fund, the Government Department of Business, Energy and Industrial Strategy, and the Wellcome Trust (Ref: SBF004|1034).

References

1. Wettschureck N, Offermanns S (2005) Mammalian G proteins and their cell type specific functions. Physiol Rev 85(4):1159–1204. https://doi.org/10.1152/physrev.00003.2005
2. Klepac K, Kilić A, Gnad T, Brown LM, Herrmann B, Wilderman A, Balkow A, Glöde A, Simon K, Lidell ME, Betz MJ, Enerbäck S, Wess J, Freichel M, Blüher M, König G, Kostenis E, Insel PA, Pfeifer A (2016) The Gq signalling pathway inhibits brown and beige adipose tissue. Nat Commun 7(1):10895. https://doi.org/10.1038/ncomms10895
3. Bazin H, Trinquet E, Mathis G (2002) Time resolved amplification of cryptate emission: a versatile technology to trace biomolecular interactions. J Biotechnol 82(3):233–250. https://doi.org/10.1016/s1389-0352(01)00040-x
4. Rhee SG (2001) Regulation of phosphoinositide-specific phospholipase C. Annu Rev Biochem 70(1):281–312.

https://doi.org/10.1146/annurev.biochem.70.1.281

5. Sun MY, Geyer M, Komarova YA (2017) IP (3) receptor signaling and endothelial barrier function. Cell Mol Life Sci 74(22): 4189–4207. https://doi.org/10.1007/s00018-017-2624-8
6. Ohanian J, Ohanian V (2001) Lipid second messenger regulation: the role of diacylglycerol kinases and their relevance to hypertension. J Hum Hypertens 15(2):93–98. https://doi.org/10.1038/sj.jhh.1001139
7. Mak DO, McBride SM, Petrenko NB, Foskett JK (2003) Novel regulation of calcium inhibition of the inositol 1,4,5-trisphosphate receptor calcium-release channel. J Gen Physiol 122(5):569–581. https://doi.org/10.1085/jgp.200308808
8. Dillon SB, Murray JJ, Uhing RJ, Snyderman R (1987) Regulation of inositol phospholipid and inositol phosphate metabolism in chemoattractant-activated human polymorphonuclear leukocytes. J Cell Biochem 35(4): 345–359. https://doi.org/10.1002/jcb.240350409
9. Liu K, Titus S, Southall N, Zhu P, Inglese J, Austin CP, Zheng W (2008) Comparison on functional assays for Gq-coupled GPCRs by measuring inositol monophospate-1 and intracellular calcium in 1536-well plate format. Curr Chem Genomics 1:70–78. https://doi.org/10.2174/1875397300801010070
10. Trinquet E, Bouhelal R, Dietz M (2011) Monitoring Gq-coupled receptor response through inositol phosphate quantification with the IP-one assay. Expert Opin Drug Discov 6(10):981–994. https://doi.org/10.1517/17460441.2011.608658
11. Garbison KE, Heinz BA, Lajiness ME, Weidner JR, Sittampalam GS (2004) Assay guidance manual. Eli Lilly & Company and the National Center for Advancing Translational Sciences
12. Jin L, Zeng X, Liu M, Deng Y, He N (2014) Current progress in gene delivery technology based on chemical methods and nano-carriers. Theranostics 4(3):240–255. https://doi.org/10.7150/thno.6914

Chapter 3

Measuring IP3 Generation in Real-Time Using a NanoBiT Luminescence Biosensor

Caroline M. Gorvin

Abstract

G protein-coupled receptors that activate Gq/11 regulate a range of physiological processes including neurotransmission, energy homeostasis, blood pressure regulation, and calcium homeostasis. Activation of Gq/11-coupled receptors stimulates the generation of inositol 1,4,5-trisphosphate (IP3), which mobilizes intracellular calcium release from the endoplasmic reticulum. This chapter describes an assay that uses a NanoBiT-IP3 luminescent biosensor to detect increases in IP3 in live cells. It describes how to perform these assays to assess signaling by the ghrelin receptor and the calcium-sensing receptor in HEK293 cells.

Key words Calcium-sensing receptor, Ghrelin receptor, Gq/11 signaling, Luminescence, Phospholipase C

1 Introduction

G protein-coupled receptors (GPCRs) comprise the largest family of receptors in humans with over 800 members and perform a wide variety of physiological functions in the sensory (e.g., taste and light perception), cardiovascular, endocrine, gastrointestinal, neuronal, and inflammatory systems [1]. GPCRs are the target of many FDA-approved drugs, and mutations in >50 GPCRs have been described to cause monogenic human diseases including bone and mineral disorders, obesity, thyroid dysfunction, reproductive disorders, and tumor development [2]. GPCRs perform their function by coupling to heterotrimeric guanine-nucleotide binding proteins (G proteins), comprising a membrane-anchored Gβγ dimer and a Gα subunit [1]. Upon ligand binding, GPCRs are activated and mediate conformational changes within Gα proteins that allow exchange of GDP for GTP, and dissociation of Gα-GTP from Gβγ. This allows both Gβγ and Gα-GTP subunits to activate diverse downstream signaling pathways. Gα subunits can be divided into four major classes based upon the effector that elicits these

Caroline M. Gorvin (ed.), *Calcium Signaling: Methods and Protocols*, Methods in Molecular Biology, vol. 2861,
https://doi.org/10.1007/978-1-0716-4164-4_3,

downstream signaling pathways. Both Gαs and Gαi/o act upon adenylate cyclase to increase and decrease cAMP, respectively. cAMP can activate a range of downstream proteins including the protein kinase A enzyme. The Gα12/13 subfamily activates Rho kinases that induce rearrangement of the cytoskeleton. The Gαq/11 subfamily activates phospholipase C (PLC), which hydrolyzes phosphatidylinositol 4,5-bisphosphate (PIP2) to generate the second messengers diacylglycerol (DAG) and inositol 1,4,5-trisphosphate (IP3). DAG activates protein kinase C (PKC) and the mitogen-activated protein kinase (MAPK) cascades, while IP3 binds to IP3 receptors on the endoplasmic reticulum to release stored intracellular calcium ($Ca^{2+}{}_i$) into the cytosol [3].

GPCRs that couple to Gq/11 to activate IP3-$Ca^{2+}{}_i$ signaling pathways include the angiotensin receptors that regulate blood pressure [4]; the calcium-sensing receptor (CaSR) that maintains serum calcium concentrations [5]; the ghrelin receptor (GHSR) which has important roles in energy homeostasis [6]; and muscarinic acetylcholine receptors [7] and metabotropic glutamate receptors [8] that regulate neurotransmitter responses. Gq/11 signaling can be assessed in a number of ways including assessment of G protein activation using resonance energy techniques such as BRET and FRET, measurement of $Ca^{2+}{}_i$ release using fluorescent calcium indicators by microscopy or plate-based assays, measurement of MAPK activation (e.g., by assessing protein phosphorylation by Western blot or AlphaScreen assays), or assessment of gene transcription by luciferase reporter assays. The measurement of IP3 proved difficult for many years as it is rapidly metabolized [9]. Traditionally, IP3 was assessed using radioactivity assays, which can be complex to perform, and is unfavorable due to additional safety precautions required when handling radioactive substances. Moreover, these assays are low throughput and are difficult to use to screen compounds. The development of the homogenous time-resolved fluorescence (HTRF) IP-one assay, which measures the IP1 metabolite of IP3, improved the ability of researchers to assess Gq/11 signaling in a high-throughput format [10]. The IP1 assay is nonradioactive and is particularly useful for measuring signaling in primary cells. However, the IP1 HTRF assay requires cell lysis and therefore cannot assess IP3 generation in real time [10]. This has led to the development of IP3 biosensors that measure IP3 generation in real time in living cells. One such IP3 biosensor uses the IP3-binding domain (IP3-binding core) of the human IP3 receptor-2 (IP3R2) fused between complementary fragments of the Nanoluciferase (NanoLuc®) [11]. This construct, which is hereafter referred to as the NanoBiT IP3 biosensor, was based on a previously described IP3-Rluc biosensor [12]. Upon binding of IP3, the IP3-binding core undergoes conformational changes which allow proximity between the LgBiT and SmBiT proteins and formation of the NanoLuc protein. The more IP3 that is

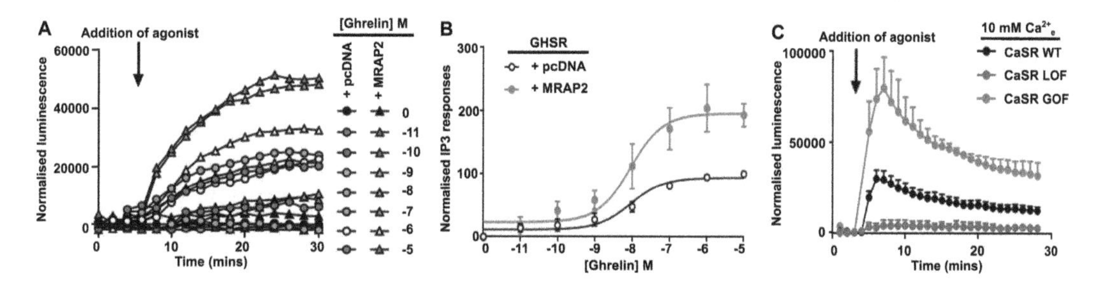

Fig. 1 Example of data from the NanoBiT IP3 biosensor assay. (**a**) Typical luminescence values for HEK293 cells transfected with GHSR and the NanoBiT IP3 biosensor and exposed to a range of ghrelin concentrations. Here the cells are co-transfected with either pcDNA control or the MRAP2 accessory protein, which is known to increase GHSR responses [14]. (**b**) Dose-response generated using the AUC values from $N = 4$ experiments. (**c**) NanoBiT IP3 responses generated by activation of the calcium-sensing receptor. The assay can accurately detect gain-of-function (GOF) and loss-of-function (LOF) mutations in the CaSR that cause human disease

generated by receptor activation, the more luciferase is produced, and this can be measured in real time using a plate reader that is capable of kinetic measurements (Fig. 1a).

The NanoBiT IP3 biosensor has been used to investigate signaling by the muscarinic acetylcholine receptor M3, the endothelin receptor type-A [4], and the angiotensin-II type-1 receptor (AT1R) [11] using coelenterazine as the substrate. These studies reported small fold-change responses (1–1.2-fold) [11]. We have subsequently adapted the assay to use furimazine as the substrate [13] and can produce more robust fold changes (Fig. 1b, c). Using these conditions we are able to measure signaling by the ghrelin receptor (Fig. 1a, b) and detect activating and inactivating variants in the calcium-sensing receptor (Fig. 1c). This chapter will describe the protocol for using the NanoBiT IP3 biosensor to detect activation of the GHSR and CaSR.

2 Materials

Prepare all solutions using ultrapure water (18 MΩ-cm resistance at 25 °C). Prepare and store all reagents at room temperature, unless otherwise stated.

2.1 Expression Plasmids

1. NanoBiT IP3 biosensor (LgBiT-IP3R2-SmBiT): pCAGGS backbone with LgBiT and SmBiT sequences fused to the N-terminus and C-terminus, respectively, of the IP3-binding core domain (IBC) of human type 2 IP3 receptor (Gene symbol ITPR2; residues 225–604), flanked by a 15-amino acid linker. DNA sequences and construction of the plasmid were previously described [11]. The plasmid is available by material transfer agreement with Asuka Inoue, Tohoku University, Japan.

2. HA-SNAP-GHSR: pRK5 backbone with a mGluR5 signal sequence (to enhance cell surface expression), then HA-SNAP N-terminal tags immediately after, and human GHSR without its initiation codon. This plasmid was made by my group and is available on request.
3. MRAP2-FLAGx3: Construction of the MRAP2 plasmid was described previously [14]. The construct is available on request from Julien Sebag, University of Iowa.
4. pcDNA3.1-FLAG-CaSR-WT: Full-length CASR with an N-terminal FLAG tag. Construction was previously described [13]. The plasmid is available on request.

2.2 Cell Culture

1. Adherent human embryonic kidney (AdHEK) 293 cells (*see* **Note 1**).
2. Complete growth medium: Dulbecco's modified Eagle medium (DMEM) with 4.5 g/L D-Glucose and L-glutamine, 10% (v/v) fetal bovine serum (FBS).
3. 0.05% Trypsin-EDTA solution.
4. Humidified 37 °C, 5% CO_2 incubator.
5. Cell culture hood.
6. Hemocytometer for cell counting.
7. Trypan blue: Make working solution of 0.4% trypan blue in PBS.
8. 75 cm^2 culture flask (*see* **Note 2**).
9. 96-well white assay plate: TC-treated, flat non-clear bottom with lid.
10. Phase-contrast microscope.

2.3 Transfection

1. Lipofectamine™ 2000 Transfection Reagent.
2. Serum-free DMEM media.
3. 0.5 mL microcentrifuge tubes.
4. Expression plasmids are added at the following concentrations: 200 ng LgBiT-IP3R2-SmBiT, 50 ng HA-SNAP-GHSR or pcDNA3.1-CaSR, 50 ng MRAP2-FLAGx3 or empty vector control.

2.4 IP3 Biosensor Assays

1. Ca^{2+} and Mg^{2+}-free Hank's buffered saline solution (HBSS).
2. Standard HBSS.
3. Phosphate-buffered saline (PBS): Dissolve one tablet in 500 mL water. Autoclave and store at room temperature.
4. IP3 assay substrate (obtained from Promega as Nano-Glo® reagent): Dilute Nano-Glo luciferase assay substrate (furimazine) in the manufacturer's supplied buffer or HBSS (either

Ca^{2+}- and Mg^{2+}-free HBSS for CaSR assays or standard HBSS for GHSR assays) at a 1:100 dilution (*see* **Note 3**).

5. Ligand (ghrelin): reconstitute to 1 mM stock solution in DMSO and store at −20 °C in single-use aliquots (*see* **Note 4**). For assays we use a range of concentrations from −12 to −5 M to obtain a full concentration-response curve. Working solutions to add to 96-well plates are made in HBSS at a 10x concentration, such that the final concentration is −12 to −5 M.
6. Ligand (calcium): Working solution of $CaCl_2$. Dilute 1 M $CaCl_2$ in water to make a working solution of 100 mM $CaCl_2$ as higher concentrations can precipitate out of solution. Adjust the solution to maintain osmotic balance. For assays we use a range of concentrations (usually 0, 1, 2, 3, 5, 7, 10, 15 mM) to obtain a full concentration-response curve. Working solutions to add to 96-well plates are made in Ca^{2+}- and Mg^{2+}-free HBSS at a 10× concentration, such that the final concentration is 0–15 mM.

2.5 Equipment

1. Multichannel pipette for eight tips (range 50–300 μL) for plating cells.
2. Multichannel pipette for eight tips (range 1–10 μL) for adding agonist.
3. Plate reader capable of performing kinetic measurements of luminescence and with a chamber able to maintain a 37 °C temperature. We predominantly use a Promega Glomax Discover plate reader but have also used a BMG Labtech PHERAstar. If possible, generate a kinetic assay template that reads luminescence (with integration time of 0.5 s) for 4 cycles each comprising 2 min during which the whole plate will be read; a pause during which the plate can be ejected, substrate added, then the protocol resumed; ≥12 cycles each comprising 2 min during which the whole plate will be read (*see* **Note 5**).

3 Methods

All cell culture work (seeding and transfection) should be performed aseptically in a cell culture hood.

3.1 Seeding Cells in 96-Well White Assay Plates

1. Remove media from the cell culture flask.
2. Add 2–3 mL Trypsin to the flask and incubate at room temperature for ~2 min. Check cells are rounding or lifting under a light microscope (*see* **Note 6**).
3. Neutralize the trypsin by adding ~5 mL media. Mix by pipetting up and down with a 10 mL stripette.

4. Count cells on a hemocytometer. Remove 100 μL of cell suspension and add to a 0.5 mL microcentrifuge tube. Add 100 μL 0.4% trypan blue. Add 100 μL of cell/trypan blue mixture to the hemocytometer and place a coverslip on top. Count the number of live and dead (i.e., unstained vs. trypan blue-stained cells) using a phase-contrast microscope at 20× magnification.
5. Seed cells at 10,000 cells per well in a total volume of 100 μL pre-warmed media with FBS using a multichannel pipette. Incubate at 37 °C, 5% CO_2 for 7–8 h (*see* **Note 7**).

3.2 Transfection of Cells

We use one column (comprising 8 wells) of a 96-well plate for each experimental condition (Fig. 2). We therefore make one stock of transfection mix for the eight wells of the column. This protocol describes a transfection setup for a column; amounts can be adjusted to suit your experiment:

1. Add 100 μL serum-free media (*see* **Note 8**) to each of two 0.5 mL microcentrifuge tubes.

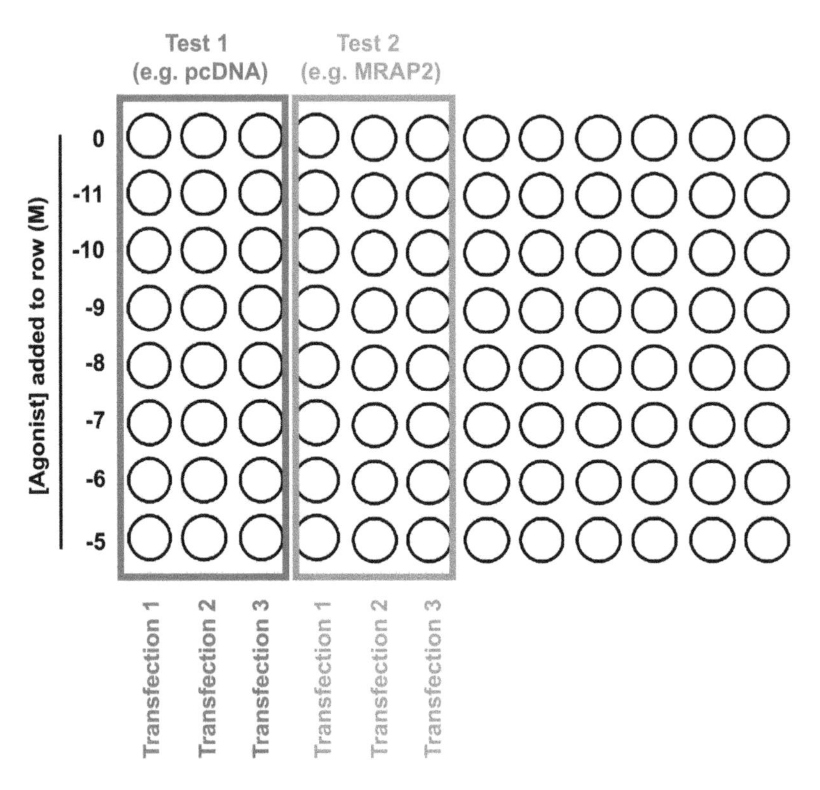

Fig. 2 Example plate setup for the IP3 biosensor assay. Cartoon showing a 96-well plate with transfections indicated for each condition. Typically, $N = 3$ transfections are performed and averaged to obtain one biological replicate. Transfection mixtures are made for one column, and agonist concentrations are aliquoted into an eight-well PCR tube to simplify the addition of agonist across the plate using a multichannel pipette

2. Add 2 μL Lipofectamine 2000 to one tube and the appropriate amounts of each plasmid (see Subeading 2.3) to the other tube. Incubate for 5 min.
3. Pipette the contents of tube 1 to tube 2 and mix gently by pipetting. Incubate for up to 20 minutes (*see* **Note 9**).
4. Pipette 24 μL of the contents to each well of a single column on the 96-well plate. Incubate at 37 °C, 5% CO_2 for 48 h.

3.3 IP3 Biosensor Assay

1. Remove all media from wells and wash once in PBS.
2. Add 50 μl Ca^{2+}- and Mg^{2+}-free HBSS for CaSR assays or standard HBSS for GHSR assays (*see* **Note 10**). Incubate cells for at least 1 h at 37 °C, 5% CO_2.
3. Add 40 μL of IP3 assay substrate to each well of the 96-well plate, and incubate for 10 min to allow stabilization of the luminescent signal.
4. Read baseline luminescence for four cycles (8 min total) or until the baseline has stabilized (*see* **Note 11**).
5. Add agonist at a 10× concentration (*see* **Note 12**) and then record for a further 25 min (*see* **Note 5**).

3.4 Analysis of Data

1. Export the data in an Excel format.
2. Subtract the vehicle response from all experimental responses.
3. For each response, calculate the mean of the baseline values and subtract this from the agonist-mediated signal (Fig. 1).
4. Calculate the AUC. We use GraphPad Prism to perform this.
5. Normalize AUC values to the control (e.g., wild-type CaSR in studies of mutant CaSR), with the top value set as 100% and the bottom as 0%.
6. Use these values to plot a concentration-response curve with a four-parameter sigmoidal fit (Fig. 1).

4 Notes

1. We used AdHEK293 cells in our studies as they are adherent, do not endogenously express CaSR or GHSR, and are routinely used for GPCR signaling studies [15–19].
2. We routinely culture cells in 75cm^3 culture flasks, but any size vessel can be used. AdHEK293 cells usually require passaging 2–3 times a week. They should not be left to grow >80% confluent and should be transfected at 50–70% confluency.
3. The original description of the IP3 biosensor described using coelenterazine as the substrate for these assays [11]. However, we were unable to obtain reliable data using coelenterazine and

instead use furimazine, which is optimized for use with NanoBiT.

4. MK0677 can be used as a ghrelin receptor agonist instead of ghrelin. Both ghrelin and MK0677 are reconstituted in DMSO.
5. A total number of cycles after agonist are added can be adjusted depending upon your experiment. A total of 30 min is usually more than enough to observe saturation as IP3 responses for GPCRs are usually rapid.
6. Cells should lift at room temperature, although if the cells are very adherent, incubation at 37 °C may be required.
7. Cells can be plated at a lower density and incubated for more than 1 day; however, cells should be between 50% and 70% confluent when transfected. The time between plating and transfecting cells can be adjusted. We have performed transfections after 1 day of incubation with similar results; however, we tend to avoid using the outer wells of the plate under these circumstances to avoid plate edge effects.
8. Manufacturer's instructions for transfection with Lipofectamine 2000 recommend using Opti-MEM rather than serum-free media. When we directly compared Opti-MEM with serum-free DMEM, we observed no noticeable difference in GFP-positive cells by microscopy or luminescence values on plate readers. We also observed no differences in transfection efficiency when mixtures were added directly to cells in serum-free media or in serum-containing media.
9. This time can be adjusted to as little as 5-min incubation with no noticeable difference in transfection efficiency.
10. Changing media to Ca^{2+}- and Mg^{2+}-free HBSS is vital for CaSR assays to ensure low levels of basal calcium.
11. Luminescence values should be over 1000; otherwise, the data is unreliable and difficult to distinguish from background values (which typically fall within the 50–500 range on our plate reader).
12. We aliquot our solutions into eight-well PCR tubes to simplify adding ligands across the plate using a multichannel pipette.

Acknowledgments

This work was supported by an Academy of Medical Sciences Springboard Award supported by the British Heart Foundation, Diabetes UK, the Global Challenges Research Fund, the Government Department of Business, Energy and Industrial Strategy, and the Wellcome Trust. Ref: SBF004|1034 (CMG) and a Sir Henry

Dale Fellowship jointly funded by the Wellcome Trust and the Royal Society (Grant Number 224155/Z/21/Z).

References

1. Weis WI, Kobilka BK (2018) The molecular basis of G protein-coupled receptor activation. Annu Rev Biochem 87:897–919. https://doi.org/10.1146/annurev-biochem-060614-033910
2. Thompson MD, Percy ME, Cole DEC, Bichet DG, Hauser AS, Gorvin CM (2024) G protein-coupled receptor (GPCR) gene variants and human genetic disease. Crit Rev Clin Lab Sci:1–30. https://doi.org/10.1080/10408363.2023.2286606
3. Gorvin CM (2022) GPCR's and endocrinology. In: Terry Kenakin GB, Michel MC (eds) Comprehensive pharmacology, vol 4, 1st edn. Elsevier, pp 3–52
4. Kawakami K, Yanagawa M, Hiratsuka S, Yoshida M, Ono Y, Hiroshima M, Ueda M, Aoki J, Sako Y, Inoue A (2022) Heterotrimeric Gq proteins act as a switch for GRK5/6 selectivity underlying beta-arrestin transducer bias. Nat Commun 13(1):487. https://doi.org/10.1038/s41467-022-28056-7
5. Chang W, Pratt S, Chen TH, Nemeth E, Huang Z, Shoback D (1998) Coupling of calcium receptors to inositol phosphate and cyclic AMP generation in mammalian cells and Xenopus laevis oocytes and immunodetection of receptor protein by region-specific antipeptide antisera. J Bone Miner Res 13(4):570–580. https://doi.org/10.1359/jbmr.1998.13.4.570
6. Wang Y, Guo S, Zhuang Y, Yun Y, Xu P, He X, Guo J, Yin W, Xu HE, Xie X, Jiang Y (2021) Molecular recognition of an acyl-peptide hormone and activation of ghrelin receptor. Nat Commun 12(1):5064. https://doi.org/10.1038/s41467-021-25364-2
7. Dell'Acqua ML, Carroll RC, Peralta EG (1993) Transfected m2 muscarinic acetylcholine receptors couple to G alpha i2 and G alpha i3 in Chinese hamster ovary cells. Activation and desensitization of the phospholipase C signaling pathway. J Biol Chem 268(8): 5676–5685
8. Niswender CM, Conn PJ (2010) Metabotropic glutamate receptors: physiology, pharmacology, and disease. Annu Rev Pharmacol Toxicol 50:295–322. https://doi.org/10.1146/annurev.pharmtox.011008.145533
9. Garbison KE, Heinz BA, Lajiness ME (2004) IP-3/IP-1 assays. In: Markossian S, Grossman A, Arkin M et al (eds) Assay guidance manual. Bethesda, Rockville
10. Pottie E, Stove CP (2022) In vitro assays for the functional characterization of (psychedelic) substances at the serotonin receptor 5-HT(2A) R. J Neurochem 162(1):39–59. https://doi.org/10.1111/jnc.15570
11. Inoue A, Raimondi F, Kadji FMN, Singh G, Kishi T, Uwamizu A, Ono Y, Shinjo Y, Ishida S, Arang N, Kawakami K, Gutkind JS, Aoki J, Russell RB (2019) Illuminating G-protein-coupling selectivity of GPCRs. Cell 177(7): 1933–1947. e1925. https://doi.org/10.1016/j.cell.2019.04.044
12. Ataei F, Torkzadeh-Mahani M, Hosseinkhani S (2013) A novel luminescent biosensor for rapid monitoring of IP3 by split-luciferase complementary assay. Biosens Bioelectron 41:642–648. https://doi.org/10.1016/j.bios.2012.09.037
13. Abid HA, Inoue A, Gorvin CM (2021) Heterogeneity of G protein activation by the calcium-sensing receptor. J Mol Endocrinol 67(2):41–53. https://doi.org/10.1530/JME-21-0058
14. Rouault AAJ, Rosselli-Murai LK, Hernandez CC, Gimenez LE, Tall GG, Sebag JA (2020) The GPCR accessory protein MRAP2 regulates both biased signaling and constitutive activity of the ghrelin receptor GHSR1a. Sci Signal 13(613). https://doi.org/10.1126/scisignal.aax4569
15. Centeno PP, Herberger A, Mun HC, Tu C, Nemeth EF, Chang W, Conigrave AD, Ward DT (2019) Phosphate acts directly on the calcium-sensing receptor to stimulate parathyroid hormone secretion. Nat Commun 10(1): 4693. https://doi.org/10.1038/s41467-019-12399-9
16. Leach K, Gregory KJ, Kufareva I, Khajehali E, Cook AE, Abagyan R, Conigrave AD, Sexton PM, Christopoulos A (2016) Towards a structural understanding of allosteric drugs at the human calcium-sensing receptor. Cell Res 26(5):574–592. https://doi.org/10.1038/cr.2016.36
17. Nesbit MA, Hannan FM, Howles SA, Babinsky VN, Head RA, Cranston T, Rust N, Hobbs MR, Heath H 3rd, Thakker RV (2013) Mutations affecting G-protein subunit alpha11 in hypercalcemia and hypocalcemia. N Engl J

Med 368(26):2476–2486. https://doi.org/10.1056/NEJMoa1300253

18. Ramirez VT, van Oeffelen W, Torres-Fuentes-C, Chruscicka B, Druelle C, Golubeva AV, van de Wouw M, Dinan TG, Cryan JF, Schellekens H (2019) Differential functional selectivity and downstream signaling bias of ghrelin receptor antagonists and inverse agonists. FASEB J 33(1):518–531. https://doi.org/10.1096/fj.201800655R

19. Holst B, Cygankiewicz A, Jensen TH, Ankersen M, Schwartz TW (2003) High constitutive signaling of the ghrelin receptor – identification of a potent inverse agonist. Mol Endocrinol 17(11):2201–2210. https://doi.org/10.1210/me.2003-0069

Chapter 4

Monitoring Calcium-Sensing Receptor (CaSR)-Induced Intracellular Calcium Flux Using an Indo-1 Flow Cytometry Assay

Caroline M. Gorvin and Sarah A. Howles

Abstract

The calcium-sensing receptor (CaSR) has a critical role in maintaining serum calcium concentrations within the normal physiological range, and mutations in the receptor, or components of its signaling and trafficking pathway, cause disorders of calcium homeostasis. Inactivating mutations cause neonatal severe hyperparathyroidism or familial hypocalciuric hypercalcemia (FHH), while gain-of-function mutations cause autosomal dominant hypocalcemia (ADH). Characterizing the functional impact of mutations of the CaSR, and components of the CaSR-signaling pathway, is clinically important to enable correct diagnoses of FHH and ADH, optimize management, and prevent inappropriate parathyroidectomy or vitamin D supplementation. CaSR signals predominantly by activating the G-alpha subunit-11 to mobilize calcium release from intracellular stores. Thus, measurement of CaSR-induced intracellular calcium ($Ca^{2+}{}_{i}$) signaling is the gold standard method to investigate the pathogenicity of CaSR genetic variants. This protocol describes a method to assess CaSR-induced $Ca^{2+}{}_{I}$ signaling using the Indo-1 calcium indicator dye and flow cytometry. This method has been used to assess multiple genetic variants in CaSR and components of its signaling and trafficking pathway in HEK293 cells.

Key words Adaptor protein-2 sigma subunit, Calcium homeostasis, Calcium flux, G-alpha protein-11, Hyper−/hypocalcemia, Parathyroid hormone

1 Introduction

Calcium is the most abundant mineral in the human body, and tight control of intra- and extracellular calcium concentrations is crucial for normal physiology [1]. Extracellular free calcium concentrations are typically within the range of 1.1–1.3 mM, and intracellular concentrations are ~100 nM. The calcium-sensing receptor (CaSR) is the master regulator of calcium homeostasis, detecting small changes in extracellular calcium concentrations and initiating intracellular signaling pathways and hormone release to alter calcium

Caroline M. Gorvin (ed.), *Calcium Signaling: Methods and Protocols*, Methods in Molecular Biology, vol. 2861,
https://doi.org/10.1007/978-1-0716-4164-4_4,

uptake, secretion, or excretion and maintain calcium concentrations within the normal physiological range [2].

The CaSR is a G-protein-coupled receptor (GPCR) that is primarily expressed on the cell surface of the chief cells of the parathyroid gland, apical and basolateral surfaces of renal tubular segments, and bone marrow stem cells, chondrocytes, and osteoblasts [3, 4]. Mutations of the CaSR result in disorders of extracellular calcium homeostasis. Thus, heterozygous CaSR loss-of-function mutations cause familial hypocalciuric hypercalcemia (FHH) [5], and biallelic loss-of-function mutations cause neonatal severe hyperparathyroidism (NSHPT) [5, 6]; gain-of-function mutations cause autosomal dominant hypocalcemia (ADH) [7]. Additionally, mutations in components of the CaSR-signaling and trafficking pathways also cause FHH and ADH. FHH and ADH type 2 are caused by inactivating and activating mutations, respectively, in G-protein alpha 11 (Gα11), by which the receptor signals and FHH type 3 is due to loss-of-function mutations of the adaptor protein 2 sigma subunit 1 (AP2σ), by which CaSR is internalized [8, 9]. FHH is characterized by a lifelong elevation in serum calcium concentration, reduced urinary calcium excretion, and inappropriately normal circulating PTH concentrations. NSHPT commonly presents in the neonatal period with extreme hypercalcemia, failure to thrive, hypotonia, anorexia, constipation, and, in some cases, respiratory distress due to hyperparathyroid-bone disease-induced rib fractures [6, 10]. ADH has the opposite phenotype to FHH, with mild to moderate hypocalcemia and PTH levels inappropriately in the low-normal range; 34% of ADH patients have hypercalciuria; the remaining patients have urinary calcium excretions that are inappropriately normal [7, 11, 12].

Characterizing the functional impact of mutations of the CaSR, and components of the CaSR-signaling pathway, is clinically important to enable correct diagnoses of FHH and ADH, optimize management, and prevent inappropriate parathyroidectomy or vitamin D supplementation [13, 14]. Furthermore, classification of the functional impact of variants can prevent misdiagnoses being made based on nonfunctional polymorphisms [15]. In addition, in vitro assays allow the potential utility of pharmacological agents to be assessed [16, 17].

Flow cytometry, using the calcium binding dye Indo-1 acetoxymethylester (Indo-1-AM), provides a means to assess the impact of genetic variants on CaSR-signaling responses. During flow cytometry, cells labeled with a fluorescent dye are placed in suspension and passed one cell at a time in front of a light source, or light sources, with wavelengths selected to stimulate fluorescence. Simultaneously, each cell is analyzed for visible light scatter and one or more fluorescence parameters. Indo-1-AM is a fluorescent dye used in calcium flux assays that is excited by ultraviolet light, and acts as a ratiometric calcium indicator, having an emission

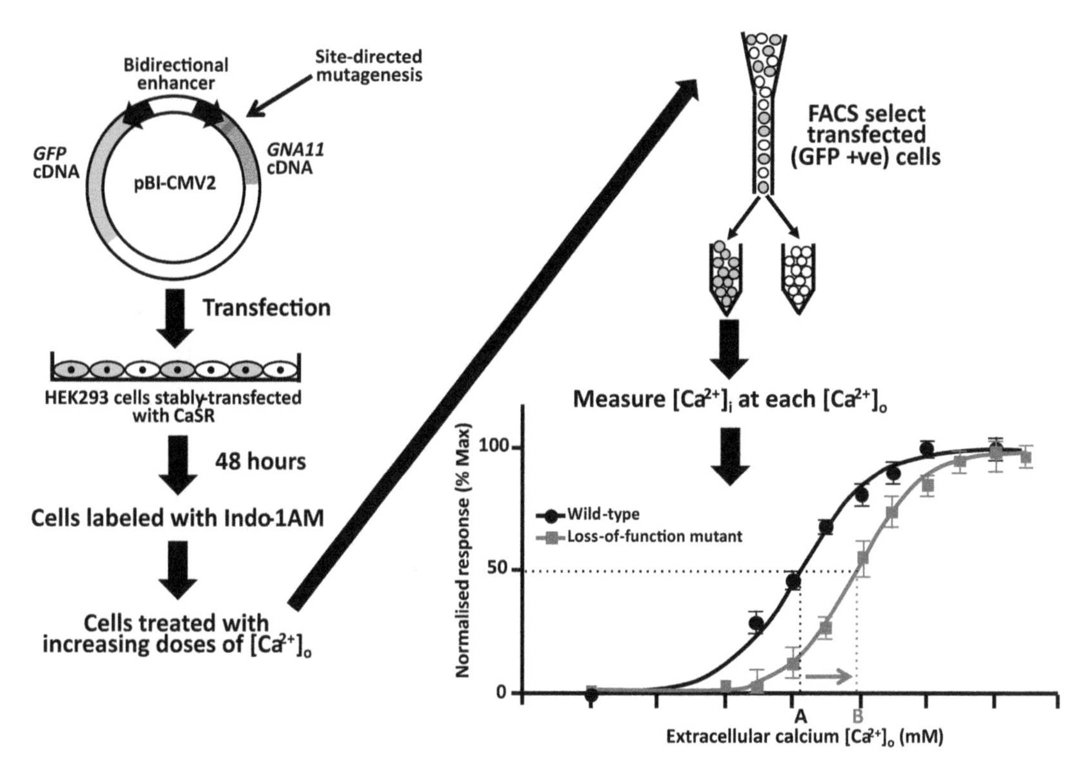

Fig. 1 Overview of the Indo-1 AM calcium flux flow cytometry assay. The gene of interest, *GNA11*, is cloned into the pBI-CMV2 expression plasmid which expresses GFP and *GNA11* on separate enhancers. Site-directed mutagenesis can be used to introduce human genetic variants identified in FHH or ADH patients into pBI-CMV2. HEK293 cells stably expressing CaSR are transiently transfected with the pBI-CMV2 construct. Two days later, cells are labeled with Indo-1 AM and taken to the flow cytometer. Cell populations are gated to select GFP-positive cells, and then exposed to increasing doses of extracellular Ca^{2+} and intracellular Ca^{2+} responses recorded. Normalized responses are plotted and EC_{50} values derived to compare wild-type and mutant responses

maximum of 475 nm when calcium-free and an emission maximum of 400 nm when calcium bound. This dye is typically used in combination with a transfected fluorescent marker to allow identification of cells transfected with a gene of interest (Fig. 1). Indo-1 AM is well suited for use in flow cytometry assays, including multicolor fluorescence applications, as a single laser, such as the 351–364 nm spectral lines of an argon-ion laser, can be used for excitation. This assay system is more time efficient than measurement of single-cell outputs, and assessment of bulk signaling responses is typically reproducible over many independent experiments.

In this article, we present a protocol for using the Indo-1-AM flow cytometry assay to assess the effects of mutations in Gα11 on CaSR signaling, and to determine whether a CaSR-positive allosteric modulator, cinacalcet, modulates impaired CaSR signal transduction.

2 Materials

2.1 Expression Plasmids (See Note 1)

1. pBI-CMV2-Gα11: full-length human *GNA11* was cloned in-frame into a pBI-CMV2 mammalian expression vector (purchased from Clontech), designed to simultaneously express Gα11 and green fluorescent protein (GFP); expression of these two proteins was driven by independent promoters.

2.2 Cell Culture

1. Human embryonic kidney (HEK) 293 cells stably expressing CaSR (hereafter designated HEK-CaSR) (*see* **Note 2**).
2. Complete growth medium: DMEM-Glutamax, 10% (v/v) fetal bovine serum (FBS), 400 μg/mL geneticin (also called G418 by some companies) (*see* **Note 3**).
3. 0.05% Trypsin-EDTA solution.
4. Phosphate-buffered saline (PBS): Dissolve one tablet in 500 mL water. Autoclave and store at room temperature.
5. Cell culture hood.
6. Humidified 37 °C, 5% CO_2 incubator.
7. 75cm^2 culture flasks.
8. Phase-contrast microscope.
9. Hemocytometer for cell counting.
10. Trypan blue.

2.3 Transient Transfection

1. Lipofectamine™ 2000 Transfection Reagent.
2. Serum-free DMEM media.
3. 1.5 mL microcentrifuge tubes.

2.4 Flow Cytometry

1. Indo-1-acetoxymethylester (Indo-1-AM): Add 50 μL DMSO to 50 μg Indo-1-AM. Protect solution from the light and vortex briefly. Indo-1 should be stored in the fridge and needs to be used within 1–2 weeks (*see* **Note 4**).
2. Cinacalcet (CaSR-positive allosteric modulator): Dissolve (2-hydroxypropyl)-β-cyclodextrin in ultrapure water to make a 20% carrier. Add cinacalcet to the carrier solution to make a 2 mM stock solution.
3. Ca^{2+}- and Mg^{2+}-free Hanks' buffered saline solution (HBSS).
4. Cell dissociation buffer (Hanks' based).
5. 15 mL centrifuge tubes.
6. 5 mL (12 × 75 mm) round bottom polypropylene tubes with snap cap.
7. Agonist solutions: Dilute 1 M $CaCl_2$ 1:10 in HBSS to make a 100 mM working solution (*see* **Note 5**).

8. Flow cytometer fitted with a 350 nm UV laser to measure Indo-1 AM (*see* **Note 6**).
9. Optional controls:

   Positive control ionomycin: Dissolve 1 mg/mL in DMSO to make a stock solution, and then add to samples at 6 μM (*see* **Note 7**).

   Negative control EGTA: Make stock of 5 mM EGTA in water and use at 1–2 mM. EGTA should be added to cells 1 h before analysis.

2.5 Software Required for Data Analysis

1. Cytomation SUMMIT (*see* **Note 8**).
2. Graphpad Prism (*see* **Note 9**).

3 Methods

All cell culture work (seeding and transfection) should be performed aseptically in a cell culture hood. This protocol describes the Indo-1 calcium flux assay in HEK-CaSR cells to examine the effect of a genetic variant in *GNA11*, the gene encoding Gα11, and to determine whether cinacalcet, a CaSR-positive allosteric modulator, affects $Ca^{2+}{}_e$-induced CaSR responses. Therefore, six flasks of HEK-CaSR will be required for the following tests:

1. pBI-CMV2 empty vector (vehicle),
2. pBI-CMV2-Gα11-WT (vehicle),
3. pBI-CMV2- Gα11-mutant (vehicle),
4. pBI-CMV2 empty vector (10 nM cinacalcet),
5. pBI-CMV2-Gα11-WT (10 nM cinacalcet),
6. pBI-CMV2- Gα11-mutant (10 nM cinacalcet).

3.1 Day 1: Cell Seeding

1. Remove media from the cell culture flasks.
2. Add 2–3 mL Trypsin to the flask and incubate at room temperature for ~2 min. Check if cells are rounding or lifting under a light microscope (*see* **Note 10**).
3. Neutralize the trypsin by adding ~5 mL media. Mix by pipetting up and down with a 10 mL stripette.
4. Count cells on a hemocytometer. Count the number of cells using a phase-contrast microscope at 20× magnification. Seed cells at 2,100,000 cells per flask in pre-warmed media with FBS. Incubate at 37 °C, 5% CO_2 overnight.

3.2 Day 2: Transfection

1. Add 750 μL serum-free media to each of two 1.5 mL microcentrifuge tubes.
2. Add 80 μL Lipofectamine 2000 to one tube and 16 μg pBI-CMV2-GNA11 construct to the other tube. Incubate for 5 min (*see* **Note 11**).
3. Pipette the contents of tube 1 to tube 2 and mix gently by pipetting. Incubate for up to 20 min (*see* **Note 12**).
4. Pipette the contents of the tube dropwise to cells on the coverslip. Incubate at 37 °C, 5% CO_2 for 24 h.

3.3 Day 3: Cell Expansion (Optional)

1. Split each T80 into a T175. Incubate at 37 °C, 5% CO_2 for 24 h (*see* **Note 13**).

3.4 Day 4: Flow Cytometry

3.4.1 Preparation of Cells for Flow Cytometry

1. Remove media from cells and add 10 mL Ca^{2+} and Mg^{2+}-free HBSS.
2. Remove HBSS and use 10 mL cell dissociation buffer (Hanks based) to dissociate cells. Incubate at 37 °C, 5% CO_2 for approximately 10 min until cells begin to lift (*see* **Note 14**).
3. Collect cells into a 15 mL centrifuge tube and spin at 400 *g* for 5 min.
4. Pour off supernatant carefully and add 10 mL Ca^{2+}- and Mg^{2+}-free HBSS. Pipette up and down 2× to wash the cells.
5. Centrifuge at 400 *g* for 5 min.
6. Pour off supernatant carefully and add 5 mL Ca^{2+}-and Mg^{2+}-free HBSS. Pipette up and down 2× to wash the cells.
7. Centrifuge at 400 *g* for 5 min.
8. Carefully remove the supernatant. Add 5 mL Ca^{2+}- and Mg^{2+}-free HBSS.
9. Add 1 μL Indo-1 AM per 1 mL of HBSS. Cover in foil and incubate for 1 h at 37 °C (*see* **Notes 15** and **16**).
10. Pellet cells at 400 *g* for 5 min.
11. Pour off supernatant carefully and add 10 mL Ca^{2+}- and Mg^{2+}-free HBSS. Pipette up and down 2× to wash the cells.
12. Centrifuge at 400 *g* for 5 min.
13. Remove supernatant and resuspend in 5 mL Ca^{2+}- and Mg^{2+}-free HBSS.
14. If using cinacalcet (or other allosteric modulator), add to the HBSS, and incubate for at least 1 h before running on the flow cytometer (*see* **Note 17**).
15. Cells can then be taken to the flow cytometer (*see* **Note 18**).

3.4.2 Calcium Flux Assay on the Flow Cytometer

1. Aliquot 1 mL of test sample to a new flow cytometry tube.
2. Set the flow cytometer to a low flow rate. Visualize forward scatter height and side scatter height of the whole cell population on a linear scale (Fig. 2a). Gate the population of interest (live cells with GFP fluorescence) and gate out debris and dead cells (*see* **Note 19**).
3. Plot the cells within this first gate to show the pulse width vs. forward scatter and form a new gate to include only single cell events (Fig. 2b).
4. Visualize single parameter histograms showing counts vs. GFP fluorescence (Fig. 2c). Form a final gate to show only the GFP-positive cells.
5. Finally plot the fluorescence ratio of Indo-1 AM vs. time (Fig. 2d).
6. Set to record 1,000,000 events. Record basal responses, then add 10 μL of 100 mM Ca^{2+} working solution (i.e., 1 mM response), and record for up to 60 s.
7. Remove flow cytometry tube and add 5 μL of 100 mM Ca^{2+} working solution. Record for up to 60 s.
8. Repeat **step 7** adding the following volumes of Ca^{2+} working solution (5, 10, 50 μL) to achieve the full dose–response curve.
9. Measure the volume of solution remaining in the sample and record the figure.
10. Repeat **steps 6–9** for all samples (*see* **Note 20**).

3.5 Analysis

1. Open the sample in Cytomation SUMMIT software, and measure the total time recorded for each calcium response (Fig. 3).

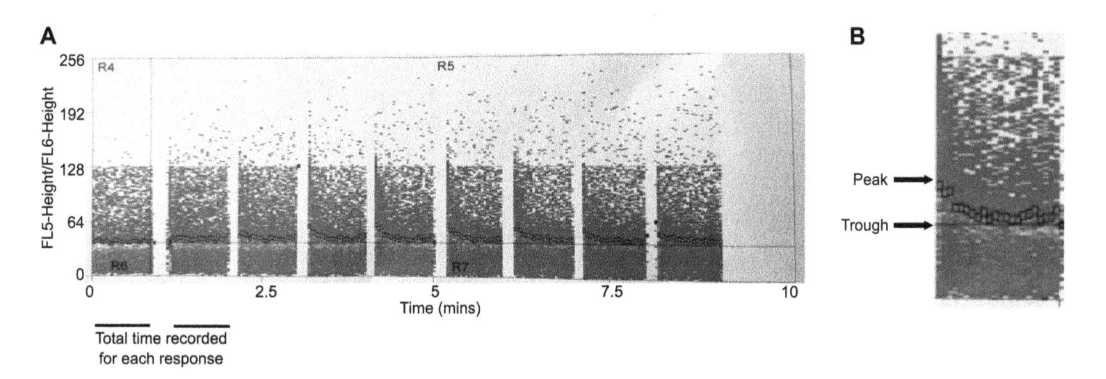

Fig. 3 Analysis of Indo-1 AM data. (**a**) Graph showing $Ca^{2+}{}_e$-induced $Ca^{2+}{}_i$ responses in cells expressing CaSR and Gα11. For analyses, the data can be opened in Cytomation SUMMIT or similar flow cytometry software. In step 1, the total time recorded for each response is measured (the first two responses are shown by the solid black line). (**b**) In step 9, the peak and trough of the transient calcium response are measured for each calcium addition and then the cumulative response plotted in Graphpad

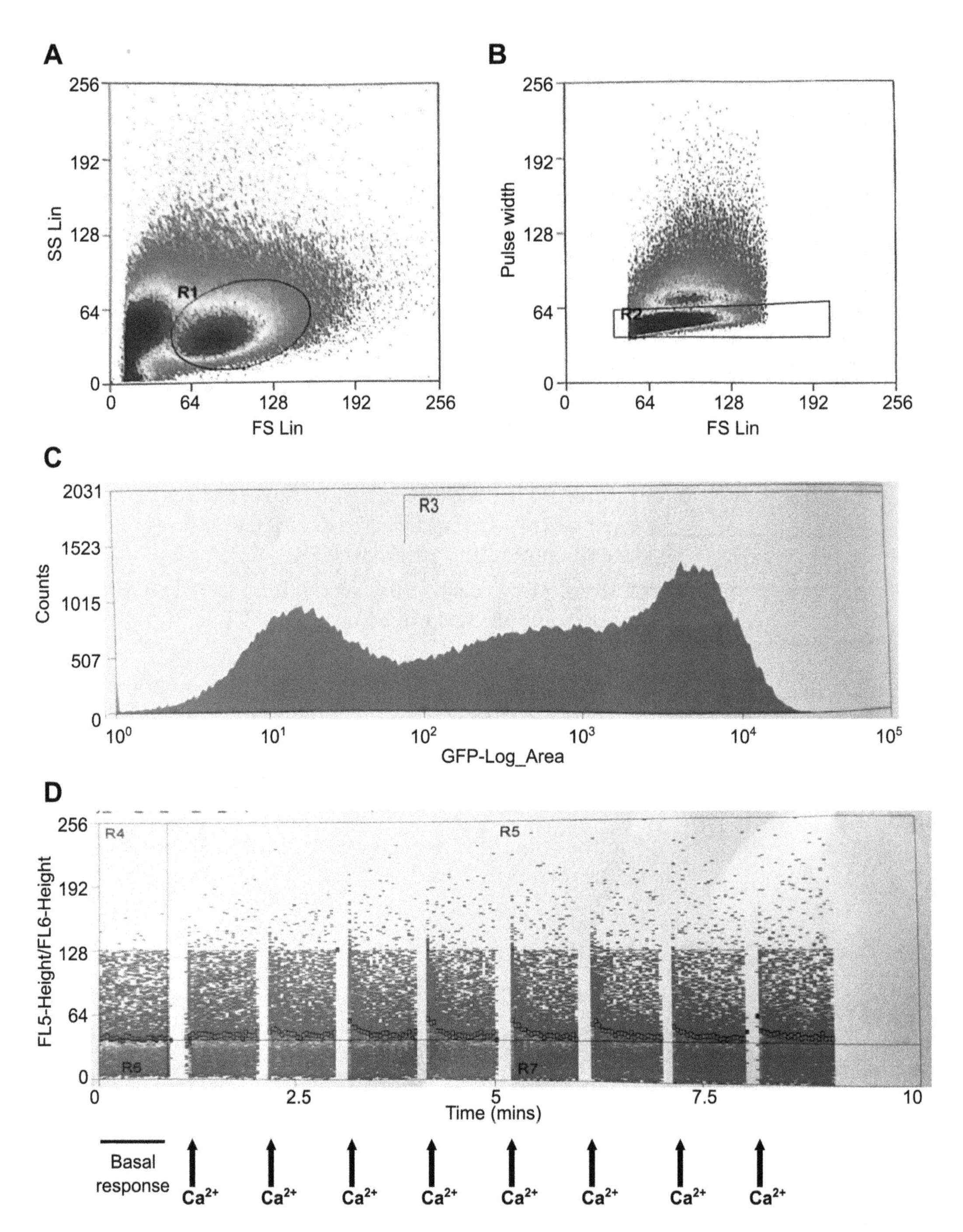

Fig. 2 Examples of gating and data produced from calcium flux flow cytometry assay. (**a**) Visualization of side scatter (SS) vs. forward scatter (FS) for the whole cell population. Live cells (R1 gate) were selected from dead cells. (**b**) Cells from R1 were further gated to select only single cells (R2 gate). (**c**) Single parameter histograms showing counts vs. GFP fluorescence. GFP-positive cells were selected in the R3 gate. (**d**) Visualization of the fluorescence ratio of Indo-1 AM vs. time. Arrows below indicate where Ca^{2+} was added. Calcium transient responses can be seen in blue

2. Subtract the volume of solution remaining in the flow cytometer tube from the volume total (1100 μL, i.e., 1 mL sample + 100 μL calcium).
3. Divide the volume of solution by the total time recorded to calculate the flow rate in μL/s.
4. Multiply the cumulative time by the flow rate to calculate the volume removed at each stage.
5. Calculate the volume remaining by subtracting the volume removed from the final volume.
6. Add the volume of calcium added to volume remaining to calculate the final volume.
7. Divide calcium added by the final volume to calculate the actual concentration of calcium added.
8. Calculate the cumulative corrected calcium.
9. In Cytomation SUMMIT, measure the peak and trough of the mean fluorescence ratio of the transient calcium response (Fig. 3). Subtract the trough from the peak.
10. Calculate the cumulative mean of transient responses (as each calcium concentration was added to the same sample).
11. Express this number as the percentage response of the top concentration.
12. In Graphpad Prism plot the corrected calcium from **step 8** on the x-axis and the responses from **step 11** on the y-axis.
13. Derive responses at the calcium concentrations desired and normalize the responses such that the basal response is 0 and the top response 100% (see graph in Fig. 1).
14. In Graphpad Prism, calculate the nonlinear regression of the concentration response curve to derive the EC_{50}.

4 Notes

1. We have used a range of plasmids to investigate human and mouse genetic variants in *CASR*, *GNA11*, and *AP2S1* [8, 9, 15–24]. To investigate CaSR variants, a pEGFP-N1 construct was transiently transfected into HEK293 cells. To investigate *GNA11* and *AP2S1* variants, HEK-CaSR cells were transiently transfected with pBI-CMV2 (GFP) or pBI-CMV4 (RFP) expression plasmids expressing Gα11 or AP2σ.
2. We chose to use cells stably expressing CaSR to ensure that each batch of cells expressed the same amount of CaSR protein and that only the test plasmid (e.g., GNA11 or AP2S1) or drug concentration (e.g., cinacalcet or NPS-2143 (a negative CaSR allosteric modulator)) varied in each experiment. To ensure

that the concentration of CaSR is the same, residual cells can be pelleted following flow cytometry and protein extracted for studies of protein expression (e.g., by Western blot analysis).

3. HEK-CaSR cells should be rederived from liquid nitrogen vials in DMEM-Glutamax with 10% FBS, without geneticin. After cells have settled and grown to ~50% confluency, media can be changed to selection media containing geneticin. We found that this improved the health of cells upon rederivation from liquid nitrogen.
4. Addition of 0.04% Pluronic F-127 can increase the aqueous solubility of Indo-1 AM.
5. We diluted to 100 mM to avoid precipitation of calcium from solution at higher concentrations and to enable addition of agonist at ≥5 μL amounts at the flow cytometer.
6. We used a MoFlo modular flow cytometer from Beckman Coulter with a JDSU Xcyte UV laser and a Coherent sapphire laser or argon laser to measure RFP or GFP, respectively. Many facilities will require training before the flow cytometer can be used. Please contact your local facility to find out their requirements.
7. Ionomycin acts as a positive control to achieve maximal intracellular calcium response. If the flow cytometer performs poorly, ionomycin can be used to ensure the machine is able to detect calcium transients. Ionomycin can be stored in small aliquots (~10 μL) at −20 °C.
8. Other flow cytometry software can be used. FlowJo is a popular alternative.
9. Other graphing software capable of plotting nonlinear regression and calculating EC_{50} can be used.
10. Cells should lift at room temperature, although if the cells are very adherent, incubation at 37 °C may be required.
11. These steps will need to be repeated for all your transfections with the appropriate plasmids. The concentration of plasmid can be adjusted if a different backbone is used. We found that the pBI-CMV2 and pBI-CMV4 plasmids had low transfection efficiency (~40–50%), compared to ~70–80% transfection efficiency achieved for pEGFP-N1-CaSR in other studies. Transfection efficiency should be optimized for your expression construct prior to flow cytometry studies.
12. This time can be adjusted. We have tested transfection at a range of incubation times between 5 and 20 min and found no difference in the efficiency of transfected cells.
13. We performed this step to separate cells out and prevent overconfluency that can result in flow cytometer blockages.

Alternatively, cells can be passed through a cell strainer on day 4 prior to flow cytometry to reduce cell clumping.

14. We used cell dissociation buffer as it is less harsh than trypsin and avoids cell damage. Alternatively, accutase or similar mild detachment buffers can be used.
15. For initial experiments additional negative controls can be performed including preparing a batch of cells with no Indo-1 AM and a batch with Indo-1 AM and no GFP as extra controls to determine background fluorescence for each condition.
16. Before proceeding to the next step, microscopic evaluation can be used to check whether cells have taken up Indo-1.
17. Several concentrations of cinacalcet should be tested to ensure a dose-dependent effect occurs. We recommended starting with 10 nM and using 2–3 other concentrations (e.g., 5 nM, 20 nM, 30 nM).
18. Cells can be stored warm in a polystyrene box with a lid for several hours without loss of fluorescence. Cells should be warmed prior to running on the machine to 37 °C.
19. This allows visualization of cells based on size and granularity. Debris and dead cells often form a cluster in the bottom left side of the SSC vs. FSC plot.
20. Once completed and the residual volumes in each tube have been calculated, samples can be pelleted and protein lysed for use in Western blot analysis. We performed such analyses in several studies of CaSR signaling [21, 25].

Acknowledgments

This work was supported by a Sir Henry Dale Fellowship jointly funded by the Wellcome Trust and the Royal Society (Grant Number 224155/Z/21/Z) to C.M.G. and a Wellcome Trust Clinical Career Development Fellowship to S.A.H (Grant number 220668/Z/20/Z).

References

1. Brown EM (2013) Role of the calcium-sensing receptor in extracellular calcium homeostasis. Best Pract Res Clin Endocrinol Metab 27(3): 333–343. https://doi.org/10.1016/j.beem.2013.02.006
2. Brown EM (1991) Extracellular Ca2+ sensing, regulation of parathyroid cell function, and role of Ca2+ and other ions as extracellular (first) messengers. Physiol Rev 71(2): 371–411. https://doi.org/10.1152/physrev.1991.71.2.371
3. Hendy GN, Canaff L, Cole DE (2013) The CASR gene: alternative splicing and transcriptional control, and calcium-sensing receptor (CaSR) protein: structure and ligand binding sites. Best Pract Res Clin Endocrinol Metab 27(3):285–301. https://doi.org/10.1016/j.beem.2013.02.009

4. Chang W, Tu C, Chen TH, Komuves L, Oda Y, Pratt SA, Miller S, Shoback D (1999) Expression and signal transduction of calcium-sensing receptors in cartilage and bone. Endocrinology 140(12):5883–5893. https://doi.org/10.1210/endo.140.12.7190
5. Pollak MR, Brown EM, Chou YH, Hebert SC, Marx SJ, Steinmann B, Levi T, Seidman CE, Seidman JG (1993) Mutations in the human Ca(2+)-sensing receptor gene cause familial hypocalciuric hypercalcemia and neonatal severe hyperparathyroidism. Cell 75(7):1297–1303. https://doi.org/10.1016/0092-8674(93)90617-y
6. Pearce SH, Trump D, Wooding C, Besser GM, Chew SL, Grant DB, Heath DA, Hughes IA, Paterson CR, Whyte MP et al (1995) Calcium-sensing receptor mutations in familial benign hypercalcemia and neonatal hyperparathyroidism. J Clin Invest 96(6):2683–2692. https://doi.org/10.1172/JCI118335
7. Pollak MR, Brown EM, Estep HL, McLaine PN, Kifor O, Park J, Hebert SC, Seidman CE, Seidman JG (1994) Autosomal dominant hypocalcaemia caused by a Ca(2+)-sensing receptor gene mutation. Nat Genet 8(3):303–307. https://doi.org/10.1038/ng1194-303
8. Nesbit MA, Hannan FM, Howles SA, Babinsky VN, Head RA, Cranston T, Rust N, Hobbs MR, Heath H 3rd, Thakker RV (2013) Mutations affecting G-protein subunit alpha11 in hypercalcemia and hypocalcemia. N Engl J Med 368(26):2476–2486. https://doi.org/10.1056/NEJMoa1300253
9. Nesbit MA, Hannan FM, Howles SA, Reed AA, Cranston T, Thakker CE, Gregory L, Rimmer AJ, Rust N, Graham U, Morrison PJ, Hunter SJ, Whyte MP, McVean G, Buck D, Thakker RV (2013) Mutations in AP2S1 cause familial hypocalciuric hypercalcemia type 3. Nat Genet 45(1):93–97. https://doi.org/10.1038/ng.2492
10. Gannon AW, Monk HM, Levine MA (2014) Cinacalcet monotherapy in neonatal severe hyperparathyroidism: a case study and review. J Clin Endocrinol Metab 99(1):7–11. https://doi.org/10.1210/jc.2013-2834
11. Yamamoto M, Akatsu T, Nagase T, Ogata E (2000) Comparison of hypocalcemic hypercalciuria between patients with idiopathic hypoparathyroidism and those with gain-of-function mutations in the calcium-sensing receptor: is it possible to differentiate the two disorders? J Clin Endocrinol Metab 85(12):4583–4591. https://doi.org/10.1210/jcem.85.12.7035
12. Roszko KL, Stapleton Smith LM, Sridhar AV, Roberts MS, Hartley IR, Gafni RI, Collins MT, Fox JC, Nemeth EF (2022) Autosomal dominant hypocalcemia type 1: a systematic review. J Bone Miner Res 37(10):1926–1935. https://doi.org/10.1002/jbmr.4659
13. Hannan FM, Thakker RV (2013) Calcium-sensing receptor (CaSR) mutations and disorders of calcium, electrolyte and water metabolism. Best Pract Res Clin Endocrinol Metab 27(3):359–371. https://doi.org/10.1016/j.beem.2013.04.007
14. Cavaco BM, Canaff L, Nolin-Lapalme A, Vieira M, Silva TN, Saramago A, Domingues R, Rutter MM, Hudon J, Gleason JL, Leite V, Hendy GN (2018) Homozygous calcium-sensing receptor polymorphism R544Q presents as hypocalcemic hypoparathyroidism. J Clin Endocrinol Metab 103(8):2879–2888. https://doi.org/10.1210/jc.2017-02407
15. Howles SA, Gorvin CM, Cranston T, Rogers A, Gluck AK, Boon H, Gibson K, Rahman M, Root A, Nesbit MA, Hannan FM, Thakker RV (2023) GNA11 variants identified in patients with hypercalcemia or hypocalcemia. J Bone Miner Res 38(6):907–917. https://doi.org/10.1002/jbmr.4803
16. Babinsky VN, Hannan FM, Gorvin CM, Howles SA, Nesbit MA, Rust N, Hanyaloglu AC, Hu J, Spiegel AM, Thakker RV (2016) Allosteric modulation of the calcium-sensing receptor rectifies signaling abnormalities associated with G-protein alpha-11 mutations causing hypercalcemic and hypocalcemic disorders. J Biol Chem 291(20):10876–10885. https://doi.org/10.1074/jbc.M115.696401
17. Howles SA, Hannan FM, Babinsky VN, Rogers A, Gorvin CM, Rust N, Richardson T, McKenna MJ, Nesbit MA, Thakker RV (2016) Cinacalcet for symptomatic hypercalcemia caused by AP2S1 mutations. N Engl J Med 374(14):1396–1398. https://doi.org/10.1056/NEJMc1511646
18. Hannan FM, Howles SA, Rogers A, Cranston T, Gorvin CM, Babinsky VN, Reed AA, Thakker CE, Bockenhauer D, Brown RS, Connell JM, Cook J, Darzy K, Ehtisham S, Graham U, Hulse T, Hunter SJ, Izatt L, Kumar D, McKenna MJ, McKnight JA, Morrison PJ, Mughal MZ, O'Halloran D, Pearce SH, Porteous ME, Rahman M, Richardson T, Robinson R, Scheers I, Siddique H, Van't Hoff WG, Wang T, Whyte MP, Nesbit MA, Thakker RV (2015) Adaptor protein-2 sigma subunit mutations causing familial hypocalciuric hypercalcaemia type 3 (FHH3) demonstrate genotype-phenotype correlations, codon bias

and dominant-negative effects. Hum Mol Genet 24(18):5079–5092. https://doi.org/10.1093/hmg/ddv226

19. Piret SE, Gorvin CM, Pagnamenta AT, Howles SA, Cranston T, Rust N, Nesbit MA, Glaser B, Taylor JC, Buchs AE, Hannan FM, Thakker RV (2016) Identification of a G-protein subunit-alpha11 gain-of-function mutation, Val340Met, in a family with autosomal dominant hypocalcemia type 2 (ADH2). J Bone Miner Res 31(6):1207–1214. https://doi.org/10.1002/jbmr.2797
20. Hannan FM, Gorvin CM, Babinsky VN, Olesen MK, Stewart M, Wells S, Cox RD, Nemeth EF, Thakker RV (2020) Calcilytic NPSP795 increases plasma calcium and PTH in an autosomal dominant hypocalcemia type 1 mouse model. JBMR Plus 4(10):e10402. https://doi.org/10.1002/jbm4.10402
21. Gorvin CM, Hannan FM, Howles SA, Babinsky VN, Piret SE, Rogers A, Freidin AJ, Stewart M, Paudyal A, Hough TA, Nesbit MA, Wells S, Vincent TL, Brown SD, Cox RD, Thakker RV (2017) Galpha(11) mutation in mice causes hypocalcemia rectifiable by calcilytic therapy. JCI Insight 2(3):e91103. https://doi.org/10.1172/jci.insight.91103
22. Roszko KL, Bi R, Gorvin CM, Brauner-Osborne H, Xiong XF, Inoue A, Thakker RV, Stromgaard K, Gardella T, Mannstadt M (2017) Knockin mouse with mutant Galpha (11) mimics human inherited hypocalcemia and is rescued by pharmacologic inhibitors. JCI Insight 2(3):e91079. https://doi.org/10.1172/jci.insight.91079
23. Gorvin CM, Metpally R, Stokes VJ, Hannan FM, Krishnamurthy SB, Overton JD, Reid JG, Breitwieser GE, Thakker RV (2018) Large-scale exome datasets reveal a new class of adaptor-related protein complex 2 sigma subunit (AP2sigma) mutations, located at the interface with the AP2 alpha subunit, that impair calcium-sensing receptor signalling. Hum Mol Genet 27(5):901–911. https://doi.org/10.1093/hmg/ddy010
24. Gorvin CM, Rogers A, Stewart M, Paudyal A, Hough TA, Teboul L, Wells S, Brown SD, Cox RD, Thakker RV (2017) N-ethyl-N-nitrosourea-induced adaptor protein 2 sigma subunit 1 (Ap2s1) mutations establish Ap2s1 loss-of-function mice. JBMR Plus 1(1):3–15. https://doi.org/10.1002/jbm4.10001
25. Gorvin CM, Cranston T, Hannan FM, Rust N, Qureshi A, Nesbit MA, Thakker RV (2016) A G-protein subunit-alpha11 loss-of-function mutation, Thr54Met, causes familial hypocalciuric hypercalcemia type 2 (FHH2). J Bone Miner Res 31(6):1200–1206. https://doi.org/10.1002/jbmr.2778

Chapter 5

Bioluminescence Resonance Energy Transfer (BRET) Assay to Measure Gq Recruitment to the Ghrelin Receptor

Aqfan Jamaluddin

Abstract

This protocol chapter outlines a guide for investigating ghrelin receptor activation of Gq signaling using Bioluminescence Resonance Energy Transfer (BRET). Focusing on the eBRET^2 and NanoBRET™ variants, the chapter covers steps from cell culture to transfection, ligand stimulation, and BRET measurements, offering a robust protocol to examine the temporal aspects of Gq signaling in live cells. This methodology facilitates a nuanced understanding of Gq-mediated cellular responses, but also provides researchers with a valuable tool for dissecting other related signaling pathways.

Key words Coelenterazine-H, Furimazine, G protein recruitment, G protein, Protein interactions, Receptor activity, Renilla luciferase

1 Introduction

G protein-coupled receptors (GPCRs) are an important class of cell surface receptors responsible for influencing various physiological processes [1, 2]. Gq/11 signaling is a specific pathway initiated by GPCRs, where, upon ligand binding, the receptor recruits and activates G proteins of the Gq/11 family leading to the activation of downstream effectors such as phospholipase C (PLC) and subsequent generation of IP3. IP3 binds to IP3-receptors at the ER, resulting in a release of intracellular calcium ($Ca^{2+}{}_i$), which subsequently activates protein kinase C signaling pathways [3–5]. The Gq/11 signaling ultimately regulates diverse cellular responses including cell differentiation, metabolism, and proliferation [1, 2].

Bioluminescence Resonance Energy Transfer (BRET) assays are a robust tool in the field of molecular biology and pharmacology, offering a sensitive and real-time method for assessing receptor activity and interactions in live cells. BRET assays leverage the principles of energy transfer between a bioluminescent donor,

Caroline M. Gorvin (ed.), *Calcium Signaling: Methods and Protocols*, Methods in Molecular Biology, vol. 2861,
https://doi.org/10.1007/978-1-0716-4164-4_5,

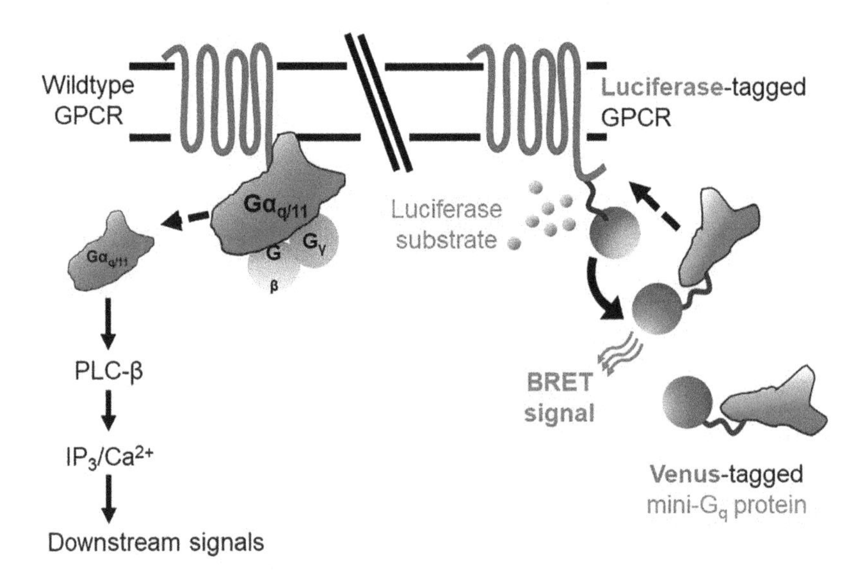

Fig. 1 GPCR-Gq activity measured by BRET. Schematic showing signaling by a Gq-coupled GPCR (left), with the same receptor modified to measure its activity by BRET (right). In the wild-type scenario, upon activation, the GPCR recruits and activates the Gq protein, initiating dissociation of Gα from Gβγ, and facilitates downstream signaling cascades such as IP3-intracellular calcium ($Ca^{2+}{}_i$) signaling. In the BRET assay, the luciferase-tagged GPCR undergoes activation, leading to the recruitment of Venus fluorophore-tagged Gq protein. The proximity and proper orientation of the luciferase (donor) and Venus (acceptor) molecules upon Venus-Gq recruitment enable BRET to occur. Utilizing the appropriate luciferase substrate, catalysis by the luciferase generates luminescence and initiates BRET

typically Renilla luciferase (Rluc) or NanoLuc® luciferase (Nluc), and a fluorescent acceptor, commonly a fluorescent protein such as green fluorescent protein (GFP) or yellow fluorescent protein (Venus/eYFP). The efficiency of this energy transfer is dependent on the distance and relative orientation of the donor and acceptor molecules [6, 7]. In addition to this, the ability to attach the fluorophores to targeted points of interest, such as membrane-bound or intracellular proteins, makes BRET an excellent tool for studying protein–protein interactions and conformational changes associated with receptor activation. Given this, BRET has proven to be a valuable tool in investigating the recruitment and interaction between GPCRs and Gq proteins. Typically, the luciferase donor is genetically fused to a GPCR of interest, and the fluorescent acceptor, usually GFP or YFP, is tagged to the G protein. Upon ligand binding and receptor activation, G protein recruitment to GPCRs promotes proximity between the donor and acceptor and thus facilitates energy transfer, resulting in a measurable BRET signal (Fig. 1).

BRET assays can assess several aspects of GPCR functionality. By manipulating the experimental conditions or introducing mutations in the GPCR or Gq protein, researchers can dissect the molecular determinants governing the specificity and efficiency of

GPCR-Gq protein coupling. Additionally, in contrast to some endpoint assays that rely on cell lysates and proxy measures of secondary messengers to probe for GPCR-Gq protein interactions, BRET allows the study of direct protein interactions within the context of live cells, providing a more physiologically relevant environment and quantitative data in real time. Such real-time data allows researchers to quantify the kinetics of Gq protein recruitment to activated GPCRs, offering insights into the temporal aspects of signal transduction. This capability is particularly valuable in drug discovery and development [8], where understanding the kinetics of ligand–receptor interactions is crucial for optimizing therapeutic interventions.

There are several varieties of BRET assays, $BRET^1$, $BRET^2$, and $BRET^3$, and preference is dependent on experimental requirements [9]. The key difference between them lies in the luciferase protein utilized and the subsequent substrate used to catalyze luminescence. $BRET^1$ uses Rluc as the donor molecule, eYFP as the acceptor, and coelenterazine-H as the substrate. $BRET^2$ uses coelenterazine 400A (also known as DeepBlue C) as the substrate, and GFP2/GFP10 as the acceptor [10]. Additionally, signal strength can be enhanced by using Rluc8 as the donor ($eBRET^2$) [6]. $BRET^2$ generally yields better signal-to-noise ratio due to the reduced acceptor–donor emission spectra overlap. The downside is that coelenterazine 400A decays faster than coelenterazine-H and therefore $BRET^2$ is not recommended for longer duration kinetic measurements [11]. $BRET^3$ utilizes luciferin as the substrate, Firefly luciferase as the donor, and DsRed as the acceptor. Firefly luciferase possesses less autofluorescence but has a larger overlap between acceptor and donor emission spectra [9]. NanoBRET™ (Nluc)-based BRET uses the 100× brighter Nluc as the donor and furimazine as the substrate. NanoBRET™ also has very good separation between donor emission and acceptor emission, resulting in improved sensitivity compared to Rluc8/YFP [12].

In this chapter, we describe methods to assess recruitment of a fluorescent-labeled G protein to a GPCR with a C-terminally tagged Rluc8 or Nluc. Fluorescent-labeled G protein mimics (named MiniG proteins) have also become popular as surrogates for full G proteins for screening purposes. MiniG proteins exhibit notable modifications compared to standard Gα subunits including a truncated N terminus, which eliminates membrane anchors and the Gβγ-binding surface, as well as the absence of the α-helical domain. Additionally, mutations have been introduced to enhance protein stability in vitro including a mutation in the C-terminal α5 helix which stabilizes receptor–miniG complexes [13]. MiniG proteins also include a nuclear export sequence (NES) to improve accessibility of mGsq to receptors at the cell membrane [13]. However, a recent study has shown that miniG proteins may affect GPCR trafficking [14]. Therefore, findings should be verified with independent assays, and the effects on trafficking of your GPCR of interest should be assessed. The Venus fluorophore has the same spectral properties as eYFP, but has a 30-fold higher relative fluorescence, matures faster, and is less sensitive to pH and chloride ion concentrations [15].

This protocol describes the activation of Gq by the ghrelin receptor (GHSR), a GPCR involved in energy homeostasis [16]. Steps required to perform these assays from cell culture to data analysis are covered in this chapter. The example protocol uses adherent HEK293 cells, but the assay can accommodate different types of cells or cells in suspension with minor modifications.

2 Materials

2.1 Expression Plasmids

1. GHSR-Rluc8 or GHSR-Nluc: Full-length human GHSR was cloned in-frame at the N-terminus of Rluc8 available from Addgene or Nluc available from Promega.
2. pEGFP-Gαq: Full-length human Gαq was cloned in-frame at the N-terminus of GFP within the pEGFP-N1 vector (Clontech). Or, if using miniG proteins, NES-Venus-mGsq: The plasmid was a gift from Nevin Lambert, University of Georgia. The construction of the plasmid has been described [13].
3. Prepare plasmid constructs of your receptor-Rluc8/Nluc and G protein-YFP/GFP in MilliQ or Ultrapure water (*see* **Note 1**). Obtain the concentration of plasmid in the solution after plasmid purification, and store at 4 °C for short–/medium-term storage, or −20 °C for longer-term storage.

2.2 Cell Culture

1. Adherent HEK293 cells.
2. 6-well clear cell culture plates: TC-treated, flat bottom with lid.
3. 96-well white assay plate: TC-treated, flat non-clear bottom with lid.
4. Growth media: High glucose Dulbecco's modified Eagle medium (DMEM) media, 10% bovine serum (*see* **Note 2**).
5. Transfection media: DMEM media without serum.
6. 0.05% Trypsin-EDTA solution.
7. 1× sterile phosphate-buffered saline (PBS).
8. Hemocytometer for cell counting and trypan blue.
9. Lipofectamine™ 2000.
10. Replating media: FluoroBrite™ DMEM, 10% bovine serum (*see* **Note 3**).
11. 1× Hank's buffered saline solution (HBSS).
12. Phase-contrast microscope.

2.3 Substrates for BRET Assay

1. Coelenterazine-H (substrate for Rluc BRET assays): Resuspend coeleterazine-H in ethanol to make 1 mM stock and store at −80 °C and in light-protected aliquots.
2. Furimazine (substrate for NanoBRET™ assays): Furimazine is obtained from Promega as Nano-Glo® Live Cell Reagent, the primary proprietor of Nluc-based technologies.
3. Ligand (ghrelin): Reconstitute to 1 mM stock solution in DMSO and store at −20 °C in single-use aliquots (*see* **Note 4**).

2.4 Equipment

1. BRET-compatible plate reader capable of maintaining physiological temperature at 37 °C in the plate chamber. The ideal filter setup will include a band-pass (BP) filter to measure the donor signal (emission 450 nm/BP 80 nm) and a long-pass (LP) filter to measure the acceptor signal (Emission 530 nm/LP) (*see* **Note 5**). We have used both BMG Labtech PHERAstar and Promega GloMax Discover plate readers and find that both plate readers gave comparable results after data are normalized (*see* **Note 6**).

3 Methods

3.1 Cell Culture

1. Gently wash flask containing HEK293 cells once with 1× PBS.
2. Detach cells from flasks with 2–3 mL of 0.05% trypsin (sufficient volume to cover flask surface) and observe cells for detachment from surface. Cells should take no longer than 5 min to detach with gentle tapping.
3. Neutralize the trypsin with an equal volume of growth media. Aspirate cells up and down carefully to remove clumps and detach any remaining cells on the flask surface.
4. Obtain a small sample of cells from the cell suspension and count cells with a hemocytometer to determine total cell amount in the suspension.
5. Plate cells into 6-well plates at approximately 2–3 × 10^5 cells per well in 2 mL of media.
6. Incubate cells overnight at 37 °C and 5% CO_2.

3.2 Transfection

1. Dilute plasmids at a GPCR-Rluc/Nluc to G protein-YFP/GFP protein ratio of 1:10 in 100 μL of the transfection media. This will become the transfection mix. We use 50 ng of GHSR-Rluc/Nluc (*see* **Note 7**).
2. Add 2 μL of Lipofectamine™ 2000 into the transfection mix (*see* **Note 8**).
3. Incubate the transfection mix for a minimum of 5 min at room temperature.

4. Add 100 μL of the transfection mix dropwise into the well.
5. Incubate cells at 37 °C and 5% CO_2 for 2 days before replating.

3.3 Replating of Cells into White 96-Well Plates

1. Aspirate out the old media and wash cells with 1× PBS.
2. Remove PBS and add 1 mL of replating media.
3. Detach and de-clump cells by pipetting up and down.
4. Pipette cells into a 96-well plate at approximately 100 μL/well (*see* **Note 9**). For a dose–response curve, we typically plate cells from a single well of a 6-well plate across a column (i.e., 8 wells) of a 96-well plate.
5. Incubate cells for a minimum of 4 h at 37 °C and 5% CO_2 before performing the assay to allow cells to settle to the well surface (*see* **Note 10**).

3.4 Substrate and Ligand Addition

1. Coelenterazine-H and furimazine should be prepared at 40 μL/well. Dilute at 1:100 in 1×HBSS or the manufacturer's recommended buffer (*see* **Notes 11** and **12**).
2. Prepare ligand dilutions at 10 μL/well. Ligand should be diluted with 1× HBSS to 5× the final concentration in the well for Rluc-BRET, or 10× the final concentration in the well for NanoBRET™.
3. Set the plate reader chamber temperature to 37 °C 10 min before starting assays to allow stabilization of the temperature across the whole plate.
4. Set the plate reader to read ratiometric luminescence at 450 nm and 530 nm every 2 min for approximately 20–30 cycles. Use an integration time of 0.25–2 s. We set our plate reader to pause after four cycles to allow addition of ligand, and then immediately resume the program to perform the additional cycles.
5. Pipette out the replating media from the 96-well to replace with the luciferase substrate (*see* **Note 13**):
   - (a) For Rluc-BRET, pipette out all media from the 96-well, and replace with 40 μL of the prepared Coelenterazine-H substrate (Fig. 2).
   - (b) For NanoBRET™, pipette out only 50 μL of the replating media from the 96-well, and replace with 40 μL of the prepared furimazine-based Nano-Glo® Live Cell Reagent (Fig. 2).
6. Read baseline luminescence for four cycles (8 min total).
7. Add 10 μL of diluted ligand to the well.
8. Immediately continue reading for the remainder of the cycle reads.

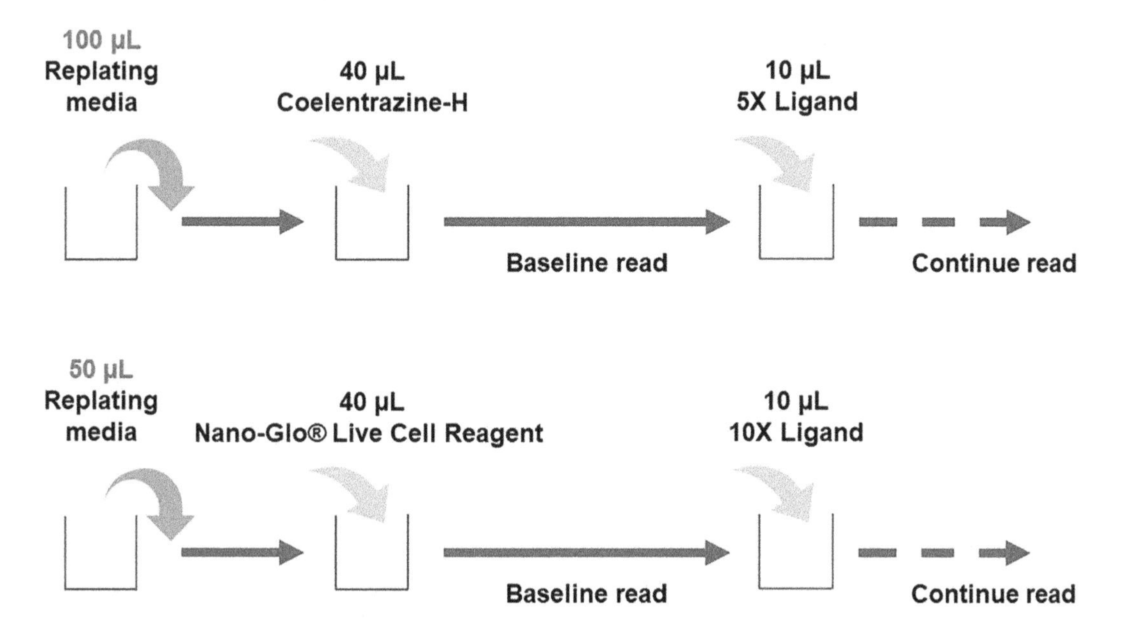

Fig. 2 Step process of substrate and ligand addition for both Rluc BRET and NanoBRET™ assays

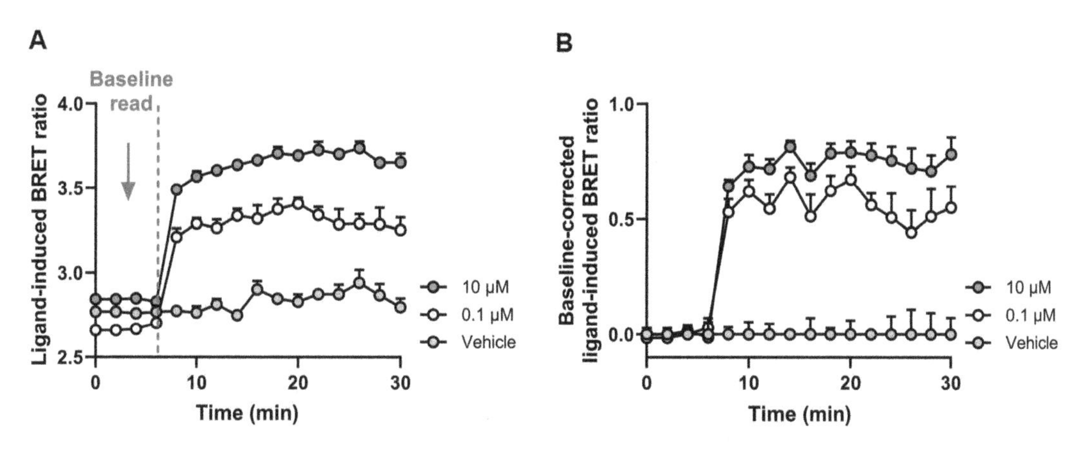

Fig. 3 Graphs of analyzed BRET signal. (**a**) The BRET ratio is plotted against time. Note that the initial baseline read for all datasets does not show similar BRET ratio signals. (**b**) The same dataset following baseline correction. First, the *baseline-corrected* ligand-induced BRET ratio was obtained by subtracting the mean vehicle-induced BRET ratio signal from all datasets. This is then followed by subtracting the mean of the four baseline reads from each response to normalize data to 0

3.5 Data Analysis

1. Obtain the BRET ratio by dividing acceptor [530] signal/ donor [450] signal. Depending on the plate reader file output, BRET ratio will automatically be calculated or will have to be done manually.
2. On a data analysis software, such as GraphPad Prism, plot the raw BRET ratio signal against time (Fig. 3a).

3. For each timepoint, subtract the mean vehicle-induced BRET ratio signal from all datasets. This is the *baseline-corrected* ligand-induced BRET ratio (Fig. 3b).
4. Again, from each timepoint of the now *baseline-corrected* datasets, subtract the average BRET ratio signal of the baseline read (mean of their corresponding first 4 pre-stimulation *baseline-corrected* BRET ratio) from all datasets. This will anchor the BRET signal to 0 from point of stimulation and remove all background BRET from the dataset (Fig. 3b) [17] (*see* **Note 14**).

4 Notes

1. There is the possibility that the BRET assay may not elicit strong signals despite all attempted technical corrections and optimization. Likely causes are unfavorable relative orientation of BRET acceptors and donors or not being in close proximity to generate a BRET signal [7]. A solution could be to switch the acceptor and donor probes either between the two proteins or otherwise between C- and N-termini of the intracellular protein of interest. The acceptor and donor probes in fusion proteins are also rotationally constricted, and therefore the BRET signal is heavily dependent on the conformation of the folded proteins and the complexes that they form. Therefore, modifying the linker region of the fusion proteins can alter the rigidity of the probes and allow for RET to occur more optimally. As a positive control, a construct encoding a fusion protein containing both donor and acceptor (e.g., Rluc-Venus) can be used to transfect the cells and to test for BRET signal.
2. Antibiotics can be supplemented into the media, but it is not recommended.
3. Alternatively, DMEM with no phenol can be substituted.
4. DMSO or ethanol can be used to reconstitute less soluble ligands, but the final DMSO or ethanol percentage that cells will be exposed to should be considered. MK0677 can be used as a ghrelin receptor agonist instead of ghrelin.
5. BRET signals are expressed as a ratio of the light coming from the acceptor (fluorescence) over the light coming from the donor (luminescence).
6. BRET ratios can range in value between 0 and 10 depending on transfection efficiency and the plate reader used. Different plate readers will have different sensitivity. Given the option, perform all experiments on one plate reader exclusively for comparable results.

7. Optimization of transfected DNA amount is highly recommended if results are not optimal. BRET ratio is dependent on the level of expression of the donor-tagged protein. Generally, the amount of donor-tagged plasmids used for transfection should be lower than the amount of acceptor-tagged plasmids. Lowering the level of donor protein expression will inflate the acceptor–donor ratio. However, the ideal acceptor–donor ratio can be further optimized. Too low a concentration of an acceptor (relative to donor) can often result in more variability in BRET output, while too high a concentration of the acceptor can lead to nonspecific BRET due to random collisions. This is a phenomenon that may arise if membranes are overcrowded with proteins, facilitating random collisions that allow RET to occur. A BRET titration curve should be able to discern the optimal ratio between acceptor and donor. Transfect cells with a constant amount of donor plasmid (Rluc or Nluc), and an increasing amount of acceptor plasmid (Venus or YFP). Subtract the BRET ratio of sample wells from the BRET ratio obtained when the donor is expressed alone, and you will obtain the $BRET_{net}$. Then, plot $BRET_{net}$ values as a function of acceptor–donor ratio:

$$BRET_{net} = \left(\frac{\text{Acceptor}}{\text{Donor}}\right)\text{Sample wells} - \left(\frac{\text{Acceptor}}{\text{Donor}}\right)\text{“Donor only” wells}$$

If BRET signals are attributed to random BRET, then $BRET_{net}$ should increase linearly to acceptor amount. Conversely, if protein–protein interactions are true, the $BRET_{net}$ increases as a hyperbolic function until a saturation point occurs where there is a limited concentration of donors to interact and induce BRET with the acceptors (Fig. 4a). Note that good judgment should also be used to interpret curves as bystander BRET can also saturate at high enough acceptor–donor ratios (Fig. 4b).

From the hyperbolic function, we can obtain both the $BRET_{max}$, the highest $BRET_{net}$ value obtained; and the $BRET_{50}$, the acceptor–donor ratio that elicited 50% of the $BRET_{max}$ value (Fig. 4a). Using an acceptor–donor ratio lower than $BRET_{50}$ in subsequent experiments runs the risk of less sensitivity and more variability in the BRET signal due to too low a concentration of acceptors. Where the linear function of bystander $BRET_{net}$ (for a protein of similar expression and localization) intersects the $BRET_{max}$ value indicates that the acceptor–donor ratio could be too high and bystander BRET is likely a contributing factor to the observed BRET signal (Fig. 4c). In an ideal scenario, the ideal acceptor–donor ratio

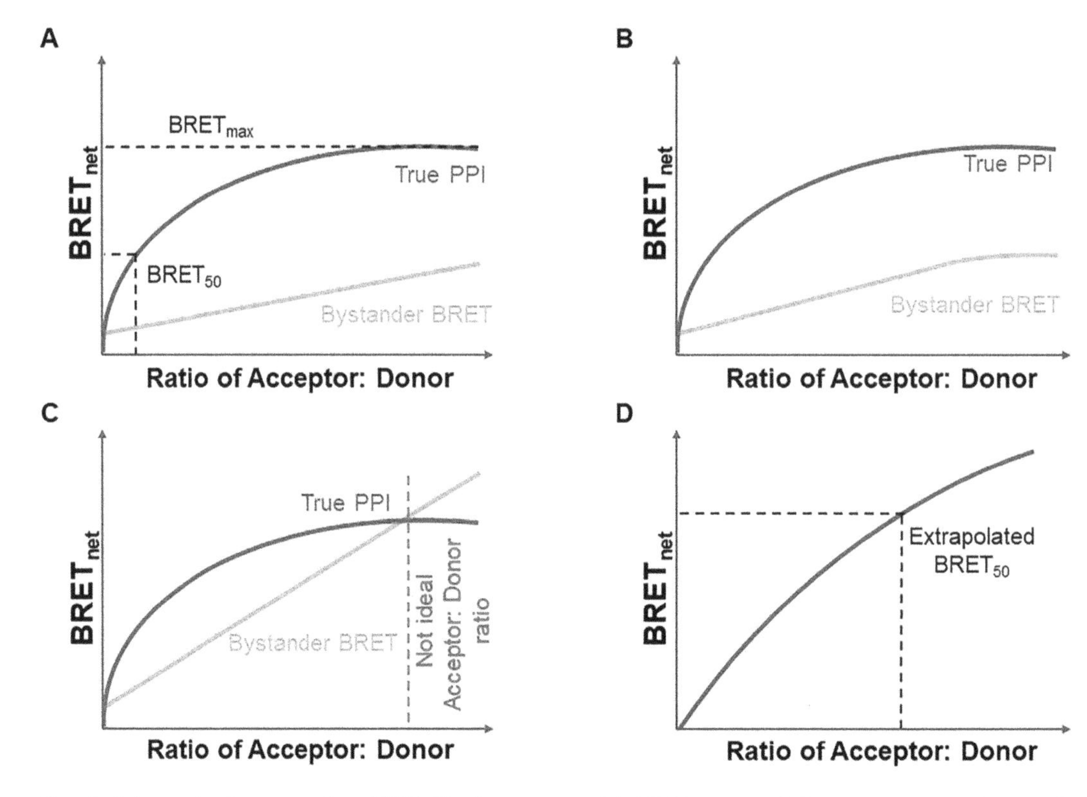

Fig. 4 Schematic demonstrating BRET titration curves. (**a**) BRET expected from a true protein–protein interaction (PPI) against one that occurs from random collision (bystander BRET). The two comparative values, $BRET_{max}$ and $BRET_{50}$, can be measured from the true PPI BRET curve. (**b**) An example of bystander BRET saturating at very high acceptor–donor range. (**c**) The recommended cutoff for acceptable acceptor–donor ratio if bystander BRET of a similar protein intersects the true PPI curve. (**d**) The extrapolated $BRET_{50}$ can be used to gauge for ideal acceptor–donor ratio if $BRET_{max}$ is not able to be measured directly from the curve

is the one that yielded $BRET_{net}$ close to the $BRET_{max}$ value. If the hyperbolic function does not reach a saturation point within the tested range, the extrapolated $BRET_{50}$ can also be a good gauge for the ideal acceptor–donor ratio (Fig. 4d).

A BRET titration curve is also a good experiment to determine if constitutive interactions occur between receptors and a protein of interest. Simply repeat the titration experiment without ligand stimulation. If there are no interactions occurring, a linear function of $BRET_{net}$ should occur (bystander BRET). If a distinct hyperbolic function of $BRET_{net}$ is observed, constitutive interaction between receptor and protein is a likely occurrence.

8. Alternative transfection methods can be used. For example, PEI can yield better transfection efficiency and is more affordable than Lipofectamine™ 2000 but can cause more cell death. Validate and optimize transfection efficiency.

9. Ideally, the number of cells replated from a 6-well plate to a 96-well plate should be consistent not only between wells but also between independent experimental repeats to reduce variability in the BRET assay. This would require a second cell counting step prior to replating. However, cell counting can be omitted provided strict consistency is maintained with the experimental schedule from the first cell counting/plating. For our protocol above, we calculated that we consistently replated cells into the 96-well plate at approximately 10^5 cells/well, and therefore chose to omit the second cell counting step due to its redundancy.
10. Alternatively, cells can be replated into 96-well plates the day before the BRET experiment is performed. In this case, consider not plating cells in the columns and rows occupying the edges of the plate to minimize plate edge effects.
11. Prepare the final diluted volumes of coelenterazine-H and Nano-Glo® Live Cell Reagent fresh for each experiment to prevent breakdown of substrates. If preparing in advance of approximately a few hours, keep the substrates on ice or refrigerated, and protected from light. Nano-Glo® Live Cell Reagent, for example, can lose up to 10% activity in approximately 3 h at 20 °C, while only 10% of its activity can be lost in approximately 7 h if kept at 4 °C [18].
12. The volume of substrates needed for either assay should ideally be tested to ensure BRET signal is robust. However, we have found that we can also lessen the amount of substrate used and still obtain relatively good signal. Particularly for Nano-BRET™ assay, we do not have to aspirate the whole replating media and replace it with the Nano-Glo® Live Cell Reagent, provided the replating media contains no components that could affect furimazine-based luminescence or the luciferase activity itself (also *see* **Note 3**).
13. If aspirating the replating media in its entirety, ensure that there are no residual volumes as this could significantly alter the BRET signal due to variability in the total final volume in the wells.
14. This step can be omitted if there are only two datasets for analysis, i.e., treatment vs vehicle. Not anchoring the BRET signal to 0 from point of stimulation can provide more information in alternative experimental designs, i.e., receptor–intracellular protein association or dissociation assays. Note that it is crucial to obtain the *baseline-corrected* ligand-induced BRET ratio first, prior to removing background BRET from the datasets, and not vice versa. This is to avoid the likelihood of normalization artifacts in subsequent data collation of the independent repeats.

Acknowledgments

This work was performed in the laboratory of Dr. Caroline Gorvin, University of Birmingham. Assay establishment was supported by a Sir Henry Dale Fellowship jointly funded by the Wellcome Trust and the Royal Society (Grant Number 224155/Z/21/Z) and an Academy of Medical Sciences Springboard Award supported by the British Heart Foundation, Diabetes UK, the Global Challenges Research Fund, the Government Department of Business, Energy and Industrial Strategy, and the Wellcome Trust (Ref: SBF004| 1034).

References

1. Wettschureck N, Offermanns S (2005) Mammalian G proteins and their cell type specific functions. Physiol Rev 85(4):1159–1204. https://doi.org/10.1152/physrev.00003.2005
2. Klepac K, Kilić A, Gnad T, Brown LM, Herrmann B, Wilderman A, Balkow A, Glöde A, Simon K, Lidell ME, Betz MJ, Enerbäck S, Wess J, Freichel M, Blüher M, König G, Kostenis E, Insel PA, Pfeifer A (2016) The Gq signalling pathway inhibits brown and beige adipose tissue. Nat Commun 7(1):10895. https://doi.org/10.1038/ncomms10895
3. Rhee SG (2001) Regulation of phosphoinositide-specific phospholipase C. Annu Rev Biochem 70(1):281–312. https://doi.org/10.1146/annurev.biochem.70.1.281
4. Ohanian J, Ohanian V (2001) Lipid second messenger regulation: the role of diacylglycerol kinases and their relevance to hypertension. J Hum Hypertens 15(2):93–98. https://doi.org/10.1038/sj.jhh.1001139
5. Mak DO, McBride SM, Petrenko NB, Foskett JK (2003) Novel regulation of calcium inhibition of the inositol 1,4,5-trisphosphate receptor calcium-release channel. J Gen Physiol 122(5):569–581. https://doi.org/10.1085/jgp.200308808
6. Dacres H, Michie M, Wang J, Pfleger KDG, Trowell SC (2012) Effect of enhanced Renilla luciferase and fluorescent protein variants on the Förster distance of bioluminescence resonance energy transfer (BRET). Biochem Biophys Res Commun 425(3):625–629. https://doi.org/10.1016/j.bbrc.2012.07.133
7. Weihs F, Wang J, Pfleger KDG, Dacres H (2020) Experimental determination of the bioluminescence resonance energy transfer (BRET) Förster distances of NanoBRET and red-shifted BRET pairs. Anal Chim Acta X 6: 100059. https://doi.org/10.1016/j.acax.2020.100059
8. Robers MB, Dart ML, Woodroofe CC, Zimprich CA, Kirkland TA, Machleidt T, Kupcho KR, Levin S, Hartnett JR, Zimmerman K, Niles AL, Ohana RF, Daniels DL, Slater M, Wood MG, Cong M, Cheng Y-Q, Wood KV (2015) Target engagement and drug residence time can be observed in living cells with BRET. Nat Commun 6(1):10091. https://doi.org/10.1038/ncomms10091
9. Bacart J, Corbel C, Jockers R, Bach S, Couturier C (2008) The BRET technology and its application to screening assays. Biotechnol J 3(3):311–324. https://doi.org/10.1002/biot.200700222
10. Leduc M, Breton B, Galés C, Gouill CL, Bouvier M, Chemtob S, Heveker N (2009) Functional selectivity of natural and synthetic prostaglandin EP₄ receptor ligands. J Pharmacol Exp Ther 331(1): 297–307. https://doi.org/10.1124/jpet.109.156398
11. Jiang T, Yang X, Zhou Y, Yampolsky I, Du L, Li M (2017) New bioluminescent coelenterazine derivatives with various C-6 substitutions. Org Biomol Chem 15(33):7008–7018. https://doi.org/10.1039/C7OB01554B
12. Machleidt T, Woodroofe CC, Schwinn MK, Méndez J, Robers MB, Zimmerman K, Otto P, Daniels DL, Kirkland TA, Wood KV (2015) NanoBRET – a novel BRET platform for the analysis of protein–protein interactions. ACS Chem Biol 10(8):1797–1804. https://doi.org/10.1021/acschembio.5b00143
13. Wan Q, Okashah N, Inoue A, Nehmé R, Carpenter B, Tate CG, Lambert NA (2018) Mini G protein probes for active G

protein-coupled receptors (GPCRs) in live cells. J Biol Chem 293(19):7466–7473. https://doi.org/10.1074/jbc.RA118.001975

14. Manchanda Y, Ramchunder Z, Shchepinova MM, Rutter GA, Inoue A, Tate EW, Jones B, Tomas A (2021) Expression of mini-G proteins specifically halt cognate GPCR trafficking and intracellular signalling. bioRxiv:2021.2011.2024.469908. doi: https://doi.org/10.1101/2021.11.24.469908
15. Okita C, Sato M, Schroeder T (2004) Generation of optimized yellow and red fluorescent proteins with distinct subcellular localization. BioTechniques 36(3):418–424. https://doi.org/10.2144/04363st01
16. Jamaluddin A, Gorvin CM (2023) RISING STARS: targeting G protein-coupled receptors to regulate energy homeostasis. J Mol Endocrinol 70(4). https://doi.org/10.1530/JME-23-0014
17. White CW, Vanyai HK, See HB, Johnstone EKM, Pfleger KDG (2017) Using nanoBRET and CRISPR/Cas9 to monitor proximity to a genome-edited protein in real-time. Sci Rep 7(1):3187. https://doi.org/10.1038/s41598-017-03486-2
18. Promega Nano-Glo® Live Cell Assay System Protocol (2024). https://www.promega.co.uk/resources/protocols/technical-bulletins/101/nano-glo-live-cell-assay-system-protocol/

Chapter 6

Analysis of Calcium-Sensing Receptor Signaling Using Dual Luciferase Assays

Franziska Lena Sörgel

Abstract

Luciferases catalyze a reaction that involves the emission of light, a phenomenon referred to as "bioluminescence". The calcium-sensing receptor (CaSR), a G protein-coupled receptor (GPCR), induces characteristic signaling pathways that stimulate extracellular signal-regulated kinase 1/2 (ERK1/2) and Ca^{2+} mobilization from the endoplasmic reticulum. ERK1/2 causes an activation of the serum response element (SRE), whereas Ca^{2+} causes an activation of the nuclear factor of activated T-cells response element (NFAT-RE). Transfection of cells with a vector containing a firefly luciferase reporter gene under the control of the SRE or NFAT-RE allows the monitoring of ERK1/2 activation and Ca^{2+} mobilization, respectively, by measuring luminescence. In a dual luciferase assay, firefly luminescence is normalized by co-transfecting an internal control vector, which contains a constitutively active promoter driving the expression of a second luciferase, namely, *Renilla* luciferase, whose activity can be quantified within the same sample. Here, a protocol for the analysis of CaSR signaling using dual luciferase assays in HEK293 cells is provided. The assays can, for example, be used to investigate functional consequences of mutations in the CaSR gene.

Key words Calcium mobilization, Calcium-sensing receptor (CaSR), Dual luciferase assay, Extracellular signal-regulated kinase (ERK), G protein-coupled receptor (GPCR), Nuclear factor of activated T-cells response element (NFAT-RE), Serum response element (SRE)

1 Introduction

Luciferases are light-producing enzymes commonly used in reporter gene assays to monitor the activity of regulatory DNA elements. They catalyze the oxygen-dependent oxidation of their substrate, a luciferin, to an electronically excited intermediate, which emits a photon of light when decaying to its ground state. This phenomenon is termed "bioluminescence". In a reporter gene assay, a luciferase reporter gene is placed downstream of the regulatory element of interest so that the element controls the expression level of the luciferase. The intensity of the light signal, which can be quantified with a luminometer, correlates with the amount

Caroline M. Gorvin (ed.), *Calcium Signaling: Methods and Protocols*, Methods in Molecular Biology, vol. 2861,
https://doi.org/10.1007/978-1-0716-4164-4_6,

of luciferase enzyme formed and therefore reflects the activity of the regulatory element [1].

Luciferase reporter gene assays can be used to analyze signaling pathways of G protein-coupled receptors (GPCRs), such as the calcium-sensing receptor (CaSR), because the pathways entail the activation of particular transcription factors that activate characteristic regulatory DNA elements [1] (Fig. 1). The CaSR is highly expressed in the parathyroid glands and in the kidneys, where it regulates parathyroid hormone secretion and urinary calcium excretion, respectively [2]. Gα-subunits from the $G\alpha_{q/11}$ family are crucial for the CaSR-mediated maintenance of a physiological extracellular Ca^{2+} concentration [3]. $G\alpha_{q/11}$ proteins activate phospholipase C, which hydrolyzes the membrane lipid phosphatidylinositol 4,5-bisphosphate, forming the second messengers diacylglycerol (DAG) and inositol 1,4,5-trisphosphate (IP_3). IP_3 binds to IP_3 receptors on the endoplasmic reticulum (ER), triggering the release of Ca^{2+} from the ER into the cytosol. Together with Ca^{2+}, DAG activates protein kinase C, which stimulates the mitogen-activated protein kinase cascade to activate extracellular signal-regulated kinase 1/2 (ERK1/2). The CaSR additionally activates ERK1/2 via $G\alpha_{i/o}$ proteins and also G protein-independently via β-arrestins [4]. $G\alpha_{i/o}$ proteins inhibit adenylyl cyclase, resulting in a decrease in cAMP levels and thus a decrease in protein kinase A activity, relieving its inhibitory effect on ERK1/2 activation [5]. Intracellular Ca^{2+} mobilization and ERK1/2 activation both result in a change in gene expression. Activated ERK1/2 translocates from the cytosol into the nucleus, where it phosphorylates the transcriptional activator Elk-1. Together with a dimer of serum response factor, phosphorylated Elk-1 binds as a ternary complex to the serum response element (SRE), activating SRE-mediated transcription [6]. Cytosolic Ca^{2+} activates the phosphatase calcineurin, which dephosphorylates nuclear factor of activated T-cells (NFAT) proteins. Dephosphorylated NFAT proteins translocate to the nucleus, where they bind to and activate the NFAT response element (NFAT-RE) [7]. Molecular cloning of the SRE or the NFAT-RE into an expression vector upstream of a luciferase reporter gene and transfer of that vector into CaSR-expressing cells allows the quantification of ERK1/2 activation and Ca^{2+} mobilization following CaSR activation. The SRE and the NFAT-RE luciferase assay can be used to analyze functional consequences of *CASR* variants, which can cause human disorders of calcium homeostasis, on CaSR signaling.

The total amount of luciferase formed is determined not only by the activity of the response element but also by other confounding variables, such as variations in cell number and transfection efficiency. These variables can be corrected for by co-transfecting two luciferase reporter genes in a single assay termed "dual luciferase assay". The expression of one luciferase, referred to as the

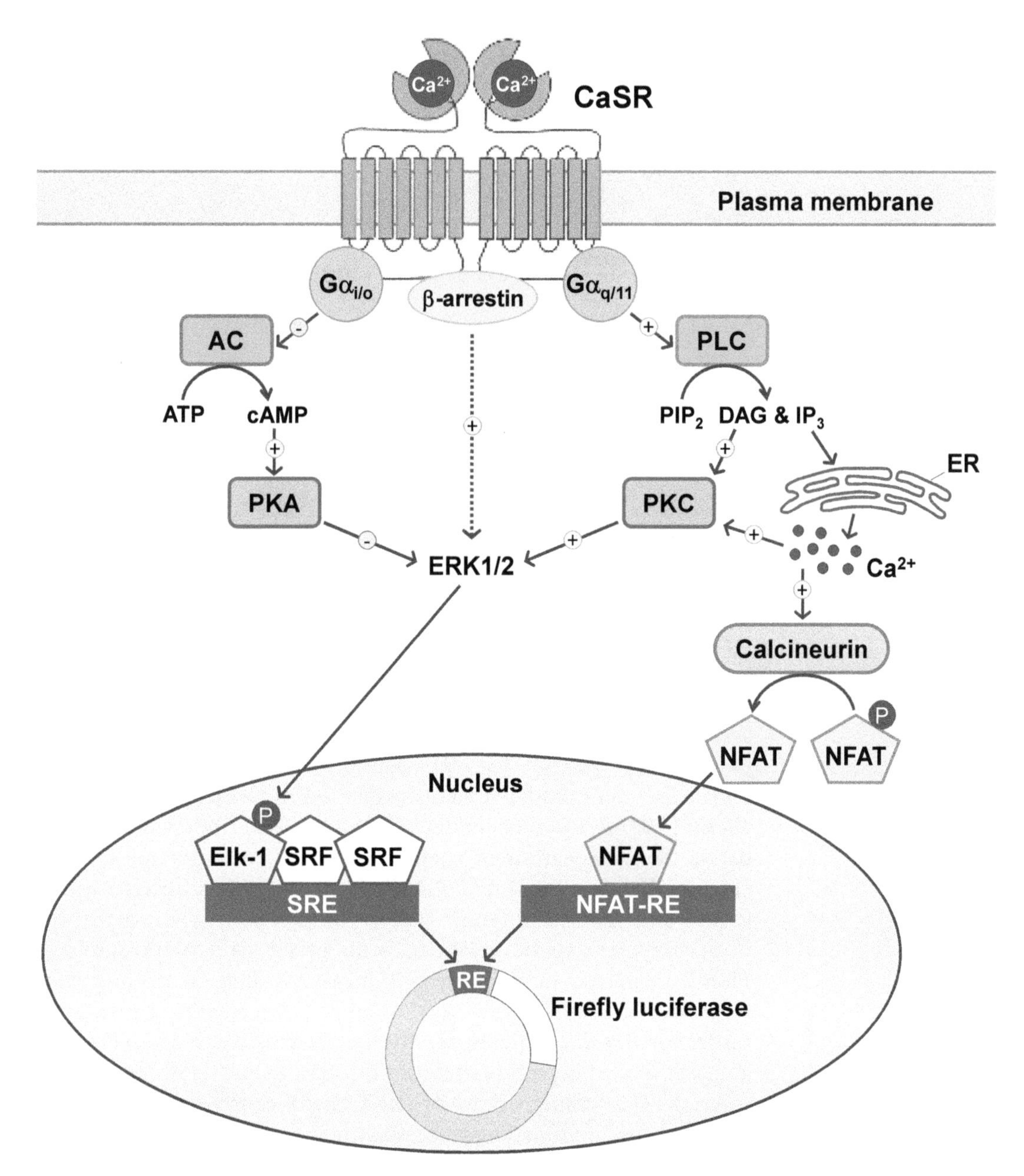

Fig. 1 Signaling pathways of the calcium-sensing receptor (CaSR). The CaSR is activated by extracellular Ca^{2+}. It couples to Gα subunits from the $G\alpha_{q/11}$ and $G\alpha_{i/o}$ families. $G\alpha_{q/11}$ stimulates phospholipase C (PLC), which hydrolyzes phosphatidylinositol 4,5-bisphosphate (PIP_2), forming the second messengers diacylglycerol (DAG) and inositol 1,4,5-trisphosphate (IP_3). IP_3 induces the release of Ca^{2+} from the endoplasmic reticulum (ER) into the cytosol. DAG and Ca^{2+} stimulate protein kinase C (PKC), which activates extracellular signal-regulated kinase 1/2 (ERK1/2). $G\alpha_{i/o}$ proteins inhibit adenylyl cyclase (AC), resulting in a decrease in cAMP levels and hence a decrease in protein kinase A (PKA) activity, relieving its inhibitory effect on ERK1/2 activation. Furthermore, β-arrestins contribute to ERK1/2 activation by acting as a signaling scaffold. Cytosolic Ca^{2+} activates calcineurin. Calcineurin dephosphorylates nuclear factor of activated T-cells (NFAT) proteins, which can now translocate to the nucleus and activate the NFAT response element (NFAT-RE). ERK1/2 activates Elk-1 in the nucleus, which together with a dimer of serum response factor (SRF), activates the serum response element (SRE). When the SRE or NFAT-RE are cloned into an expression vector upstream of a firefly luciferase reporter gene, ERK1/2 activation and intracellular Ca^{2+} mobilization can be monitored by measuring luminescence

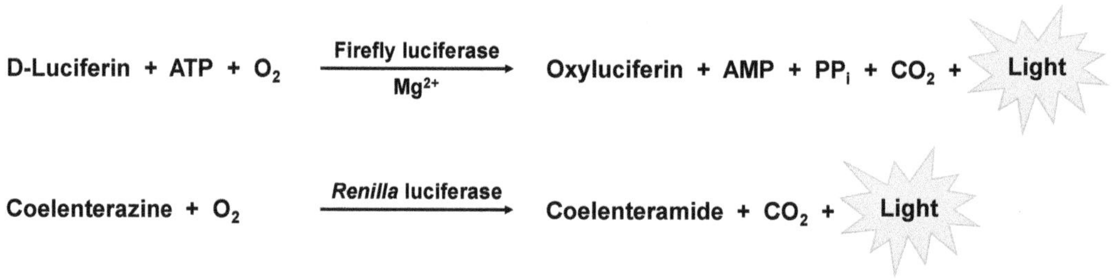

D-Luciferin + ATP + O_2 —Firefly luciferase, Mg^{2+}→ Oxyluciferin + AMP + PP_i + CO_2 + Light

Coelenterazine + O_2 —*Renilla* luciferase→ Coelenteramide + CO_2 + Light

Fig. 2 Bioluminescent reactions catalyzed by firefly and *Renilla* luciferase. Firefly luciferase catalyzes the ATP- and oxygen-dependent oxidative decarboxylation of D-luciferin to oxyluciferin. *Renilla* luciferase catalyzes the oxygen-dependent oxidative decarboxylation of coelenterazine to coelenteramide. Oxyluciferin and coelenteramide are both produced in an electronically excited state. Relaxation to their ground state is accompanied by the emission of a photon of light

experimental reporter, is driven by the regulatory element of interest. The expression of the second luciferase is controlled by a constitutively active promoter (e.g., a housekeeping or viral promoter). It can therefore be used as an internal control to normalize the activity of the reporter luciferase [8, 9].

Two luciferase reporter genes commonly used in dual luciferase assays are the ones from *Photinus pyralis* (firefly) and *Renilla reniformis* (sea pansy) [8, 9]. Firefly luciferase catalyzes the ATP-dependent oxidative decarboxylation of D-luciferin to oxyluciferin [10]. *Renilla* luciferase catalyzes the ATP-independent oxidative decarboxylation of coelenterazine to coelenteramide [11] (Fig. 2). Because firefly and *Renilla* luciferase utilize different substrates and require different buffer compositions, the activity of both enzymes can be measured sequentially in a single sample. Firefly luciferase luminescence is measured first by adding the respective substrate buffer to the sample. Afterwards, a substrate buffer for *Renilla* luciferase is added, activating the *Renilla* luciferase reaction while simultaneously quenching the firefly luciferase reaction. Quenching occurs by the high concentration (1.1 M) of NaCl in the *Renilla* luciferase substrate buffer, which prohibits firefly luciferase activity, and the increased EDTA concentration, which chelates Mg^{2+} ions required as a cofactor for firefly luciferase [12]. In most cases, *Renilla* luciferase is used as an internal control, and firefly luciferase is used as an experimental reporter. Normalization is performed by calculating the ratio of firefly luminescence to *Renilla* luminescence [8, 9].

Here, a detailed protocol for the analysis of CaSR-mediated SRE activation using a dual luciferase assay in Human Embryonic Kidney (HEK) 293 cells in a 96-well plate format is described. HEK293 cells are an established and widely used model system to characterize CaSR signaling [13–18]. This chapter comprises a 4-day protocol to transiently express wild-type or mutant CaSR along with the luciferase reporter genes (see Subheading 3.1). On

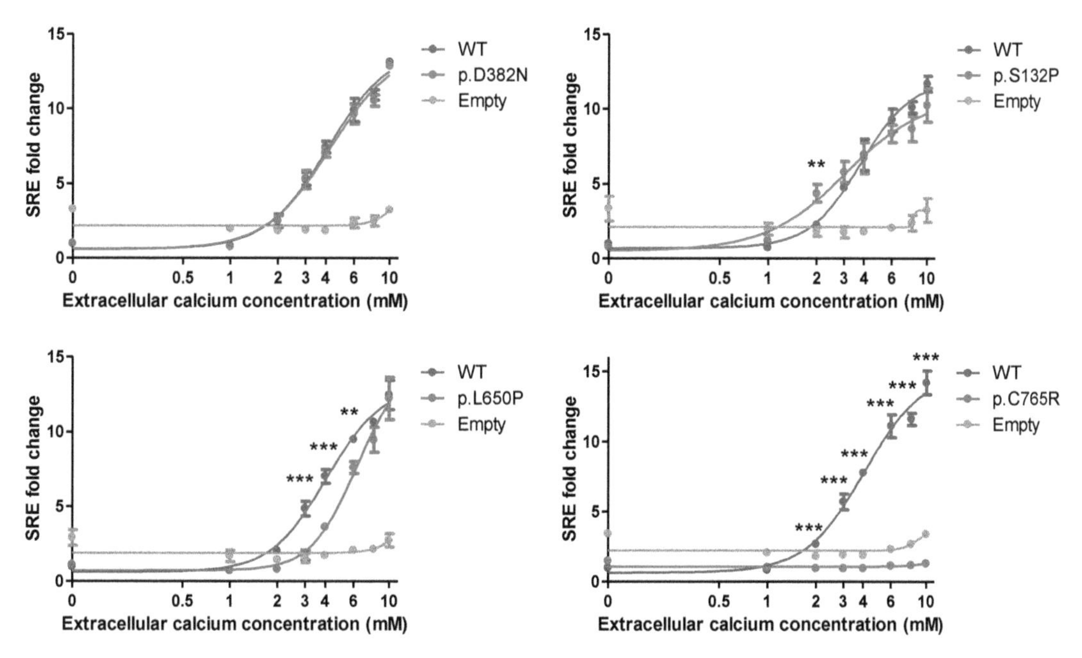

Fig. 3 CaSR-mediated serum response element (SRE) activation after stimulation with different concentrations of extracellular Ca^{2+} for exemplary mutants in comparison with the wild-type (WT) receptor. SRE activation is expressed as fold change relative to the normalized SRE activity of WT CaSR at a concentration of 0.1 mM $CaCl_2$. Curve fitting was performed with GraphPad Prism using the four-parameter model. The curves for p.L650P and p.C765R are typical for loss-of-function mutations that can be found in patients with familial hypocalciuric hypercalcemia type 1 (FHH1). In contrast, the curve for p.S132P is typical for gain-of-function mutations that can be found in patients with autosomal dominant hypocalcemia type 1 (ADH1). p.D382N is an example for a mutant that does not affect receptor activity (i.e., unlikely to be pathogenic). Data are derived from three independent experiments ($n = 3$). Data are presented as mean ± SD. ** $P \leq 0.01$, *** $P \leq 0.001$. Empty: Cells were transfected with an empty backbone vector

day 4, the CaSR is activated with different concentrations of extracellular Ca^{2+} over a couple of hours. Afterwards, the cells are lyzed to release the luciferases. The amount of firefly luciferase that was formed, driven by CaSR signaling, and the amount of *Renilla* luciferase as internal control are quantified using self-made substrate buffers (see Subheading 3.2). With self-made substrate buffers, the costs for 100 dual luciferase assays amount to approximately 5 euros and are therefore >30-fold less than with commercially available assay kits. The use of a microplate reader with automated injection of the luciferase substrate buffers enables high-throughput analysis of *CASR* variants with highly reproducible results. Figure 3 shows exemplary dose–response curves for the activation of the SRE by selected CaSR mutants that differentially affect CaSR signaling in comparison with the wild-type receptor. In general, loss-of-function mutations cause hypercalcemic disorders such as familial hypocalciuric hypercalcemia type 1 (FHH1), and gain-of-function mutations cause hypocalcemic disorders such as autosomal dominant hypocalcemia type 1 (ADH1) [3].

2 Materials

Prepare all solutions using ultrapure water (resistivity of 18.2 MΩ·cm at 25 °C).

2.1 Vectors

1. Experimental reporter vector containing a SRE driving the expression of the firefly luciferase reporter gene (*luc2P*) (pGL4.33[*luc2P*/SRE/Hygro] vector (Promega)) (*see* **Note 1**).
2. Internal control vector containing a constitutively active promoter driving the expression of the *Renilla* luciferase reporter gene. We use a pRL vector that contains a ubiquitin C (UbC) promoter.
3. *CASR* expression vector. Our *CASR* expression vector is based on a pcDNA3.1(+) vector backbone. We express the CaSR with an N-terminal c-myc tag.

2.2 Cell Culture

1. Human Embryonic Kidney (HEK) 293 cells.
2. Complete growth medium: High glucose Dulbecco's Modified Eagle's Medium (DMEM) supplemented with 10% (v/v) fetal bovine serum (FBS) (and antibiotics).
3. Phosphate-buffered saline (PBS).
4. Trypsin-EDTA solution.
5. 37 °C water bath.
6. Humidified 37 °C, 5% CO_2 incubator.
7. Cell culture hood.
8. Cell counter of choice.
9. 15 mL centrifuge tubes and centrifuge.
10. 10 cm cell culture dishes.
11. White-walled 96-well cell culture plates (*see* **Note 2**).
12. 0.1 mg/mL poly-D-lysine solution. Add 50 mL sterile water to 5 mg poly-D-lysine powder. Mix well. Store at 4 °C.

2.3 Transfection

1. Lipofectamine™ 2000 Transfection Reagent (Invitrogen).
2. Opti-MEM™ I Reduced Serum Medium (Gibco).
3. 1.5 mL or 2 mL reaction vials.

2.4 CaSR Activation with Extracellular Ca^{2+}

1. 1 M $CaCl_2$ stock solution (purchased).
2. 100 mM $CaCl_2$ working stock solution: Add 1 volume of the 1 M $CaCl_2$ solution to 9 volumes of water. Mix well.
3. $CaCl_2$-free DMEM.

2.5 Cell Lysis

1. PBS.
2. 1× Passive Lysis Buffer (*see* **Note 3**): Add 1 volume of 5× Passive Lysis Buffer (Promega) to 4 volumes of water. Mix well.

2.6 Luciferase Substrate Buffers

Luciferase substrate buffers are prepared based on stock solutions of the individual components:

1. 530 mM ATP solution: Weigh out 2.92 g ATP disodium salt. Add water to a final volume of 10 mL and mix. Dispense into aliquots that are stored at −20 °C.
2. Bovine serum albumin (BSA). Store at 4 °C.
3. 14.3 mM coelenterazine solution: Add 826 μL ethanol to 5 mg coelenterazine. Mix and dispense into aliquots that are stored at −20 °C.
4. 270 mM coenzyme A solution: Add water to a final volume of 4.8 mL to 1 g coenzyme A free acid and mix. Dispense into aliquots that are stored at −20 °C.
5. 1 M dithiothreitol solution (purchased). Dispense into aliquots that are stored at −20 °C.
6. 117.5 mM D-luciferin solution: Add 28.153 mL water to 1 g D-luciferin sodium salt. Mix and dispense into aliquots that are stored at −20 °C.
7. 500 mM EDTA disodium salt solution: Weigh out 93.06 g EDTA disodium salt dihydrate. Add about 400 mL water. Place on a magnetic stirrer. Add NaOH pellets until EDTA starts to dissolve. Adjust pH to 8.0 using a NaOH solution. Add water to a final volume of 500 mL. Store at room temperature.
8. 1 M ?K_2HPO_4 solution (pH 5.1): Weigh out 174.18 g ? K_2HPO_4. Add about 900 mL water. Place on a magnetic stirrer and adjust pH to 5.1 using HCl. Add water to a final volume of 1 L. Store at room temperature.
9. 100 mM $MgSO_4$ solution: Weigh out 6.02 g $MgSO_4$. Add water to a final volume of 500 mL and mix. Store at room temperature.
10. 2 M NaCl solution: Weigh out 116.88 g NaCl. Add water to a final volume of 1 L and mix. Store at room temperature.
11. 4.6 M NaN_3 solution (*see* **Note 4**): Weigh out 11.96 g NaN_3. Add water to a final volume of 40 mL and mix. Store at room temperature.
12. 200 mM tricine solution: Weigh out 17.92 g tricine. Add water to a final volume of 500 mL and mix. Dispense into aliquots that are stored at −20 °C.

13. *Firefly luciferase substrate buffer* [19]: 470 μM D-luciferin, 530 μM ATP, 2.67 mM $MgSO_4$, 270 μM coenzyme A (*see* **Note 5**), 33.3 mM dithiothreitol, 0.1 mM EDTA, 20 mM tricine, pH 7.8.

    For 200 mL substrate buffer, add 166.76 mL water, 200 μL of a 530 mM ATP solution, 5.34 mL of a 100 mM $MgSO_4$ solution, 200 μL of a 270 mM coenzyme A solution, 6.66 mL of a 1 M dithiothreitol solution, 40 μL of a 500 mM EDTA solution, and 20 mL of a 200 mM tricine solution to a glass beaker. Place on a magnetic stirrer and adjust pH to 7.8 using NaOH. Add 800 μL of a 117.5 mM D-luciferin solution (*see* **Note 6**). Mix and dispense into aliquots that are stored at −20 °C (*see* **Note 7**). Avoid repeated freeze-thaw cycles.

14. *Renilla luciferase substrate buffer* [20]: 1.43 μM coelenterazine, 0.44 mg/mL BSA, 1.1 M NaCl, 2.2 mM EDTA, 1.3 mM NaN_3, 220 mM ?K_2HPO_4, pH 5.0.

    For 200 mL substrate buffer, weigh 88 mg BSA in a glass beaker. Add 45.04 mL water, 110 mL of a 2 M NaCl solution, 880 μL of a 500 mM EDTA solution, 56.52 μL of a 4.6 M NaN_3 solution, and 44 mL of a 1 M ?K_2HPO_4 solution (pH 5.1). Place on a magnetic stirrer and adjust pH to 5.0 using HCl. Add 20 μL of a 14.3 mM coelenterazine solution (*see* **Note 6**). Mix and dispense into aliquots that are stored at −20 °C (*see* **Note 7**). Avoid repeated freeze-thaw cycles.

2.7 Equipment for Luciferase Activity Quantification

1. Microplate luminometer or multimode microplate reader with at least two automated injectors (e.g., a Tecan infinite® M200 microplate reader) (*see* **Note 8**).
2. Orbital shaker or rocking platform for microplates.

3 Methods

3.1 Cell Culture-Transient Expression and Activation of the CaSR

Perform all cell culture work (e.g., cell seeding, transfection, replating, and stimulation) aseptically under a cell culture hood and ensure sterility of all equipment, reagents, buffers, and media. The cell culture part of the method comprises a 4-day protocol. All volumes and amounts are given on a per-well or per-dish basis.

3.1.1 Day 1: Cell Seeding

1. Seed 3×10^6 HEK293 cells in a 10 cm cell culture dish in a total volume of 10 mL pre-warmed (37 °C) complete growth medium.
2. Grow the cells in a 37 °C, 5% CO_2 incubator for 24 h.

3.1.2 Day 2: Transfection

1. Replace the cell culture medium with 10 mL fresh pre-warmed (37 °C) complete growth medium.

2. Dilute 6 μg of the firefly luciferase vector (pGL4.33[*luc2P*/SRE/Hygro], 3 μg of the *Renilla* luciferase internal control vector, and 3 μg of the *CASR* expression vector in Opti-MEM™ I Reduced Serum Medium in a total volume of 500 μL in a 1.5 or 2 mL reaction vial (*see* **Notes 9** and **10**). Mix well.
3. Add 20 μL of the Lipofectamine™ 2000 Transfection Reagent to 480 μL Opti-MEM™ I Reduced Serum Medium in a 1.5 or 2 mL reaction vial. Mix well.
4. Incubate at room temperature for 5 min.
5. Add the diluted vectors (see **step 2**) to the diluted Lipofectamine™ 2000 Reagent (see **step 3**).
6. Mix gently (e.g., by flicking the reaction vial).
7. Spin down briefly with a benchtop microcentrifuge to collect the transfection mix at the bottom of the reaction vial.
8. Incubate at room temperature for 20 min.
9. Carefully add the entire volume (1 mL) to the cells in a drop-wise manner. Gently sway the dish to ensure an even distribution of the transfection mix.
10. Grow the cells in a 37 °C, 5% CO_2 incubator for 24 h.

3.1.3 Day 3: Replating and Serum Starvation

1. Aspirate the cell culture medium from the 10 cm cell culture dishes.
2. Carefully rinse the cells with 8 mL pre-warmed (37 °C) PBS. Aspirate PBS.
3. Add 2 mL of a pre-warmed (37 °C) trypsin-EDTA solution.
4. Incubate at room temperature for 4 min.
5. Stop trypsinization by adding 8 mL pre-warmed (37 °C) complete growth medium (*see* **Note 11**).
6. Transfer the detached cells to a 15 mL centrifuge tube.
7. Centrifuge at 300 × *g* for 3 min.
8. Aspirate the supernatant.
9. Resuspend cell pellets in 10 mL pre-warmed (37 °C) complete growth medium (*see* **Note 12**).
10. Transfer each 100 μL of the cell suspension to the wells of a poly-D-lysine-coated (see Subheading 3.1.5) white-walled 96-well plate (*see* **Notes 13** and **14**).
11. Maintain the cells in a 37 °C, 5% CO_2 incubator for 8 h (*see* **Note 15**).
12. Aspirate the cell culture medium completely and add 150 μL pre-warmed (37 °C) serum-free DMEM containing 0.5 mM

$CaCl_2$ (prepare using the 100 mM $CaCl_2$ working stock solution and the $CaCl_2$-free DMEM and mix well) (*see* **Note 16**).

13. Maintain the cells in a 37 °C, 5% CO_2 incubator for 16 h (*see* **Note 17**).

3.1.4 Day 4: CaSR Activation and Cell Lysis

1. Prepare serum-free DMEM containing different concentrations of $CaCl_2$ ranging from 0.1 mM to at least 10 mM (stimulation medium) (*see* **Note 18**) using the 100 mM $CaCl_2$ working stock solution (*see* **Note 19**) and pre-warmed (37 °C) $CaCl_2$-free DMEM. Mix well.
2. Aspirate the serum-free DMEM containing 0.5 mM $CaCl_2$ completely, and replace it with 150 μL of the stimulation medium with a given $CaCl_2$ concentration (*see* **Notes 20** and **21**).
3. Maintain the cells in a 37 °C, 5% CO_2 incubator for 3 h (*see* **Note 22**).
4. Aspirate the stimulation medium.
5. Carefully rinse the cells with 100 μL PBS (*see* **Note 23**). Aspirate PBS completely.
6. Add 20 μL 1× Passive Lysis Buffer (*see* **Note 23**).
7. Store the plates with lyzed cells at −20 °C (*see* **Note 24**).

3.1.5 Poly-D-Lysine Coating

1. Add 50 μL of a 0.1 mg/mL poly-D-lysine solution per well. Make sure that the entire cell culture surface is covered by rocking the plate.
2. Incubate at room temperature for 1 h.
3. Remove the poly-D-lysine solution (*see* **Note 25**).
4. Rinse the culture surface with 200 μL ultrapure water. Aspirate the water completely.
5. Leave the plate uncovered to dry for 2 h under a cell culture hood.
6. Use coated plates immediately or wrap them in a plastic bag and store them at 4 °C for a few days.

3.2 Quantification of Luciferase Activities

1. Thaw aliquots of the luciferase substrate buffers at room temperature while protecting them from light (*see* **Note 26**).
2. Thaw the 96-well plate with cell lysates at room temperature for 10 min.
3. Place the plate on an orbital shaker or rocking platform at room temperature for 20 min.
4. Depending on the luminometer or microplate reader, clean and prime the injectors with luciferase substrate buffers following the manufacturer's instructions for the device.

5. Perform the measurements at room temperature.
6. Inject 100 μL firefly luciferase substrate buffer into the first well (rate: 150 μL/s), and immediately quantify firefly luciferase activity (settle time: 1.000 ms; integration time: 10 s).
7. Inject 100 μL *Renilla* luciferase substrate buffer into the same well (rate: 150 μL/s), and immediately quantify *Renilla* luciferase activity (settle time: 1.000 ms; integration time: 10 s).
8. Repeat **steps 6** and 7 for the other wells.

4 Notes

1. To analyze the activation of the NFAT-RE, use the pGL4.30 [*luc2P*/NFAT-RE/Hygro] vector (Promega) instead.
2. It is recommended to use plates with a solid bottom. A white, solid bottom will reflect the light signal and maximize the luminescence signal output. However, if the microplate luminometer or multimode microplate reader (see Subheading 2.7) allows only a bottom read mode, use plates with a clear bottom.
3. It is recommended to make the 1× Passive Lysis Buffer fresh every time.
4. Sodium azide (NaN_3) is highly toxic. Only handle with gloves under a fume hood while wearing a lab coat.
5. Adding coenzyme A to the firefly luciferase substrate buffer results in a more intense and sustained light signal. In the absence of coenzyme A, dehydroluciferyl-adenylate, which is formed as a side product, acts as a potent inhibitor of the reaction. In the presence of coenzyme A, dehydroluciferyl-coenzyme A is formed instead, which is a much weaker inhibitor of the reaction [21].
6. Add the luciferase substrates (D-luciferin or coelenterazine) after adjusting the pH to minimize exposure of the substrates to light and oxygen.
7. We found that the substrate buffers can be stored at −20 °C for a few months without a decrease in effectiveness. For more long-term storage, a temperature of −80 °C is recommended.
8. If no plate-reading luminometer is available, luciferase activities can also be measured manually in each well using a tube luminometer. However, the reproducibility of results may be lower when adding the substrate buffers manually compared to when using automated injection with standardized time spans between adding the substrate buffers and measuring luminescence. Moreover, manual measurements are more time-consuming, lowering the throughput of samples.

9. It is recommended to use aliquots of the same working stock solution of a vector for a series of biological replicates to minimize experimental variation.
10. The activity of the NFAT-RE is much lower compared to the activity of the SRE, resulting in much lower levels of firefly luciferase activity. This may be explained by the fact that ERK1/2 activation is a convergence point of many signaling pathways ($G\alpha_{q/11}$, $G\alpha_{i/o}$, and β-arrestin). In contrast, the activation of the NFAT-RE is determined only by cytosolic Ca^{2+} levels ($G\alpha_{q/11}$) [4]. Varying the amount of *CASR* expression vector and pGL4.30[*luc2P*/NFAT-RE/Hygro] vector may help to get reproducible data.
11. Treat the dishes in the same order in each step, i.e., add complete growth medium in the same order to the individual dishes as trypsin was added to ensure approximately equal times of trypsinization in each dish.
12. Right after aspirating the supernatant, immediately resuspend all pellets in complete growth medium to avoid the cells existing as pellets for too long. We always resuspend the pellets in initially 1 mL complete growth medium and add the remaining 9 mL complete growth medium immediately before transferring the cell suspension to the 96-well plate.
13. It is not necessary to count the cells at this point. Well-to-well variability in cell number will be normalized by co-transfection of the internal control vector [9]. However, cells should be collected carefully from the 10 cm dishes to avoid major variations in cell number that may affect cell behavior (e.g., add PBS carefully to the dishes to prevent cell loss while washing and transfer the entire volume of cell suspension to the 15 mL centrifuge tube after trypsinization).
14. For convenience, transfer the cell suspension to a sterile reagent reservoir and dispense it in the 96-well plate using a multichannel pipette.
15. After 8 h, the cells have stably attached to the surface. To be able to monitor cell attachment when using assay plates with solid bottom, cells could additionally be replated in a poly-D-lysine coated 96-well cell culture plate with clear bottom.
16. For convenience, aspirate the cell culture medium from multiple wells simultaneously using a multichannel adapter for a vacuum aspirator, transfer the serum-free DMEM with 0.5 mM $CaCl_2$ to a sterile reagent reservoir, and dispense it into the 96-well plate using a multichannel pipette.
17. Before starting the stimulation, CaSR signaling pathways are minimized for a couple of hours by maintaining the cells in a reduced extracellular Ca^{2+} concentration (0.5 mM). This is

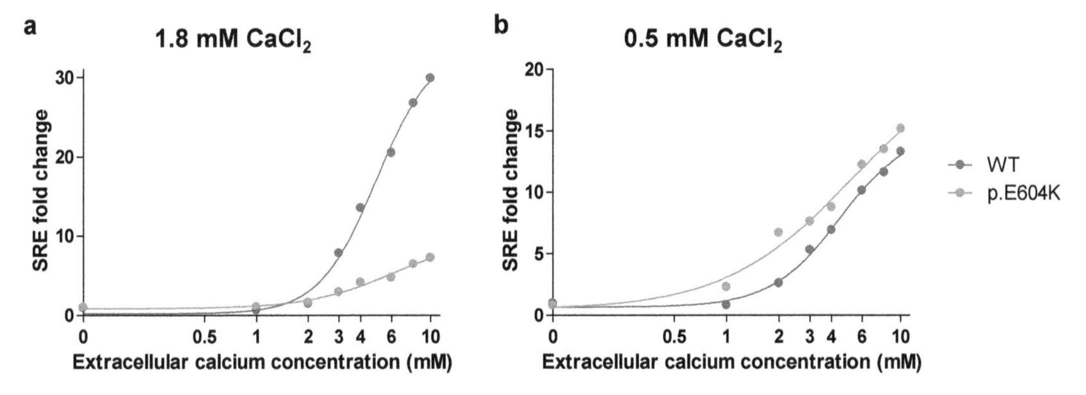

Fig. 4 CaSR-mediated serum response element (SRE) activation after stimulation with different concentrations of extracellular Ca^{2+} for the gain-of-function mutant p.E604K [24, 25] in comparison with the wild-type (WT) receptor. Before stimulation, the cells were either maintained in (**a**) standard serum-free DMEM with 1.8 mM $CaCl_2$ or in (**b**) serum-free DMEM with 0.5 mM $CaCl_2$ for 16 h. SRE activation is expressed as fold change relative to the normalized SRE activity of WT CaSR at a concentration of 0.1 mM $CaCl_2$. Curve fitting was performed with GraphPad Prism using the four-parameter model. Data is derived from one experiment

necessary to be able to identify *CASR* mutations that cause a gain of function. If the cells are maintained in standard DMEM with 1.8 mM $CaCl_2$ until starting the stimulation, gain-of-function mutations may show reduced CaSR signaling in the luciferase assay (see Fig. 4). They might therefore be misclassified as loss-of-function mutations. Of note, Ca^{2+} in the cell culture medium is required for the adhesion of cells to the culture surface and to one another [22]. Thus, Ca^{2+} cannot be completely omitted from the cell culture medium and a concentration of 0.5 mM is used. Serum starvation is required because FBS contains high levels of growth factors and hormones that would result in ERK1/2 activation. In addition, FBS contains amino acids and cations that would also activate the CaSR [23]. Moreover, proteins like albumin bind Ca^{2+} so that the concentration of free, ionized Ca^{2+} cannot be controlled. Due to the latter two factors, serum starvation is also required in the NFAT-RE luciferase assay.

18. We use the following concentrations of $CaCl_2$: 0.1, 1, 2, 3, 4, 6, 8, and 10 mM.
19. When diluting $CaCl_2$ in serum-free DMEM, insoluble Ca^{2+} salts may form at higher concentrations (e.g., 10 mM). This can be minimized by using a 100 mM working stock solution instead of the 1 M $CaCl_2$ stock solution. In addition, it is recommended to purchase a 1 M $CaCl_2$ solution instead of making it from powder. Dilute $CaCl_2$ in $CaCl_2$-free DMEM fresh every time immediately before adding the stimulation medium to the cells.

20. Three technical replicates for each $CaCl_2$ concentration are recommended. The technical replicates may also be on separate 96-well plates.
21. For convenience, remove the medium from multiple wells simultaneously using a multichannel adapter for a vacuum aspirator. The different stimulation media with the different $CaCl_2$ concentrations can be added well-wise with a single-channel pipette or by transferring the media to a sterile reagent reservoir and using a multichannel pipette.
22. For the NFAT-RE dual luciferase assay, a different stimulation time may be required.
23. For convenience, remove the stimulation medium/PBS from multiple wells simultaneously using a multichannel adapter for a vacuum aspirator and add PBS/1x Passive Lysis Buffer using a reagent reservoir and a multichannel pipette.
24. We found that cell lysates can be stored at −20 °C for up to 2 weeks without a decrease in luciferase activity. For more long-term storage, a temperature of −80 °C is recommended.
25. The poly-D-lysine solution can be reused at least four times.
26. Using the same batch of firefly and *Renilla* luciferase substrate buffer in a series of biological replicates allows virtually identical ratios of firefly activity to *Renilla* activity in each experiment.

References

1. Fan F, Wood KV (2007) Bioluminescent assays for high-throughput screening. Assay Drug Dev Technol 5(1):127–136. https://doi.org/10.1089/adt.2006.053
2. Riccardi D, Brown EM (2010) Physiology and pathophysiology of the calcium-sensing receptor in the kidney. Am J Physiol Renal Physiol 298(3):F485–F499. https://doi.org/10.1152/ajprenal.00608.2009
3. Hannan FM, Babinsky VN, Thakker RV (2016) Disorders of the calcium-sensing receptor and partner proteins: insights into the molecular basis of calcium homeostasis. J Mol Endocrinol 57(3):R127–R142. https://doi.org/10.1530/jme-16-0124
4. Gorvin CM (2018) Insights into calcium-sensing receptor trafficking and biased signalling by studies of calcium homeostasis. J Mol Endocrinol 61(1):R1–R12. https://doi.org/10.1530/jme-18-0049
5. Goldsmith ZG, Dhanasekaran DN (2007) G protein regulation of MAPK networks. Oncogene 26(22):3122–3142. https://doi.org/10.1038/sj.onc.1210407
6. Wasylyk B, Hagman J, Gutierrez-Hartmann A (1998) Ets transcription factors: nuclear effectors of the Ras-MAP-kinase signaling pathway. Trends Biochem Sci 23(6):213–216. https://doi.org/10.1016/s0968-0004(98)01211-0
7. Rao A, Luo C, Hogan PG (1997) Transcription factors of the NFAT family: regulation and function. Annu Rev Immunol 15:707–747. https://doi.org/10.1146/annurev.immunol.15.1.707
8. Allard ST (2008) Bioluminescent reporter genes. https://www.promega.de/resources/pubhub/enotes/bioluminescent-reporter-genes/. Accessed 7 Nov 2023
9. Schagat T, Paguio A, Kopish K (2007) Normalizing genetic reporter assays: approaches and considerations for increasing consistency and statistical significance. Cell Notes 17:9–12
10. Dewet JR, Wood KV, Deluca M, Helinski DR, Subramani S (1987) Firefly luciferase gene: structure and expression in mammalian cells. Mol Cell Biol 7(2):725–737. https://doi.org/10.1128/mcb.7.2.725

11. Woo JC, Howell MH, Von Arnim AG (2008) Structure-function studies on the active site of the coelenterazine-dependent luciferase from Renilla. Protein Sci 17(4):725–735. https://doi.org/10.1110/ps.073355508
12. Sherf BA, Schenborn ET, Leary LN (1998) Quenching reagents and assays for enzyme-mediated luminescence. US Patent 5,744,320, 7 June 1995
13. Gorvin CM, Frost M, Malinauskas T, Cranston T, Boon H, Siebold C, Jones EY, Hannan FM, Thakker RV (2018) Calcium-sensing receptor residues with loss- and gain-of-function mutations are located in regions of conformational change and cause signalling bias. Hum Mol Genet 27(21):3720–3733. https://doi.org/10.1093/hmg/ddy263
14. Guarnieri V, D'Elia AV, Baorda F, Pazienza V, Benegiamo G, Stanziale P, Copetti M, Battista C, Grimaldi F, Damante G, Pellegrini F, D'Agruma L, Zelante L, Carella M, Scillitani A (2012) CASR gene activating mutations in two families with autosomal dominant hypocalcemia. Mol Genet Metab 107(3):548–552. https://doi.org/10.1016/j.ymgme.2012.06.012
15. Leach K, Wen A, Davey AE, Sexton PM, Conigrave AD, Christopoulos A (2012) Identification of molecular phenotypes and biased signaling induced by naturally occurring mutations of the human calcium-sensing receptor. Endocrinology 153(9):4304–4316. https://doi.org/10.1210/en.2012-1449
16. Li D, Opas EE, Tuluc F, Metzger DL, Hou CP, Hakonarson H, Levine MA (2014) Autosomal dominant hypoparathyroidism caused by germline mutation in GNA11: phenotypic and molecular characterization. J Clin Endocrinol Metab 99(9):E1774–E1783. https://doi.org/10.1210/jc.2014-1029
17. Magno AL, Leatherbarrow KM, Brown SJ, Wilson SG, Walsh JP, Ward BK (2020) Functional analysis of calcium-sensing receptor variants identified in families provisionally diagnosed with familial hypocalciuric hypercalcaemia. Calcif Tissue Int 107(3):230–239. https://doi.org/10.1007/s00223-020-00715-1
18. Nakamura A, Hotsubo T, Kobayashi K, Mochizuki H, Ishizu K, Tajima T (2013) Loss-of-function and gain-of-function mutations of calcium-sensing receptor: functional analysis and the effect of allosteric modulators NPS R-568 and NPS 2143. J Clin Endocrinol Metab 98(10):E1692–E1701. https://doi.org/10.1210/jc.2013-1974
19. Gaunitz F, Papke M (1998) Gene transfer and expression. Methods Mol Biol 107:361–370. https://doi.org/10.1385/0-89603-519-0:361
20. Dyer BW, Ferrer FA, Klinedinst DK, Rodriguez R (2000) A noncommercial dual luciferase enzyme assay system for reporter gene analysis. Anal Biochem 282(1):158–161. https://doi.org/10.1006/abio.2000.4605
21. Fraga H, Fernandes D, Fontes R, Esteves da Silva JC (2005) Coenzyme A affects firefly luciferase luminescence because it acts as a substrate and not as an allosteric effector. FEBS J 272(20):5206–5216. https://doi.org/10.1111/j.1742-4658.2005.04895.x
22. Kemler R, Ozawa M, Ringwald M (1989) Calcium-dependent cell adhesion molecules. Curr Opin Cell Biol 1(5):892–897. https://doi.org/10.1016/0955-0674(89)90055-0
23. Hannan FM, Kallay E, Chang W, Brandi ML, Thakker RV (2018) The calcium-sensing receptor in physiology and in calcitropic and noncalcitropic diseases. Nat Rev Endocrinol 15(1):33–51. https://doi.org/10.1038/s41574-018-0115-0
24. Cavaco BM, Canaff L, Nolin-Lapalme A, Vieira M, Silva TN, Saramago A, Domingues R, Rutter MM, Hudon J, Gleason JL, Leite V, Hendy GN (2018) Homozygous calcium-sensing receptor polymorphism R544Q presents as hypocalcemic hypoparathyroidism. J Clin Endocrinol Metab 103(8):2879–2888. https://doi.org/10.1210/jc.2017-02407
25. Tan YM, Cardinal J, Franks AH, Mun HC, Lewis N, Harris LB, Prins JB, Conigrave AD (2003) Autosomal dominant hypocalcemia: a novel activating mutation (E604K) in the cysteine-rich domain of the calcium-sensing receptor. J Clin Endocrinol Metab 88(2):605–610. https://doi.org/10.1210/jc.2002-020081

Part II

Microscopy Methods to Assess Calcium Signaling

Chapter 7

Calcium Imaging in Brain Tissue Slices

Orsolya Kékesi, Nisal Keembiyage, and Yossi Buskila

Abstract

Calcium imaging is a method that was first developed in the mid-1970s yet kept developing until current days to allow accurate measurement of free calcium ions in tissues. This widely used method has provided significant advances to our understanding of cellular signal transduction, including the discovery of subcellular compartmentalization of neurons and astrocytes, the identification of multiple signaling pathways, and mapping the functional connectivity between astrocytes and neuronal networks. Here we describe a method for the loading and imaging of cell-permeable AM ester calcium-sensitive dyes for the in vitro measurement of free intracellular Ca^{2+} ions in acute brain slices.

Key words AM dyes, Braincubator, Brain slices, Calcium imaging, Neurophysiology

1 Introduction

Calcium (Ca^{2+}) ions are among the most important secondary messengers in cellular signaling, involved in a myriad of physiological processes including neurotransmitter vesicular release, synaptic plasticity, muscle contraction, and K^+ homeostasis [1, 2]. Due to the numerous processes Ca^{2+} is involved in, its intracellular concentration is tightly regulated primarily by calcium-binding proteins and ion pumps [3]. In non-neuronal cells, such as astrocytes, transient fluctuations in the intracellular calcium concentration are associated with the release of signaling molecules, regulation of blood flow, and ion homeostasis that affects both cellular and pan-glial activities due to their ability to travel through intercellular channels [1]. Indeed, close monitoring of intracellular Ca^{2+} fluctuations has led to major advances in our understanding of neuronal processes at the subcellular level, including the release of gliotransmitters, such as glutamate [4], adenosine triphosphate [5], and D-serine [6] that affects neuronal activity [1].

Calcium-sensitive fluorescent indicators have been widely used to monitor signal transduction in both in vitro and in vivo studies. Two major types of chemical calcium indicators are currently

Caroline M. Gorvin (ed.), *Calcium Signaling: Methods and Protocols*, Methods in Molecular Biology, vol. 2861, https://doi.org/10.1007/978-1-0716-4164-4_7,

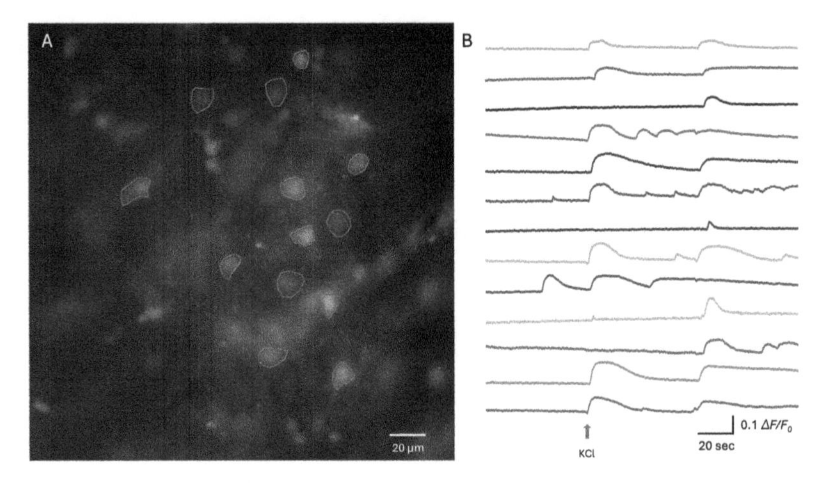

Fig. 1 Calcium imaging in acute brain slice. (**a**) Fluorescence image of neocortical slice loaded with the calcium indicator Fluo-4 AM. Blue circles outlining selected cells to record Ca^{2+} signals. (**b**) Time-lapse traces of Ca^{2+} signals recorded from the outlined cells in (**a**) before and following local application of 30 mM KCl (red arrow)

commercially available: ratiometric and non-ratiometric indicators. The ratiometric dyes, such as Fura-2 and Fura-Red, are a type of dual excitation-wavelength dyes that shift their maximum excitation wavelength in the presence of the target ion while maintaining a fixed emission wavelength. Hence, calculating the ratio between the emission/excitation^1 and emission/excitation^2 provides an accurate measurement of free calcium levels [7]. These distinctive spectral properties allow robust and reproducible measurements that decrease the impact of inaccuracies associated with dye leakage, photobleaching, and unequal dye loading. The non-ratiometric, or intensity dyes (e.g., Fluo-4 and Calcium Green) increase the intensity of fluorescence without changing the excitation or emission spectrum upon binding the calcium ion (see Fig. 1). While ratiometric indicators have the advantage of being more precise in determining the intracellular concentration, due to their excitation and emission spectra, their excitation maxima fall within the ultraviolet range which has a potential phototoxic effect on the slices. The intensity dyes have the advantage of ease of applicability, as they don't require equipment that facilitate a fast switching between excitation wavelengths; however, photobleaching and uneven dye loading can affect the quality of the measurements (extensively reviewed in [7]).

Here we describe the method our group uses to perform calcium imaging experiments in acute brain slices. To maintain the viability of brain slices, our group uses the Braincubator™, an incubation system that prolongs the lifespan of acute neuronal tissue [8–10]. Following dye loading, slices can be kept in the

Braincubator™ for an extended time period (>24 h post dissection) until they are used for imaging [11].

2 Materials

2.1 Chemicals

- Fluo-4-AM (Molecular weight (MW) 1096.95, 50 μg vials).
- Fluo-8-AM (MW 1046.94 g/mol, 50 μg vials).
- Pluronic acid-127 (MW 12500 Da).
- Fura-2-AM (MW 1001.9, 50 μg vials).
- DMSO (MW 78.13, Anhydrous).
- Carbogen (5% CO_2 mixed in 95% O_2).
- KCl (MW 74.55).
- NaH_2PO_4 monobasic monohydrate (MW 137.99).
- $NaHCO_3$ (MW 84.01).
- D-glucose (MW 180.16).
- NaCl (MW 58.44).
- $MgCl_2$ (MW 95.21).
- $CaCl_2$ (MW 110.98).

2.2 Equipment

1. Sonicator.
2. Pasteur pipette.
3. Glass vial.
4. Eppendorf tubes (3 mL).
5. Gas dispersion tube.
6. Slice carrier-custom made (*see* **Note 1**).
7. Water bath set to 37 °C.
8. Braincubator™ (PaYo Scientific Ltd. Pty; http://braincubator.com.au)—optional.
9. BX51W Olympus Microscope.

2.3 Solutions

Prepare all solutions using MilliQ water (prepared by purifying deionized water with sensitivity of 18.2 MΩ-cm at 25 °C), store all reagents at room temperature (unless stated otherwise), and dispose of waste appropriately according to waste disposal regulations:

1. Ca^{2+}-free aCSF (125 mM NaCl, 2.5 mM KCl, 1 mM $MgCl_2$, 1.25 mM NaH_2PO_4, 25 mM $NaHCO_3$, 25 mM D-glucose): pH 7.4, osmolarity 300–310 mOsm. Add 800 mL of MilliQ water to a 1 L glass beaker and add a magnetic stirrer. Place the beaker on a stirrer at 450 revolutions/minute. Weigh 7.305 g

NaCl, 0.186 g, KCl, 0.095 g $MgCl_2$, 0.149 g NaH_2PO_4, 2.1 g $NaHCO_3$, 4.5 g D-glucose, and add to the beaker while stirring. Add 200 mL of water to the 1 L beaker and measure pH and osmolarity. Adjust the osmolarity with sucrose if it is lower than the recommended range and adjust the pH with HCl if higher than required range or NaOH if lower than required range.

2. 10% Pluronic-127: Dissolve 10 mg powder of Pluronic-127 (MW 12,500) into 100 μL MilliQ water or DMSO and sonicate for 10 min. Pluronic F-127 is soluble in water up to 10% (weight/volume) or in DMSO up to 20%. Heating to 70 °C may be necessary to achieve these concentrations.
3. Stock solution: Fluo-4 AM, Fluo-8 AM, and Fura-2 AM are sold in 50 μg vials. To prepare a stock solution (~1 mM), add to each vial 45 μL of DMSO and 5 μL of 10% Pluronic acid-127 (for a final concentration of 1% to facilitate AM dye penetration to the cell). Sonicate the solution for 10–15 min (*see* **Note 2**).

3 Methods

This method is optimized for rodent brain slices (300 μm). Prepare brain slices according to resident protocols and animal ethics regulations, or alternatively use the method we previously described in detail [8, 10]. The following methods can be used to load brain slices with cell-permeable AM esters of calcium-sensitive fluorescent dyes, such as Fluo-4 AM, Fluo-8 AM, and Fura-2 AM. Fluo-4 AM is predominantly loaded into astrocytes [12]; hence, it is recommended to use an additional specific indicator, such as sulforhodamine 101 (SR101, selective marker for astrocytes) to confirm the identity of the cells. Carry out all procedures at room temperature unless otherwise stated.

3.1 Loading the Slices

1. Transport the slices using a slice carrier into a ~ 10 mL covered glass vial (2 slices/vial) and remove the excess fluid from the slices (*see* **Note 3**). Quickly and carefully add Ca^{2+}-free aCSF to the slices to a final volume of 2.5 mL.
2. Pipette 25 μL of the stock solution (1 mM) directly on top of the slices to get a final dye concentration of 10 μM. This step ensures that each slice is initially exposed to a high concentration of the AM dye and provides better penetration into deep layers of the tissue (*see* **Note 4**).
3. Place the carbogen tube right above the surface of the solution to incorporate oxygen and CO_2 into the system and close the lid to prevent light getting in. Incubate slices for 45 min at 37 °

C, or 1 h at RT (*see* **Note 5** for dye loading into brain slices from aged animals >12 weeks).

4. Following incubation, transfer the slices back to the holding chamber (at RT) or into the Braincubator™ (at 16 °C), and wait for 30 min to let the intracellular enzymes cleave the acetoxymethylester groups and for excess dye to be washed from the extracellular environment (*see* **Note 6**).
5. While waiting, the microscope and the filters can be set up according to the dye's specifications, and the imaging parameters at the discretion of the researcher (*see* **Note 7**).

3.2 Imaging

1. Following incubation, transfer the slice into a submerged bath chamber positioned under an upright microscope (*see* **Note 8**). While in DIC (differential interference contrast) mode, lower the immersible objective into the chamber until the surface of the slice is in focus. Adjust the position of the slice to the required area under investigation (*see* **Notes 8** and **9**).
2. In the image acquisition software, increase the exposure time to allow acquisition of low-light images. Make sure that the correct filter-set is in position and open the shutter to acquire images (*see* **Notes 10** and **11**).

3.3 Image Analysis

Analysis of Ca^{2+} signals can be done in various commercial and free open-source software. In our lab, we use FIJI, which is an image analysis software developed by the NIH as an open-source platform for biological image analysis [13]. The acquired image files can be imported into FIJI by using the "Plugins" → BioFormats → Bio-formats Importer function (or drag and drop).

3.3.1 Analysis of Fluorescence Signals

Once the file appears in a new window (as a stack of images), the brightness/contrast can be adjusted within Image → Adjust → Brightness/contrast. To evaluate dynamic changes in Ca^{2+} fluorescent signals during the recording, the user needs to preprocess the images to compensate for photobleaching occurring during the recording. This can be done using the TopoJ tool by going to Analyse → TopoJ → Remove slope (*see* **Note 12**).

Dynamic changes in the fluorescence intensity are usually normalized to reduce variations in fluorescence intensity that are due to fluorophore loading and tissue characteristics. A typical normalization method is to measure the signal to baseline ratio, termed $\Delta F/F_0$, where (F_0) is the background baseline value at resting state and ΔF is the difference in fluorescence intensity from baseline levels. We obtained F_0 by averaging the intensity of the first five images acquired in the stack:

- To get F_0: Go to Image → Stacks → Z project (average 1–5 images). The result (F_0) will appear in a new window; save it as a .tif file.
- To get the ΔF: The F_0 value needs to be deducted from the original set of images (F). Go to Process → Image Calculator → Subtract → select the appropriate set of frames. The ΔF file will appear in a new window; save the file.
- To get the $\Delta F/F_0$: Divide the saved ΔF file by the F_0 file. Go to Process → Image Calculator → Divide → select the appropriate set of frames. The normalized file $\Delta F/F_0$ will appear in a new window; save it as a .tif file.
- The regions of interest can be chosen by using the Analyze → Tools → ROI Manager function. Use the oval selection tool in the "Tools" menu to outline the cell contours. Click "Add" on the ROI Manager window. The selected ROIs can be saved in a compressed format in the ROI manager window by choosing "More" → Save function. The fluorescence intensity can be measured in the ROI Manager window by choosing More → Multi measure. The results will appear in a new window and can be saved as a .csv file, in which graphical charts can be generated.

4 Notes

1. A slice carrier can be made by cutting off the lower third of a 3 mL plastic Pasteur pipette and fire-polishing the edges to prevent the sharp edges from damaging the slices.
2. Cover the vial and DMSO with aluminum foil to avoid photobleaching. As the Pluronic acid forms foam during sonication, it is recommended to add the Pluronic-127 at the end of the sonication and mix the solution using the pipette tip.
3. Pipette the solution as soon as possible and try not to cause the slices to float.
4. Be very careful not to poke a hole in the slice with the pipette.
5. For slices from adult animals (>12 weeks), the loading should be done directly onto the surface of the slices to allow the dye to penetrate into deeper layers of the tissue, and incubation time should be increased to 75 min.
6. When transferring the slices one by one in the slice carrier, wait until the slice sinks down to the bottom of the carrier to use a lower amount of liquid for the transfer.
7. Turn on the fluorescent lamp at least 10 min before imaging to let the light intensity reach a steady state. The excitation maximum for Fluo-4 AM and Fluo-8 AM dyes is at 480 nm; hence,

use a band-pass filter between 460 and 490 nm and a dichroic mirror that reflects light <490 nm and transmits light >510 nm. To capture at the emission maximum (530 nm), use a band-pass filter of 515–550 nm. For the ratiometric dye Fura-2, the excitation maximum has two peaks at 340 and 380 nm, which is excited with a high-speed wavelength switcher, such as Lambda DG-4. The emitted light can be captured with an emission filter of 510 ± 20 nm.

8. Try to work in as low a light condition as possible. Make sure that the microscope is well covered with black curtains and the shutter of the fluorescent light is on to avoid unnecessary photobleaching. Long-term and high-intensity excitation causes photobleaching and degeneration of the signal. Keep the shutter closed while not in acquisition mode!
9. The top 30 μm of an acute neuronal slice contains mostly dead/damaged cells. Make sure you are acquiring images from deeper levels >30 μm.
10. The acquisition camera should be a high-speed digital camera, and the acquisition rate can be adjusted according to the researcher's discretion.
11. Ensure there is enough storage space on the computer as the acquired images have large sizes.
12. Might need to save the file as a 32 bit to get this plug-in to work.

References

1. Khakh BS, McCarthy KD (2015) Astrocyte calcium signaling: from observations to functions and the challenges therein. Cold Spring Harb Perspect Biol 7(4):a020404. https://doi.org/10.1101/cshperspect.a020404
2. Bellot-Saez A, Kekesi O, Morley JW, Buskila Y (2017) Astrocytic modulation of neuronal excitability through K(+) spatial buffering. Neurosci Biobehav Rev 77:87–97. https://doi.org/10.1016/j.neubiorev.2017.03.002
3. Bagur R, Hajnoczky G (2017) Intracellular Ca (2+) sensing: its role in calcium homeostasis and signaling. Mol Cell 66(6):780–788. https://doi.org/10.1016/j.molcel.2017.05.028
4. Perea G, Gomez R, Mederos S, Covelo A, Ballesteros JJ, Schlosser L, Hernandez-Vivanco A, Martin-Fernandez M, Quintana R, Rayan A, Diez A, Fuenzalida M, Agarwal A, Bergles DE, Bettler B, Manahan-Vaughan D, Martin ED, Kirchhoff F, Araque A (2016) Activity-dependent switch of GABAergic inhibition into glutamatergic excitation in astrocyte-neuron networks. elife 5. https://doi.org/10.7554/eLife.20362
5. Lalo U, Palygin O, Rasooli-Nejad S, Andrew J, Haydon PG, Pankratov Y (2014) Exocytosis of ATP from astrocytes modulates phasic and tonic inhibition in the neocortex. PLoS Biol 12(1):e1001747. https://doi.org/10.1371/journal.pbio.1001747
6. Mothet JP, Pollegioni L, Ouanounou G, Martineau M, Fossier P, Baux G (2005) Glutamate receptor activation triggers a calcium-dependent and SNARE protein-dependent release of the gliotransmitter D-serine. Proc Natl Acad Sci USA 102(15):5606–5611. https://doi.org/10.1073/pnas.0408483102
7. Paredes RM, Etzler JC, Watts LT, Zheng W, Lechleiter JD (2008) Chemical calcium indicators. Methods 46(3):143–151. https://doi.org/10.1016/j.ymeth.2008.09.025
8. Buskila Y, Bellot-Saez A, Kékesi O, Cameron M, Morley J (2020) Extending the life span of acute neuronal tissue for imaging and electrophysiological studies. In: Wright N

(ed) Basic neurobiology techniques, vol 152. Springer, New York, pp 235–259. https://doi.org/10.1007/978-1-4939-9944-6_10

9. Cameron M, Kékesi O, Morley JW, Tapson J (2016) Calcium imaging of AM dyes following prolonged incubation in acute neuronal tissue. PLoS One 0155468. https://doi.org/10.1371/journal.pone.0155468
10. Kekesi O, Buskila Y (2020) Method for prolonged incubation of brain slices. Bio Protoc 10(14):e3683. https://doi.org/10.21769/BioProtoc.3683
11. Cameron MA, Kekesi O, Morley JW, Bellot-Saez A, Kueh S, Breen P, van Schaik A, Tapson J, Buskila Y (2017) Prolonged incubation of acute neuronal tissue for electrophysiology and calcium-imaging. J Vis Exp 120. https://doi.org/10.3791/55396
12. Wang X, Takano T, Nedergaard M (2009) Astrocytic calcium signaling: mechanism and implications for functional brain imaging. Methods Mol Biol 489:93–109. https://doi.org/10.1007/978-1-59745-543-5_5
13. Schindelin J, Arganda-Carreras I, Frise E, Kaynig V, Longair M, Pietzsch T, Preibisch S, Rueden C, Saalfeld S, Schmid B, Tinevez JY, White DJ, Hartenstein V, Eliceiri K, Tomancak P, Cardona A (2012) Fiji: an open-source platform for biological-image analysis. Nat Methods 9(7):676–682. https://doi.org/10.1038/nmeth.2019

Chapter 8

Two-Photon Microscopy to Measure Calcium Signaling in the Living Brain

Kendall A. Curtis, Tenzin Kunkhyen, and Claire E. J. Cheetham

Abstract

Two-photon microscopy enables imaging of calcium signaling at cellular or subcellular resolution up to hundreds of microns deep in the living brain. Changes in the brightness of fluorescent calcium indicators provide a readout of calcium levels over time, affording information about neuronal activity and/or calcium-dependent subcellular signaling. Here, we describe a protocol for repeated two-photon imaging of calcium signals in mice expressing a genetically encoded calcium indicator that have been implanted with a chronic cranial window.

Key words Brain, Microscopy, Imaging, In vivo, Genetically encoded calcium indicator, Two-photon, Neuron

1 Introduction

Two-photon microscopy [1] provides several experimental advantages that enable repeated imaging of both structure and function deep in the living brain. Two-photon excitation requires near-simultaneous absorption of two photons of approximately half the energy (double the wavelength) required for single-photon excitation. This longer wavelength light undergoes less scattering within tissue, enabling deeper imaging. The requirement for a very high photon density also limits two-photon excitation to the focal plane. This provides optical sectioning without the need for a pinhole in front of the detector(s), hence maximizing the collection of emitted photons (which also undergo scattering in living tissue) and reducing out-of-plane photodamage. Two-photon microscopy has therefore become widely used in the study of mammalian brain structure and function, with most studies employing mice as the model organism due to their small size, short generation time, and genetic tractability.

Caroline M. Gorvin (ed.), *Calcium Signaling: Methods and Protocols*, Methods in Molecular Biology, vol. 2861,
https://doi.org/10.1007/978-1-0716-4164-4_8,

Calcium signaling provides a readout of neuronal activity as membrane depolarization increases intracellular calcium concentration via calcium influx through voltage-gated calcium channels, which may also trigger calcium release from internal stores. With calcium signaling being a ubiquitous feature of all cells, changes in calcium concentration can also provide useful information about the function of non-excitable cells in the brain, including astrocytes and oligodendrocytes. In vivo two-photon calcium imaging in the brain initially employed organic calcium indicator dyes, which were either loaded into individual neurons via recording pipettes to enable monitoring of dendritic calcium dynamics or bulk-loaded to enable simultaneous imaging of calcium signals in tens to hundreds of neurons [2, 3]. Longitudinal imaging of calcium signals at either the cellular or subcellular level became feasible with the advent of genetically encoded calcium indicators (GECIs) [4], which also permit cell type-specific expression and, in some cases, subcellular targeting and temporal control of expression. Cell type-specific expression of GECIs can be achieved using transgenic or knock-in lines, viral transduction (typically using adeno-associated virus), or in utero electroporation.

GCaMP indicators [5–10] are the most widely used family of GECIs. Based on circularly permuted green fluorescent protein fused to the calcium-binding protein calmodulin and the M13 peptide, GCaMPs have low-resting fluorescence due to protonation of the chromophore, but a conformational change elicited by calcium binding eliminates solvent access to the chromophore, with resultant deprotonation resulting in bright fluorescence. Many different GCaMP variants have been engineered, to optimize sensitivity, kinetics, and resting fluorescence for different applications. Red-emitting GECIs [11–14], and more recently developed far-red-emitting GECIs [15–18], while less widely used, offer the advantages of greater tissue penetration and reduced scattering of longer excitation and emission wavelengths, and lack of spectral overlap with channelrhodopsins to permit all-optical stimulation and recording. The choice of GECI depends on the experimental goals, and multiple GECIs can also be combined in a single experiment to monitor calcium activity in multiple neuronal types or subcellular compartments.

Here, we describe a protocol for two-photon calcium imaging using GECIs in mice implanted with a chronic cranial window to provide long-term optical access to the brain region of interest.

2 Materials

2.1 Mouse Preparation

1. Mouse expressing a genetically encoded calcium indicator implanted with a chronic cranial window and headbar.

2. Isoflurane anesthesia system including induction box.
3. Isoflurane.
4. Medical-grade oxygen.
5. Surgical microscope equipped with camera.
6. Stereotaxic frame suitable for mice equipped with cuff-style ear bars (*see* **Note 1**).
7. Feedback-controlled heat pad.
8. Fine-pointed cotton swabs.
9. 70% ethanol.
10. Sterile 26G needle.
11. Eye ointment.
12. Vacuum line with attached pipette tip and/or air duster.
13. Balance with 0.1 g resolution and weighing container for mouse.
14. 0.05 mg/mL solution of dexmedetomidine diluted in sterile saline (stock solution in a sterile vial can be stored at room temperature).
15. Sterile syringe and needle (*see* **Note 2**).

2.2 Two-Photon Microscopy

1. Two-photon microscope (e.g., Bergamo II, ThorLabs) equipped with high numerical aperture, long-working distance water-immersion objective lens (e.g., 20x XLUMPLFLN 1.0 NA 2.0 mm working distance, Olympus), femtosecond pulsed laser (e.g., Insight X3, Spectra-Physics), visible light illumination, and camera.
2. Microscope stage equipped with feedback-controlled heat pad and frame for attachment of mouse in repeatable location using headbar (*see* **Note 3**).
3. Anesthesia system (*see* **Note 4**), anesthetic (isoflurane or sevoflurane), and medical-grade oxygen.
4. Suitable image acquisition software (e.g., ThorImage).
5. (Optional) hardware and software for stimulus delivery, synchronized to image acquisition.
6. Screws to attach headbar to frame (*see* **Note 5**).
7. Deionized distilled water as immersion media.

2.3 Recovery for Subsequent Imaging Sessions

1. 0.1 mg/mL solution of atipamezole diluted in sterile saline (stock solution in a sterile vial can be stored at room temperature).
2. Sterile syringe and needle.
3. Recovery cage placed partway on a feedback-controlled heat pad.

3 Methods

This protocol is for imaging of a lightly anesthetized mouse that expresses a calcium indicator in the cell type of interest and has already been implanted with a cranial window. It is designed to be used for multiple imaging sessions in a mouse with a chronically implanted window but can also be adapted for a single imaging session immediately after completion of acute window implantation (in which case only Subheading 3.2 is required). Procedures for window implantation have already been described in detail, e.g., [19–23]. Before beginning, ensure that you have institutional and/or national approval for all animal procedures, and follow institutional guidelines for personal protective equipment.

3.1 Preparation of Mouse for Imaging

1. Set up all necessary equipment and materials. Turn on the laser and any microscope components that require warmup time.
2. Remove mouse from the home cage and anesthetize with 4% isoflurane in 1 L/min O_2 in an induction chamber. Monitor breathing carefully; once it has slowed and become deeper, check for absence of the pedal reflex, and weigh mouse. Place mouse on feedback-controlled heat pad on the stereotaxic frame, and maintain anesthesia using 1.5–2% isoflurane in 1 L/min O_2. Secure the head using the tooth bar and ear cuffs (*see* **Note 1**).
3. Ensure that anesthesia is stable and apply eye ointment to both eyes using cotton swabs (*see* **Notes 4** and **6**).
4. Center and focus surgical microscope on the surface of the window. Use vacuum line and/or air duster to remove any loose debris from the window. Clean window with 70% ethanol using a pointed cotton swab, taking care not to put downward pressure on the window. Remove any residual debris very carefully using the edge of the bevel of a 26G needle, ensuring that downward pressure on the window and scratching are avoided. It is crucial not to remove the glue and/or dental cement that seal the window to the skull.
5. Inspect the window for skull regrowth, which is evident because surface vasculature will no longer be visible. For the initial imaging session, ensure that there are sufficient clear areas to enable collection of the required images. For subsequent imaging sessions, check whether previously imaged areas are still free of skull regrowth.
6. Collect an image of the window using the camera and white light illumination, ensuring that the pattern of surface vasculature is clearly visible.

7. Administer 0.5 mg/kg (10 μL/g of 0.05 mg/mL solution) subcutaneous dexmedetomidine and reduce isoflurane to 0.75–1%. Wait 5 min for dexmedetomidine to take effect before transferring mouse to heat pad on two-photon microscope stage.

3.2 Two-Photon Microscopy

1. Attach headbar to frame on microscope stage, taking care to maintain the mouse's airway and not to exert forces that could detach the headbar from the skull or shift the position of the frame (see Fig. 1 and **Note 3**).
2. Position anesthetic delivery tube in front of mouse's nose and maintain anesthesia (*see* **Note 4**).
3. Carefully place a large drop of water on the cranial window, ensuring that none runs away from the window (*see* **Note** 7).
4. Center the objective over the window, inspecting the position from the side as a well as the front. Carefully lower the objective into the water, using fine control once you are close to the level of the water (*see* **Note 8**).
5. Once the objective is immersed, ensure that the light path is set for widefield fluorescence, turn on the camera and visible light source, and increase illumination until an image is visible. Carefully focus downward toward the window, reducing the illumination intensity as you focus down to prevent image

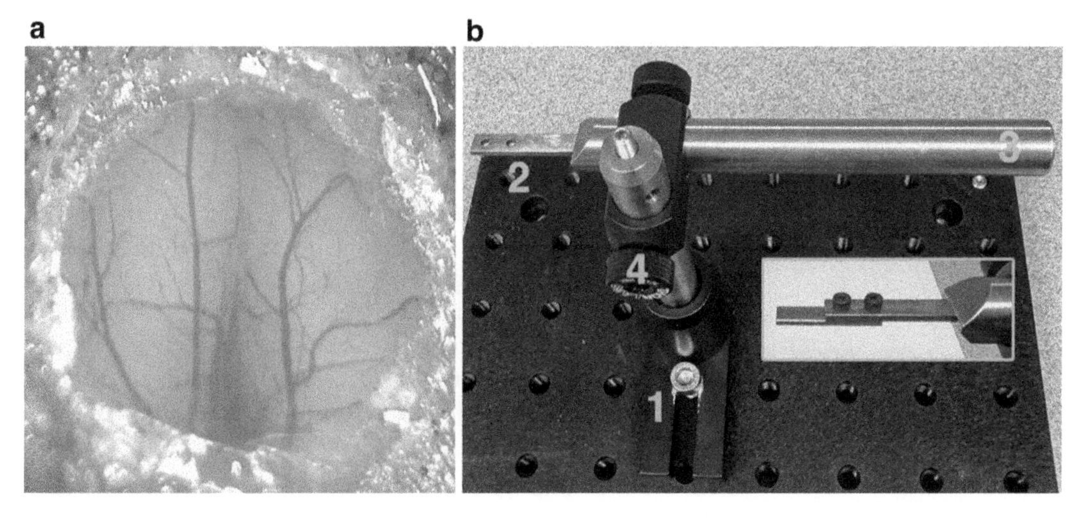

Fig. 1 Cranial window, headbar, and frame for in vivo two-photon microscopy. (**a**) Example of a 3-mm-diameter glass cranial window implanted over the mouse olfactory bulb. (**b**) Frame for reproducible positioning of mice under the two-photon microscope. The main image shows the frame that is attached to the microscope stage. Key features are the following: 1) screw for attachment to microscope stage; 2) custom portion for attachment to headbar; 3) rod can be rotated and then locked in place to optimize imaging angle; and 4) screw for height adjustment. Inset shows headbar attached to the custom piece of the frame using two screws

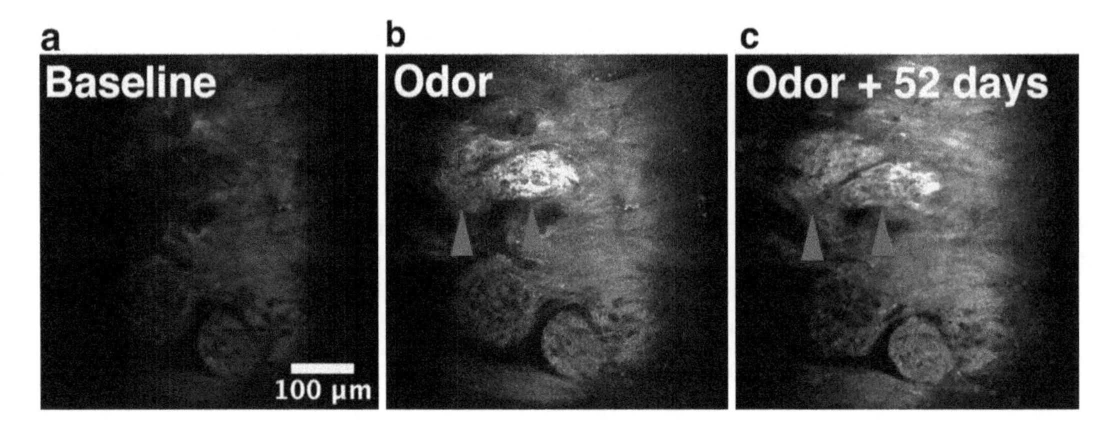

Fig. 2 Examples of two-photon calcium imaging. Images are of the glomerular layer of the olfactory bulb and were collected from an OMP-tTA;tetO-GCaMP6s mouse expressing GCaMP6s in olfactory sensory neurons, which coalesce to form glomeruli in the olfactory bulb. (**a**) Baseline fluorescence during first imaging session. (**b**) Response to odorant stimulation with 1% ethyl butyrate. (**c**) Same FOV showing the response to odorant stimulation with 1% ethyl butyrate 52 days later. Magenta arrowheads indicate odor-responsive glomeruli

saturation. Focus on the blood vessels on the surface of the brain (*see* **Note 9**).

6. Inspect the pattern of surface vasculature, and compare this to the camera image collected on the surgical microscope to determine where the blood vessels are located within the window.
7. For the initial imaging session, you will need to select a field of view (FOV) to image first. This may be possible from inspection of baseline calcium indicator fluorescence and/or fluorescence from a structural marker using widefield fluorescence. Typically, a potential FOV is selected under the camera, but this may be fine-tuned or repositioned using the two-photon image.
8. Once a potential FOV has been identified, turn off the widefield fluorescence source and switch to two-photon imaging mode in the software. Begin imaging with low laser power and detector gain settings and increase these carefully. If your software includes a pixel intensity histogram for each detection channel, use this to ensure that laser power and gain settings are set appropriately for image acquisition (*see* **Notes 10** and **11**).
9. Select a FOV. The best approach to this will depend on the goals of the experiment, but the density of GECI-expressing cells and the presence of spontaneous and/or stimulus-evoked changes in GECI fluorescence will likely be important. It may be necessary to assess multiple potential areas under the window. It may also be necessary to avoid FOVs close to large blood vessels to avoid heartbeat-related motion artifacts.

10. Once the FOV has been selected, set up acquisition settings (resolution, zoom, frame rate, etc.) and collect a z-stack containing multiple optical sections above and below the plane (s) to be collected during calcium imaging.
11. Set up all acquisition settings for calcium imaging. Set up any additional hardware and software required for stimulus delivery. If available, make use of software features that enable synchronization of external triggers with image acquisition, and automated collection of multiple trials. See Fig. 2 for example images.
12. After data collection is complete for the first FOV, record the xy coordinates and acquire a camera image of the surface vasculature. Mark the position of the FOV on a copy of the image showing the entire window.
13. Check anesthetic plane and ensure that there is sufficient water under the objective, as it can evaporate over time; add more if necessary. Check the eyes and reapply eye ointment if necessary. Select a second FOV and collect a z-stack and then the required data set of stimulus-evoked responses. Then repeat **step 12**.
14. Repeat **step 13** until data acquisition is complete for all required FOVs (*see* **Note 12**).

3.3 Recovery

1. Close shutters and set detector gain to zero and/or move out of the light path.
2. Turn off anesthetic vaporizer and oxygen.
3. Raise the objective above the mouse's head to provide sufficient clearance to remove headbar screws. Carefully wick remaining water from the window using a dry lab wipe. Remove headbar screws while supporting the mouse's head and/or holding the scruff of the neck to avoid sudden motion when the last screw is removed.
4. Remove mouse from microscope stage, administer 1 mg/kg (10 μL/g of 0.01 mg/mL solution) subcutaneous atipamezole, and place the mouse in the recovery cage. Maintain careful observation until the mouse recovers ambulatory behavior. The mouse can then be returned to regular housing.
5. Clean objective using water and lens paper only.
6. Shut down software and switch off all microscope components that require daily shutdown. Record laser operational parameters (*see* **Note 13**) and put laser into standby mode.
7. Back up acquired data to a local hard drive, server, and/or cloud storage.

3.4 Subsequent Imaging Sessions

1. Complete all steps in Subheading 3.1.
2. Complete **steps 1–3** in Subheading 3.2.
3. Set the x and y coordinates to the values recorded for the first FOV during the previous imaging session, and also perform a visual inspection to ensure that the objective is located above the window. Carefully lower the objective into the water (*see* **Note 8**).
4. Focus on the blood vessels on the surface of the brain. Adjust the FOV to match the pattern of surface vasculature from the previous imaging session (*see* **Note 14**).
5. Switch to two-photon mode, and use the fluorescent structures themselves to align the acquired FOV as carefully as possible with that collected during the previous imaging session (*see* **Note 15**).
6. Identify and collect data for all subsequent FOVs (*see* **Note 16**).
7. Recover mouse by following all steps in Subheading 3.3. Once all imaging sessions have been completed, the mouse may undergo transcardial perfusion if subsequent histology and/or immunohistochemical staining are planned or be euthanized according to institutional and/or national guidelines (*see* **Note 17**).

4 Notes

1. Cuff-style ear bars are recommended for mice as they avoid damage to the inner ear and prevent inward skull compression, which could impede the airway. The ear cuffs should be positioned symmetrically over the ears and should be tight enough to firmly hold the skull without exerting excessive pressure.
2. We use insulin syringes with permanently attached 29G needles. These syringes have virtually no dead volume (solution remaining in the syringe and/or needle after administration), 0.1 mL markings for administration of accurate drug volumes in small animals, and a short fine-gauge needle, well suited to mouse subcutaneous injection.
3. It is essential that the frame to which the headbar is attached is not moved between imaging sessions, to enable the microscope coordinate system to be used to relocate previously imaged regions. Note that even small rotational shifts will result in changes in the population of cells or structures that are visible within a FOV.
4. In combination with dexmedetomidine, light anesthesia can be stably maintained using 0.5% isoflurane in 1 L/min O_2. We

instead use sevoflurane anesthesia ($\leq$1% in 1 L/min O_2) as we image in the olfactory system, and isoflurane has a stronger background odor. Note also that anesthesia depth can affect calcium responses in the living brain. Injectable anesthesia can also be used for two-photon imaging but keep in mind that anesthetic agent(s) will gradually be metabolized so anesthesia will become lighter over time. Motion artifacts may become apparent if anesthesia is too light and the mouse starts moving, or if anesthesia is too deep, resulting in gasping.

5. It is preferable not to use a single screw to attach the headbar to the frame, as this can permit rotation between imaging sessions. Using two screws prevents this and ensures reproducible positioning.
6. Eye ointment protects the eyes from drying while the mouse is anesthetized. If not used, it is common for one or both eyes to form cataracts after one or more imaging sessions.
7. Ensure that there is sufficient glue and/or dental cement surrounding the window to support the volume of water required to image with your objective lens. We find that surrounding the window with a narrow ring of glue covered by dental cement provides sufficient surface tension to maintain a water column between the window and the objective. Another approach is to build a larger well using dental cement; whether there is sufficient space for this depends on the location of the window. Keep in mind that water immersion objectives used for two-photon imaging are often wide, and unimpeded access to the entire window requires that no part of the objective contacts the surrounding dental cement. If water finds a path away from the window, typically running off the head, the water column will gradually be lost; this can be fast or slow. If this occurs, raise the objective away from the window, remove water from the window by wicking with a dry lab wipe, and completely dry the glue/dental cement surrounding the window as well as surrounding fur that formed the path of the leaking water. It may be possible to apply a small amount of vacuum grease to prevent a subsequent leak in the same location; however, it may be necessary to return the mouse to the surgical microscope setup and apply additional glue and/or dental cement surrounding the window.
8. It is essential that the objective does not contact the window. If the objective is getting too close to the window without dipping in the water, stop, raise the objective, and add more water before continuing. Ensure that there is sufficient water to maintain a "column" between the window and the objective.
9. If blood vessels on the surface of the brain cannot be seen, the objective may not be centered over the window, but is instead

over the surrounding dental cement. Inspect the skull under the microscope to determine which direction the objective needs to be moved in to center over the window. As the dental cement is higher than the window, it should be safe to move the objective in this plane but watch the camera image continuously while adjusting. If a lower power objective is available on the microscope, this may also be useful in locating the window.

10. If there is no signal visible in the two-photon image, these are troubleshooting tips to try. Note that this list is generic and may need to be optimized for individual microscopes. It is helpful to save a troubleshooting list for your microscope in an accessible location.

    (a) With laser power and detector gain set above zero, check whether there are any nonblack pixels in the image. If so, this tells you that everything is working but your signal is very faint, whereas if all pixels have a value of zero, no light at all is reaching the detector(s). This is most easily accomplished if your software provides a pixel intensity histogram and/or enables the brightness and contrast of the displayed image to be adjusted. If the image is very dim, it may be difficult to discern especially if room lights are on.

    (b) Check that the laser is tuned to an appropriate excitation wavelength for the fluorophore(s) you are using. Ensure that the laser is operating normally

    (c) Check that the detection channel(s) and light path are selected correctly. When switching between widefield and two-photon fluorescence light paths, audible changes usually occur as the light path is switched. Occasionally, automated light path changes do not occur correctly but can be triggered by switching imaging modes again.

    (d) Confirm that all shutters are open. There is typically a shutter on the laser itself, a shutter in front of each detector, and there may be additional safety shutters to prevent PMT damage.

    (e) If available in software, check that the system is calibrated to the correct laser wavelength.

    (f) Switch to widefield imaging mode and ensure that the expected structures remain in focus.

    (g) Check for any error messages or other system settings that could be incorrect. It is helpful to save lists and/or screenshots of relevant settings on the image acquisition computer.

    (h) Restart computer, microscope controller, and any other relevant aspects of the system, e.g., shutter controllers.

11. Excessive laser excitation power (greater than ~50 mW at the back focal plane of the objective) can cause photodamage, depending on the scan speed used. Therefore, it is important to determine the laser power at the back aperture of the objective at various laser excitation power settings. Furthermore, many two-photon microscopes are equipped with automatic protection circuits that shut down detectors to prevent damage if light intensity is too high. Therefore, it is typically prudent to keep pixel values within the lower 70% of the intensity range. Note that even a single very bright pixel in an image is sufficient to trigger detector shutoff, so small adjustments in the FOV, or selection of a new FOV, can prevent this from occurring. Otherwise, reducing laser power and/or detector gain is the best approach. While structures will typically be brightest close to the surface of the brain, this depends on calcium indicator expression level and changes in calcium concentration during spontaneous or evoked activity. GECI fluorescence typically increases, sometimes by orders of magnitude, in the presence of calcium, so ensure that there is sufficient "headroom" in the intensity range to accommodate the expected magnitude of fluorescence changes.
12. The number of FOVs that is acquired depends on multiple factors:
    (a) The size of the window.
    (b) The area under the window that is clear of skull regrowth (ongoing skull regrowth will likely encroach on adjacent areas of the brain especially if imaging sessions are not closely spaced in time).
    (c) The quantity of data that is required for the planned analyses.
    (d) Whether different parts of the window represent different experimental conditions, e.g., left and right hemisphere, which may have undergone different experimental manipulations.
    (e) The length of time required for data acquisition for each FOV, relative to the total time that it is desirable to keep the mouse anesthetized (we typically aim for <3 h per imaging session).
13. A record of laser operating parameters helps to monitor for early-warning signs of problems with the laser, as parameters such as diode temperature and output power may start to drift. Early intervention can often prevent the laser from becoming inoperable and requiring repair, which results in loss of chronic imaging time points.

14. We use three complementary methods to repeatedly image FOVs over multiple days and weeks. First, our microscope provides a digital readout of the x and y coordinates. Note that while we also have digital z co-ordinates, our system has multiple z motors, making it impractical to use the z co-ordinate to return to the same imaging plane. Second, the pattern of surface vasculature, visualized on a camera using white light illumination in widefield fluorescence mode, provides reliable fiducial markers. Note that there may be small changes in surface vasculature over weeks. Finally, the pattern of neurons and/or their processes enables fine-scale adjustment of the boundaries of the FOV.
15. Some GECIs, including the widely used GCaMP6 indicators, are engineered to have low-resting fluorescence. Careful selection of the GECI to be used (e.g., jGCaMP7b has bright baseline fluorescence) [6]) or expression of a structural marker of a different wavelength with calcium-independent fluorescence may enable precise alignment of the FOV. Alternatively, it may be possible to use spontaneous or evoked changes in GECI fluorescence to align the FOV.
16. Note that if imaging continues across many weeks, fluorescence intensity in some FOVs may become dimmer over time or become invisible due to skull regrowth or dural thickening.
17. The headbar can be removed from the skull, cleaned, and autoclaved for future reuse.

References

1. Denk W, Strickler J, Webb W (1990) Two-photon laser scanning fluorescence microscopy. Sci 248:73–76
2. Stosiek C, Garaschuk O, Holthoff K et al (2003) In vivo two-photon calcium imaging of neuronal networks. Proc Natl Acad Sci USA 100:7319–7324
3. Svoboda K, Denk W, Kleinfeld D, et al (1997) In vivo dendritic calcium dynamics in neocortical pyramidal neurons
4. Miyawaki A, Llopis J, Heim R et al (1997) Fluorescent indicators for Ca2+based on green fluorescent proteins and calmodulin. Nature 388:42264
5. Nakai J, Ohkura M, Imoto K (2001) A high signal-to-noise Ca(2+) probe composed of a single green fluorescent protein. Nat Biotechnol 19:137–141
6. Dana H, Sun Y, Mohar B et al (2019) High-performance calcium sensors for imaging activity in neuronal populations and microcompartments. Nat Methods 16:649–657
7. Chen T-W, Wardill TJ, Sun Y et al (2013) Ultrasensitive fluorescent proteins for imaging neuronal activity. Nature 499:295
8. Akerboom J, Chen T-W, Wardill TJ et al (2012) Optimization of a GCaMP calcium indicator for neural activity imaging. J Neurosci 32:13819–13840
9. Tian L, Hires SA, Mao T et al (2009) Imaging neural activity in worms, flies and mice with improved GCaMP calcium indicators. Nat Methods 6:875–881
10. Tallini YN, Ohkura M, Choi B-R et al (2006) Imaging cellular signals in the heart in vivo: cardiac expression of the high-signal Ca2+ indicator GCaMP2. Proc Natl Acad Sci 103:4753–4758
11. Dana H, Mohar B, Sun Y et al (2016) Sensitive red protein calcium indicators for imaging neural activity. Elife 5:e12727
12. Inoue M, Takeuchi A, Horigane S et al (2014) Rational design of a high-affinity, fast, red

calcium indicator R-CaMP2. Nat Methods 12: nmeth.3185
13. Molina RS, Qian Y, Wu J et al (2019) Understanding the fluorescence change in red genetically encoded calcium ion indicators. Biophys J 116:1873–1886
14. Inoue M, Takeuchi A, Manita S et al (2019) Rational engineering of XCaMPs, a multicolor GECI suite for in vivo imaging of complex brain circuit dynamics. Cell 177:1346–1360. e24
15. Dalangin R, Drobizhev M, Molina RS et al (2020) Far-red fluorescent genetically encoded calcium ion indicators. bioRxiv. https://doi.org/10.1101/2020.11.12.380089
16. Hashizume R, Fujii H, Mehta S et al (2022) A genetically encoded far-red fluorescent calcium ion biosensor derived from a biliverdin-binding protein. Protein Sci 31:e4440
17. Deo C, Abdelfattah AS, Bhargava HK et al (2021) The HaloTag as a general scaffold for far-red tunable chemigenetic indicators. Nat Chem Biol 17:718–723
18. Subach OM, Barykina NV, Anokhin KV et al (2019) Near-infrared genetically encoded positive calcium indicator based on GAF-FP bacterial Phytochrome. Int J Mol Sci 20:3488
19. Cheetham CE (2018) Multiphoton intravital calcium imaging. Curr Protoc Cytom e40. https://doi.org/10.1002/cpcy.40
20. Holtmaat A, Bonhoeffer T, Chow DK et al (2009) Long-term, high-resolution imaging in the mouse neocortex through a chronic cranial window. Nat Protoc 4:1128–1144
21. Zhao Y-J, Yu T-T, Zhang C et al (2018) Skull optical clearing window for in vivo imaging of the mouse cortex at synaptic resolution. Light Sci Appl 7:17153–17153
22. Kelly EA, Majewska AK (2010) Chronic imaging of mouse visual cortex using a thinned-skull preparation. J Vis Exp 44. https://doi.org/10.3791/2060
23. Drew PJ, Shih AY, Driscoll JD et al (2010) Chronic optical access through a polished and reinforced thinned skull. Nat Methods 7:981–984

Chapter 9

Measuring Calcium-Sensing Receptor Agonist-Driven Insertional Signaling (ADIS) and Trafficking by TIRF Microscopy

Caroline M. Gorvin

Abstract

The calcium-sensing receptor (CaSR), which regulates parathyroid hormone secretion by sensing serum calcium concentrations, has developed unique trafficking mechanisms to respond to constant exposure to its orthosteric ligand calcium. CaSR rapidly responds to fluctuations in serum calcium by driving forward trafficking of the receptor to cell surfaces in a mechanism known as agonist-driven insertional signaling (ADIS). This increase in CaSR at cell surfaces is counterbalanced by both constitutive and agonist-driven internalization of the receptor. Deciphering these mechanisms is important to understand how mutations in the CaSR and components of its signaling and trafficking pathways cause human disorders of calcium homeostasis.

This chapter describes a protocol to measure CaSR ADIS and endocytosis in parallel using total internal reflection fluorescence (TIRF) microscopy. This utilizes a mammalian expression construct comprising the full-length human CaSR with an N-terminal bungarotoxin minimal-binding site that can be labeled with commercially available fluorescent ligands to measure endocytosis, and a super-ecliptic pHluorin (SEP) to measure total cell surface expression and exocytic events. This protocol could easily be adapted to simultaneously assess forward trafficking and endocytosis of other membrane proteins by TIRF microscopy.

Key words Adaptor protein-2, Calcium homeostasis, Endocytosis, Hyper−/hypocalcaemia, Parathyroid hormone

1 Introduction

The homeostatic control of serum calcium is critical for maintenance of the skeleton, regulation of neuronal excitability, muscle contraction, hormone secretion, and immune responses. The calcium-sensing receptor (CaSR) is a G protein-coupled receptor (GPCR) that performs a fundamental role in calcium homeostasis by regulating parathyroid hormone (PTH) release from the parathyroids and urinary calcium excretion at the kidneys [1]. PTH stimulates the release of calcium from bone and increases calcium reabsorption and activates vitamin D at the kidneys [2]. CaSR has a

Caroline M. Gorvin (ed.), *Calcium Signaling: Methods and Protocols*, Methods in Molecular Biology, vol. 2861,
https://doi.org/10.1007/978-1-0716-4164-4_9,

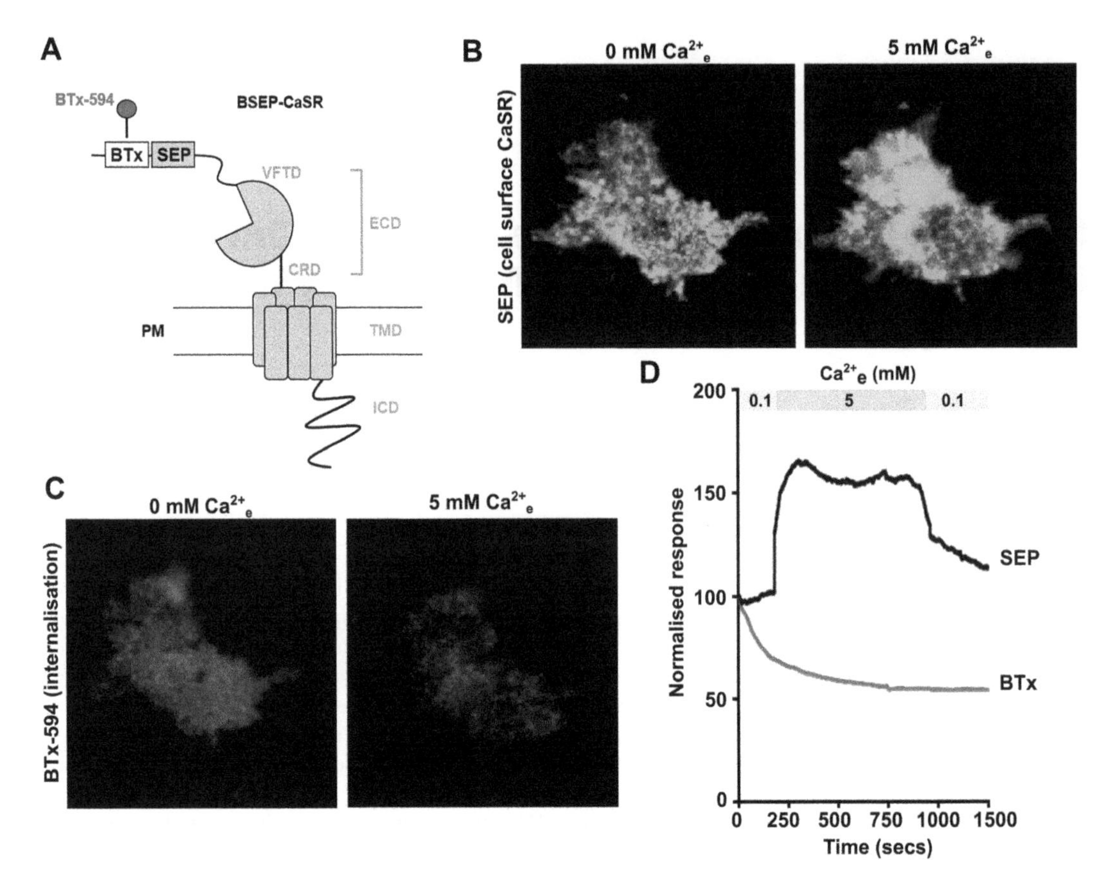

Fig. 1 Schematic showing the CaSR structure and BSEP-CaSR expression plasmid. (**a**) Cartoon of the BSEP-CaSR expression plasmid used in the TIRF studies. Each CaSR subunit comprises a large bilobed extracellular ligand-binding domain (the venus fly-trap domain, VFTD), a seven-transmembrane domain (TMD), and a long intracellular domain (ICD). The BSEP-CaSR construct consists of the full-length human CaSR with an N-terminal BTx-binding site to measure endocytosis, and a super-ecliptic pHluorin (SEP) to measure total cell surface expression and exocytic events. (**b**) Image showing cell surface CaSR-SEP. Upon addition of agonist (5 mM $Ca^{2+}{}_e$), CaSR is exported to the cell surface by ADIS and GFP fluorescence increases. (**c**) Image showing cell surface CaSR labeled with non-permeable BTx-594. Receptor internalization reduces BTx-594 at the cell surface. (**d**) Data from a recording of a single cell expressing BSEP-CaSR. Under basal conditions (0.1 mM $Ca^{2+}{}_e$), CaSR undergoes some constitutive internalization (monitored by BTx-594, red line), although total cell surface expression of the receptor is relatively stable (SEP, black line). Upon agonist addition, SEP fluorescence increases as CaSR is exported to the cell surface by ADIS

canonical structure of a class C GPCR, existing as a homodimer on cell surfaces, with a large extracellular ligand-binding domain (amino acids 1–610), a 7-transmembrane domain (amino acids 611–863), and a long intracellular domain (residues 864–1078) that harbors several phosphorylation sites [3] (Fig. 1a). The ligand-binding domain comprises a bilobed structure, known as the venus fly-trap domain (VFTD), so-called as binding of ligand to the receptor activates conformational changes that lead to closure of the lobes, and transmits further structural changes resulting in G protein activation [4]. The major ligand for CaSR is extracellular calcium ions ($Ca^{2+}{}_{e}$), and its name derives from its ability to bind calcium from blood (thus sensing serum calcium levels). Therefore, unlike many other GPCRs, CaSR is constantly exposed to its cognate ligand. For this reason, the receptor has developed a unique, sophisticated system to ensure that there is sufficient active (i.e., non-desensitized) receptor at cell surfaces to maintain serum calcium levels within a very narrow range (2.2–2.7 mmol/L or 8.5–10.5 mg/dL) [5].

CaSR cell surface expression is maintained by a balance between forward trafficking of newly synthesized receptor from the endoplasmic reticulum (ER) and Golgi, where CaSR is glycosylated, and removal from cell surfaces by endocytosis [5, 6]. However, to allow CaSR to rapidly respond to fluctuations in serum calcium concentrations, a pool of mature, fully glycosylated receptor exists in a pre-plasma membrane location, which responds rapidly to rises in serum calcium to increase cell surface expression of the receptor [5]. This phenomenon is known as agonist-driven insertional signaling (ADIS). CaSR is removed from cell surfaces by both constitutive endocytosis and agonist-mediated internalization, the latter of which may be a compensatory response to the ADIS-driven increase in receptor at the cell surface. Therefore, when assessing CaSR cell surface expression and internalization, one must assess both ADIS and endocytosis.

An elegant solution to this problem was first described in 2011 [5]. This involved live cell imaging by total internal reflection fluorescence microscopy (TIRF-M), a high-resolution technique that permits visualization of only 100–200 nm of the cell adjacent to the coverslip, and thus restricts imaging to plasma membrane events, without interference from molecules expressed in other cellular locations. To measure both ADIS and endocytic events, the Breitwieser group developed an expression plasmid (BSEP-CaSR) expressing CaSR with an N-terminal modification (Fig. 1b–d) that allows labeling of cell surface CaSR in two ways [5]. Firstly, the plasmid expresses a super-ecliptic pHluorin (SEP), which is a modified GFP that is highly sensitive to pH and fluoresces maximally upon transition from acidic to neutral pH environments. Thus, SEP is ideal for measuring exocytic events, and was used to study total CaSR expression at cell surfaces and to monitor ADIS events in live cells. Immediately upstream of SEP, the BSEP construct has a 13-amino acid bungarotoxin (BTx) binding site that can be labeled with bungarotoxin conjugated to an Alexa Fluor (commercially available from Molecular Probes) to monitor endocytosis. When this non-permeable fluorescent label is added immediately prior to imaging, it will label only CaSR at the cell surface, and upon receptor internalization, a reduction in fluorescence can be observed (Fig. 1c, d).

Understanding the forward trafficking, endocytosis, and cell surface expression of CaSR is important as mutations in the receptor cause human disorders of calcium homeostasis [7]. Inactivating mutations in CaSR cause two disorders, neonatal severe hyperparathyroidism (NSHPT) and familial hypocalciuric hypercalcemia (FHH). Infants with NSHPT present with marked elevations in serum calcium, skeletal demineralization, and failure to thrive, which are caused by homozygous or compound heterozygous CaSR mutations [7, 8]. In contrast, FHH is characterized by lifelong hypercalcemia, high or inappropriately normal PTH concentrations, and low kidney calcium excretion and is caused by heterozygous mutations in most cases [7, 8]. Activating mutations of the CaSR cause autosomal dominant hypocalcemia (ADH), associated with mild-to-moderate reductions in serum calcium concentrations, inappropriately low or normal PTH, and >70% present with symptoms of hypocalcaemia including seizures, carpopedal spasms, tetany, and paresthesia [7, 9]. Many CaSR loss-of-function mutations are associated with reduced cell surface expression [10], possibly due to intracellular retention (e.g., at the ER) [11]. The TIRF technique described within this chapter was used to show that six inactivating CaSR mutations proportionately reduce ADIS forward trafficking and calcium signaling [12]. Co-transfection with wild-type CaSR rescued ADIS and receptor signaling, indicating that drug treatments (e.g., allosteric modulators) that can rescue CaSR expression will likely restore signaling and cell surface expression of CaSR [12].

The technique described in this protocol has also been used to investigate how CaSR trafficking is affected by mutations in other proteins that are mutated in FHH and ADH. While mutations in CaSR are responsible for the majority of FHH (~65%) and ADH (~70%) cases [13], mutations in the G-alpha subunit-11, by which CaSR signals, also cause FHH and ADH [14], and mutations in the sigma subunit of the adaptor protein-2 (AP2σ), which has an important role in clathrin-mediated endocytosis, cause FHH type-3 [15]. Using the TIRF ADIS technique, AP2σ mutations were shown to increase CaSR cell surface expression and impair receptor internalization [16]. This led to the hypothesis that CaSR can continue signaling from intracellular sites (e.g., early endosomes) following internalization, and subsequent studies showed that AP2σ mutations impair this sustained signaling pathway [16]. Thus, the TIRF ADIS technique has been instrumental in identifying novel mechanisms by which CaSR signaling is regulated and has revealed new ways in which receptor expression could be exploited pharmacologically to improve treatments for disorders of calcium homeostasis. A number of other proteins have been shown to regulate CaSR cell surface expression and/or signaling [17–21], and further studies using the TIRF ADIS technique could reveal further insights into the mechanisms that govern CaSR signaling and trafficking.

This chapter will explain how to assess CaSR ADIS and endocytosis using the BSEP-CaSR expression construct and TIRF microscopy. Examples of data obtained will be shown and analysis using freely available software will be explained.

2 Materials

Prepare all solutions using ultrapure water (18 MΩ-cm resistance at 25 °C). Prepare and store all reagents at room temperature, unless otherwise stated.

2.1 Expression Plasmids

1. BSEP-CaSR: The construction of BSEP-CaSR has been described previously [5]. Briefly, the 13-amino acid α-bungarotoxin binding site (BTx) [22] was inserted in-frame after amino acid 23 of the human CaSR signal sequence. SEP was PCR amplified from the N-terminal SEP-CaSR (rat) construct [18], and cloned between the BTx site and human CaSR (Fig. 1a).

2.2 Cell Culture

1. Adherent human embryonic kidney (AdHEK) 293 cells (*see* **Note 1**).
2. Complete growth medium: Dulbecco's modified Eagle medium (DMEM) with 4.5 g/L D-Glucose and L-glutamine, 10% (v/v) fetal bovine serum (FBS).
3. 0.05% Trypsin-EDTA solution.
4. Phosphate-buffered saline (PBS): Dissolve one tablet in 500 mL water. Autoclave and store at room temperature.
5. Humidified 37 °C, 5% CO_2 incubator.
6. Cell culture hood.
7. Hemocytometer for cell counting.
8. Trypan blue: Make working solution of 0.4% trypan blue in PBS.
9. 35 mm imaging dishes.
10. 75cm^2 culture flask (*see* **Note 2**).
11. Phase-contrast microscope.

2.3 Transfection

1. Lipofectamine™ 2000 Transfection Reagent.
2. Serum-free DMEM media.
3. 0.5 mL microcentrifuge tubes.

2.4 TIRF Imaging

1. Ca^{2+}- and Mg^{2+}- free Hank's buffered saline solution (HBSS).
2. TetraSpeck microsphere standards. Use the 0.1 μm or 0.2 μm particle size.
3. Extracellular medium for imaging (hereafter called imaging buffer): 140 mM NaCl, 5 mM KCl, 0.55 $MgCl_2$, 10 mM HEPES, and 10 mM D-glucose, pH 7.4. Can be stored on the bench at room temperature for several weeks.
4. Working solution of $CaCl_2$. Dilute 1 M $CaCl_2$ in water to make a working solution of 100 mM $CaCl_2$ as higher concentrations can precipitate out of solution.
5. Activation imaging solution (i.e., agonist): Add $CaCl_2$ working solution to imaging buffer to obtain the desired final concentration (e.g., 5 mM Ca^{2+}). Adjust the solution to maintain osmotic balance.
6. α-Bungarotoxin conjugated to Alexa Fluor™ 594 (BTx-594): Reconstitute in PBS to a 1 mg/mL stock solution. Store in aliquots frozen at −20 °C.
7. 24 mm glass coverslips.
8. Chambers for holding coverslips and stage inserts (Fig. 2a, b).
9. Immersion oil-type LDF.
10. Microscope capable of TIRF imaging with a 60× or 100× objective and laser lines for measuring 488 nm and 561 nm fluorescence (*see* **Note 3**).

2.5 Software Required for Data Analysis

1. VLC Player: Available as a free download at https://www.videolan.org/vlc/
2. ImageJ: available as a free download at https://imagej.net/ij/

3 Methods

All cell culture work (seeding and transfection) should be performed aseptically in a cell culture hood. This protocol describes culture and imaging of BSEP-CaSR in adherent human embryonic kidney 293 (AdHEK293) cells (*see* **Note 1**).

3.1 Preparation of Coverslips for Imaging

1. Lift coverslips with tweezers and place in an imaging dish.
2. Add 70% ethanol to the coverslip (150–200 μL should cover), and incubate at room temperature for 10 mins.
3. Remove ethanol. Wash coverslips in ~1 mL PBS 2–3 times.
4. Coverslips are now ready to use for cell seeding or can be stored in PBS at 4 °C for 2–3 weeks (*see* **Note 4**).

A Chambers for coverslips

B Stage inserts for coverslip chambers

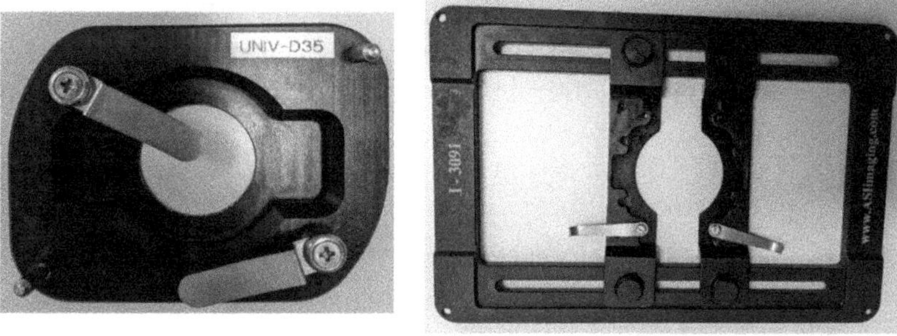

C Laser system

D Protective laser lock

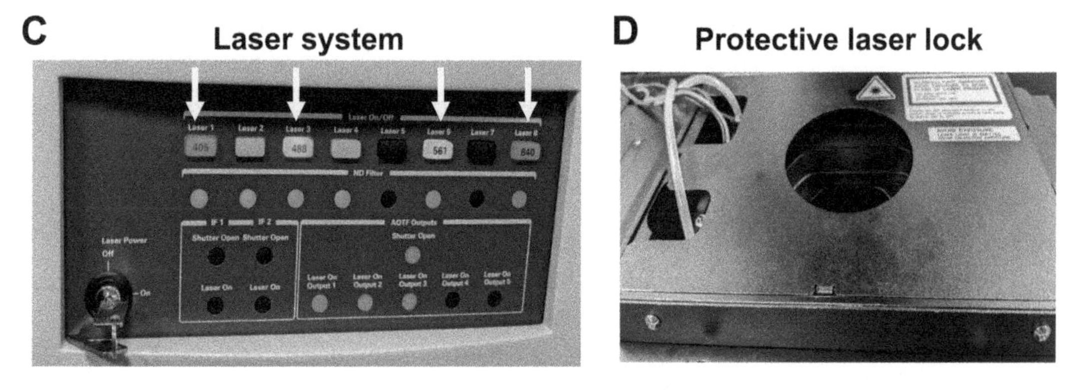

Fig. 2 Images showing different types of chambers and stage inserts used to image BSEP-CaSR by TIRF microscopy. (**a**) Images showing two types of chambers used to hold TIRF coverslips. In the chamber at the left, the coverslip is screwed into place, while in the chamber to the right, the coverslip is held in place with a red rubber seal. Imaging solution can be added on top. (**b**) Chambers can be fixed into place with a range of stage inserts to prevent movement during imaging. Images show two types used for TIRF imaging. (**c**) Image of the laser system used on one TIRF microscope that has stable lights when the laser is ready to use (as in the image indicated by arrows) and flashing lights when the system is first switched on. (**d**) Most TIRF microscopes have a protective cover that prevents the user from exposure to the lasers. Some systems have laser override systems that prevent lasers switching on if the protective cover is not placed over the stage correctly

3.2 Day 1: Seeding Cells on Coverslips

1. Remove media from the cell culture flask.
2. Add 2–3 mL Trypsin to the flask and incubate at room temperature for ~2 min. Check cells are rounding or lifting under a light microscope (*see* **Note 5**).
3. Neutralize the trypsin by adding ~5 ml media. Mix by pipetting up and down with a 10 mL stripette.
4. Count cells on a hemocytometer. Remove 100 μL of cell suspension and add to a 0.5 mL microcentrifuge tube. Add 100 μL 0.4% trypan blue. Add 100 μL of cell/trypan blue mixture to the hemocytometer and place a coverslip on top. Count the number of live and dead (i.e., unstained vs. trypan blue-stained cells) using a phase-contrast microscope at 20× magnification. Seed cells at 100,000 cells per well in a total volume of 1 mL pre-warmed media with FBS. Incubate at 37 ° C, 5% CO_2 overnight (*see* **Note 6**).

3.3 Day 2: Transfection of Cells

1. Add 100 μL serum-free media to each of two 0.5 mL microcentrifuge tubes.
2. Add 2 μL Lipofectamine 2000 (*see* **Note 7**) to one tube and 200 ng BSEP-CaSR construct to the other tube. Incubate for 5 min.
3. Pipette the contents of tube 1 to tube 2 and mix gently by pipetting. Incubate for up to 20 min (see **Note 8**).
4. Pipette the contents of the tube dropwise to cells on the coverslip. Incubate at 37 °C, 5% CO_2 for 12–24 h (*see* **Note 9**).

3.4 Day 3: TIRF Imaging

3.4.1 Preparation Steps for Imaging

1. Change media on cells to Ca^{2+}- and Mg^{2+}-free HBSS at least 1 h before imaging.
2. Switch on microscope system in advance. Typically lasers take ~15 min to warm up before they can be used. Depending on the system, there may be lights that come on when the lasers are ready to use (*see* **Note 10** and Fig. 2c).
3. Switch on humidity chamber to set the temperature to 37 °C (*see* **Note 11**).
4. Switch on the perfusion system (*see* **Note 12**), and check that the imaging chamber fills with imaging solution and that solutions are suctioned into the waste pot (*see* **Note 13**).

3.4.2 Calibration of TIRF Setup

Prior to any recordings being made, the camera alignment must be checked and calibrated with TetraSpeck beads (*see* **Note 14**).

1. Lift a clean coverslip with no cells into an imaging chamber.
2. Seal the chamber and add 1 μL TetraSpeck beads to the center of the coverslip.

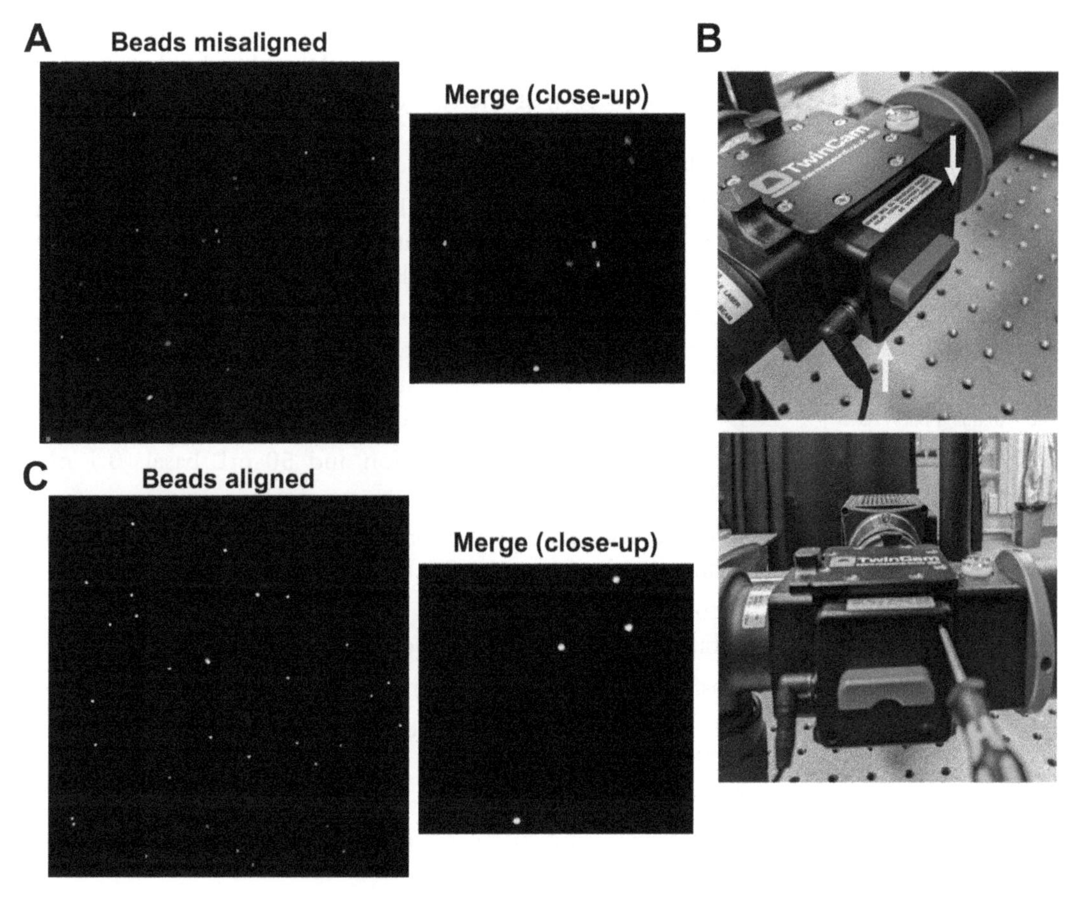

Fig. 3 Fluorophore alignment using TetraSpeck beads. (**a**) Daily calibration is required to ensure that particles are co-localized in the same optical plane. TetraSpeck microspheres are stained with fluorescent dyes for four well-separated excitation/emission peaks: 360/430 nm (blue), 505/515 nm (green), 560/580 nm (orange), 660/680 nm (dark red). (**a**) Image of TetraSpeck beads showing misalignment of the two particles. (**b**) Beads can be aligned by adjusting the two screws on the TwinCam (indicated by arrows in the top image and shown with a screwdriver in situ in the bottom image). (**c**) Following adjustment, beads should co-localize as shown

3. Add 0.5 mL imaging buffer and mix in the TetraSpeck beads with pipetting. Check there is no leakage from the chamber, and wipe the underside with tissue prior to loading on microscope.
4. Place a couple of drops of immersion oil on the objective and load the chamber onto the microscope stage and lock in place (Fig. 2b). If any oil leaks, clean up immediately.
5. Ensure any protective covers (Fig. 2d) that prevent exposure to lasers are in place prior to switching on lasers (*see* **Note 15**).
6. Switch the 488 and 594 lasers on and set at ≤50% laser power.
7. Lift the objective up toward the coverslip with the coarse focus (*see* **Note 16**). Once the objective oil touches the coverslip, use the fine focus to lift the objective. To find the focus, visualize

the specimen on the computer software using the camera view. Set the software to show the green channel and red channel in overlay view so that the two colors can be seen together and separation can be easily seen. TetraSpeck beads can be seen as small dots on screen (Fig. 3a).

8. If the pixels are misaligned (Fig. 3a), adjust the xy positions with a screwdriver in the panels at the back of the camera (Fig. 3b) until the beads overlay (Fig. 3c) (*see* **Note 17**).
9. Take several images in different locations to check the alignment is similar across the coverslip (*see* **Note 18**).

3.4.3 Imaging CaSR ADIS and Endocytosis

1. Dilute BTx-594 at 5 μg/mL in imaging buffer.
2. Prepare 50 mL agonist solution and 50 mL basal (0.1 nM $CaCl_2$) imaging buffer.
3. Remove cells from incubator and load the coverslip into the chamber using tweezers.
4. Incubate cells with BTx-594 in imaging buffer at room temperature for 3 minutes to label cells.
5. Wash coverslip two times in imaging buffer without dye, wipe the underside of the coverslip, and then load onto the microscope (*see* **Note 19**).
6. Add 1 mL imaging buffer to the chamber.
7. Locate cells using coarse focus, and then fine focus as described in Subheading 3.4.2.
8. Select cells that have similar levels of fluorescence in both channels (*see* **Note 20**).
9. Take an image of fluorescence in each channel before commencing movie acquisition.
10. Record images continuously at 10 frames/second. Image cells in basal imaging buffer for 60 s at the start of the recording, and then switch the perfusion system to Activation Solution (*see* **Note 21**).
11. After 10 min, switch the perfusion system to basal imaging buffer for 9 min (*see* **Note 22**).
12. Take a snapshot image of the cells at the end of the recording (*see* **Note 23**).

3.4.4 Analysis of Data

This will be dependent on the microscope system and software that you use. The following describes analysis when data was exported as .avi files that could not be opened in the AVI-reader plugin (*see* **Note 24**):

1. Download the VLC media player and ImageJ.
2. Open the VLC media player and drag in the file you want to open.
3. To convert the movie to images for each frame open "Tools," then "Preferences." In the "Interface Settings" tab, check the box at the bottom left corner labeled "All" under "Show settings." Then, in the "Video" drop-down menu, click "Select filters," followed by "Scene filter." In the "Scene filter" box, choose how you want the data to export e.g. png, tif, jpg, and "Recording ratio" as "1" to export every frame in the movie to a single image. Save these choices and check that the box next to "Scene Video Filter" is ticked. When you play the video, the individual image files should then export to your chosen folder.
4. Open all image files simultaneously in Image J.
5. Convert individual images to a stack by clicking on Image>Stacks>Images to Stack.
6. Draw a shape around each cell and name it as a region of interest (ROI) in the ROI manager by clicking T on the keyboard.
7. To measure fluorescence in each ROI for all frames of the stack, select More>Multi measure in the ROI manager and ensure "measure all slices" is selected. The same ROI can be applied to both the green and red channel images.
8. Ensure one ROI measures a region of similar size to the cell with no fluorescence to act as a background control.
9. Normalize the fluorescence in all of the images by first subtracting the background fluorescence, then expressing each frame relative to the fluorescence in frame one.
10. These datasets can then be plotted in software such as GraphPad Prism. The 488 fluorescence should increase upon agonist addition due to ADIS, and 594 fluorescence reduces due to receptor internalization. Upon washout the 488 fluorescence will begin to reduce while 594 will likely remain unchanged (Fig. 1).

4 Notes

1. We used AdHEK293 cells in our studies as they are adherent, do not endogenously express CaSR, and are routinely used for CaSR signaling studies [14, 23, 24] and for TIRF imaging [5, 25, 26]. We have also performed ADIS studies in HEK293 cells stably transfected with mutant AP2σ protein. Other cells such as Chinese hamster ovary (CHO) and human

retinal pigment epithelial (RPE-1) cells are also often used for TIRF as they are flat and relatively easy to transfect.

2. We routinely culture cells in 75cm^3 culture flasks, but any size vessel can be used. AdHEK293 cells usually require passaging 2–3 times a week. They should not be left to grow >80% confluent and should be transfected at 50–70% confluency.
3. We have predominantly performed this imaging on an Olympus IX-81 TIRF microscope equipped with a 60×/1.45 Apo lens. An argon ion laser was used to excite SEP and a 561 nm line of a steady-state diode laser used to excite BTx-594. The emission pathway for both fluorescent reporters was imaged simultaneously with an image splitter generating side-by-side images on the chip of an electron multiplying charge-coupled device (EMCCD) camera. We have also performed similar imaging on a custom-built microscope with a Nikon Eclipse Ti2 base, an EMCCD camera, a 488 nm argon laser, 561 nm diode laser, and a × 100/1.49 numerical aperture oil-immersion objective.
4. If using cells that do not adhere well to plastic, poly-D-lysine may be required. Following removal of ethanol, coverslips can be washed in 1× PBS two times, then poly-D-lysine either: (1) applied to coverslips directly for 10 min, followed by 2× washes in PBS, and storage at 4 °C, or (2) applied to coverslips and stored at 4 °C for later use.
5. Cells should lift at room temperature, although if the cells are very adherent, incubation at 37 °C may be required.
6. Cells can be plated at a lower density and incubated for more than 1 day.
7. Cells should only be transfected if they look flat, adherent, and stellate. We obtained the best results when cells were transfected at ~50% confluency and were not touching other cells. The time between transfection and imaging can be adjusted to as little as 12 h to ensure an optimal transfection efficiency of many GFP-positive cells without oversaturation.
8. This time can be adjusted. We have tested transfection at a range of incubation times between 5 and 20 min and found no difference in the efficiency of transfected cells.
9. Manufacturer's instructions for transfection with Lipofectamine 2000 state that Opti-MEM should be used, and cells should be incubated in serum-free media before transfection reactions are added to the cells. When we directly compared Opti-MEM with serum-free DMEM, we observed no noticeable difference in GFP-positive cells by microscopy. Additionally, we observed no differences in transfection efficiency when

transfection mixtures were added directly to cells in serum-free media or in serum-containing media.

10. The order in which each component is switched on will differ depending on the system used and the microscopy facility rules and regulations. A typical system will involve switching on a laser bank (switch on and wait for several seconds before turning key), followed by switching on cameras. Usually the final component to be switched on is the computer, and software is opened after lasers and cameras are active.
11. Imaging must be performed at 37 °C as changes in temperature have been shown to affect receptor trafficking.
12. Not all microscopes will be fitted with a perfusion system (especially in shared facilities). Although the perfusion system is useful for changing solutions, imaging of CaSR can still be performed without it, although washout cannot be observed. Instead, basal images can be collected in a chamber containing 500 μL Imaging Buffer, and then the movie paused while Activation Solution is added (at a 2× concentration). If this solution is added carefully, then the cells should not move.
13. If the perfusion moves during imaging or the suction that clears waste from the chamber is inadequate, the chamber may overfill with Imaging Solution. This can be observed as images becoming rapidly blurred during image acquisition. This must be quickly resolved to prevent solution leaking onto the objective and will require switching off the perfusion system, cancellation of image acquisition, complete removal of all buffer from the chamber, and possible cleaning of the objective. If agonist has not been added, the coverslip can be reimaged after the perfusion system problems are fixed and immersion oil is reapplied to the objective. Movement of the suction can be reduced by using adhesive putty to hold the perfusion in place.
14. Daily calibration is required to ensure that particles are co-localized in the same optical plane. TetraSpeck particles, which can be purchased as a solution or loaded onto a microscope slide, are stained with fluorescent dyes for four well-separated excitation/emission peaks: 360/430 nm (blue), 505/515 nm (green), 560/580 nm (orange), and 660/680 nm (dark red).
15. Most microscopy facilities will require that you undergo some laser training prior to microscope use. Please check local rules before starting any experiments.
16. Some microscopes have a perfect focus that helps locate the specimen. If the system you are using has perfect focus, switch this on when the oil touches the coverslip. The perfect focus

light will flash, then become a stationary light, and make a noise when the correct focus is found.

17. Some microscopy facilities may not allow users to adjust the camera alignment themselves and may require assistance from a microscopy officer or other trained member of staff. Users should check with the owner of the microscope or the facilities' manager.
18. These calibration images should be stored with any movies captured on this day as they may need to be referred to later for analysis. Calibration with TetraSpeck beads is particularly important on microscopes in core facilities where different users may use different wavelengths or camera settings.
19. Labeling can be checked on a standard microscope with fluorescence prior to TIRF imaging. If labeling is inadequate, incubate the coverslip for a longer period of time with BTx-594.
20. It is important to avoid oversaturated fluorescence. This is particularly important for the 488 channel as this fluorescence should increase upon ADIS, and oversaturation will prevent this being observed.
21. When analyzing data, the time taken for solutions to move through the perfusion system should be taken into account. In our system, solutions took 45 seconds from switching the perfusion to reaching the chamber. The movie acquisition or agonist addition timings may need adjusting depending on your perfusion system.
22. Common problems encountered during image acquisition include blurry images, or profound changes in cell shape. If adjustment of the fine focus does not improve resolution, then blurred images may be due to other factors including insufficient oil on the objective or small leaks of solution from the imaging chamber. In this case, image acquisition should be stopped and chambers removed to check oil and whether leaking from the chamber has occurred. Changes in cell shape may occur due to experimental errors (e.g., if imaging solutions have been made incorrectly, usually due to too little glucose) or due to biological changes (e.g., membrane ruffling). To avoid experimental errors, make solutions fresh on the day of imaging and re-prepare solutions if cell changes are consistently observed.
23. We only imaged one viewpoint at a time. As CaSR will be activated after calcium is added, we discarded coverslips after imaging. Therefore, several coverslips may need to be prepared on the same day if you plan to perform repeats on the same day.

24. This VLC media player step may be avoidable if your imaging software allows export of each movie frame as a .tif, or if you are able to export in a movie format that can be opened in ImageJ (e.g., using the Olympus Viewer).

Acknowledgments

This work was supported by a Sir Henry Dale Fellowship jointly funded by the Wellcome Trust and the Royal Society (Grant Number 224155/Z/21/Z). Images of microscope equipment were taken in the Imaging Facility at the Centre of Membrane Proteins and Receptors (COMPARE) at the University of Birmingham, UK.

References

1. Brown EM, Gamba G, Riccardi D, Lombardi M, Butters R, Kifor O, Sun A, Hediger MA, Lytton J, Hebert SC (1993) Cloning and characterization of an extracellular Ca(2+)-sensing receptor from bovine parathyroid. Nature 366(6455):575–580. https://doi.org/10.1038/366575a0
2. Gorvin CM (2022) Genetic causes of neonatal and infantile hypercalcaemia. Pediatr Nephrol 37(2):289–301. https://doi.org/10.1007/s00467-021-05082-z
3. Ward DT, Brown EM, Harris HW (1998) Disulfide bonds in the extracellular calcium-polyvalent cation-sensing receptor correlate with dimer formation and its response to divalent cations in vitro. J Biol Chem 273(23):14476–14483. https://doi.org/10.1074/jbc.273.23.14476
4. He F, Wu CG, Gao Y, Rahman SN, Zaoralova M, Papasergi-Scott MM, Gu TJ, Robertson MJ, Seven AB, Li L, Mathiesen JM, Skiniotis G (2024) Allosteric modulation and G-protein selectivity of the Ca(2+)-sensing receptor. Nature 626:1141. https://doi.org/10.1038/s41586-024-07055-2
5. Grant MP, Stepanchick A, Cavanaugh A, Breitwieser GE (2011) Agonist-driven maturation and plasma membrane insertion of calcium-sensing receptors dynamically control signal amplitude. Sci Signal 4(200):ra78. https://doi.org/10.1126/scisignal.2002208
6. Cavanaugh A, McKenna J, Stepanchick A, Breitwieser GE (2010) Calcium-sensing receptor biosynthesis includes a cotranslational conformational checkpoint and endoplasmic reticulum retention. J Biol Chem 285(26):19854–19864. https://doi.org/10.1074/jbc.M110.124792
7. Gorvin CM (2019) Molecular and clinical insights from studies of calcium-sensing receptor mutations. J Mol Endocrinol 63(2):R1–R16. https://doi.org/10.1530/JME-19-0104
8. Pollak MR, Brown EM, Chou YH, Hebert SC, Marx SJ, Steinmann B, Levi T, Seidman CE, Seidman JG (1993) Mutations in the human Ca(2+)-sensing receptor gene cause familial hypocalciuric hypercalcemia and neonatal severe hyperparathyroidism. Cell 75(7):1297–1303. https://doi.org/10.1016/0092-8674(93)90617-y
9. Roszko KL, Stapleton Smith LM, Sridhar AV, Roberts MS, Hartley IR, Gafni RI, Collins MT, Fox JC, Nemeth EF (2022) Autosomal dominant hypocalcemia type 1: a systematic review. J Bone Miner Res 37(10):1926–1935. https://doi.org/10.1002/jbmr.4659
10. White E, McKenna J, Cavanaugh A, Breitwieser GE (2009) Pharmacochaperone-mediated rescue of calcium-sensing receptor loss-of-function mutants. Mol Endocrinol 23(7):1115–1123. https://doi.org/10.1210/me.2009-0041
11. Mullin BH, Pavlos NJ, Brown SJ, Walsh JP, McKellar RA, Wilson SG, Ward BK (2022) Functional assessment of calcium-sensing receptor variants confirms familial hypocalciuric hypercalcemia. J Endocr Soc 6(5):bvac025. https://doi.org/10.1210/jendso/bvac025
12. Grant MP, Stepanchick A, Breitwieser GE (2012) Calcium signaling regulates trafficking of familial hypocalciuric hypercalcemia (FHH) mutants of the calcium sensing receptor. Mol Endocrinol 26(12):2081–2091. https://doi.org/10.1210/me.2012-1232

13. Dershem R, Gorvin CM, Metpally RPR, Krishnamurthy S, Smelser DT, Hannan FM, Carey DJ, Thakker RV, Breitwieser GE, Regeneron Genetics C (2020) Familial hypocalciuric hypercalcemia type 1 and autosomal-dominant hypocalcemia type 1: prevalence in a large healthcare population. Am J Hum Genet 106(6):734–747. https://doi.org/10.1016/j.ajhg.2020.04.006
14. Nesbit MA, Hannan FM, Howles SA, Babinsky VN, Head RA, Cranston T, Rust N, Hobbs MR, Heath H 3rd, Thakker RV (2013) Mutations affecting G-protein subunit alpha11 in hypercalcemia and hypocalcemia. N Engl J Med 368(26):2476–2486. https://doi.org/10.1056/NEJMoa1300253
15. Nesbit MA, Hannan FM, Howles SA, Reed AA, Cranston T, Thakker CE, Gregory L, Rimmer AJ, Rust N, Graham U, Morrison PJ, Hunter SJ, Whyte MP, McVean G, Buck D, Thakker RV (2013) Mutations in AP2S1 cause familial hypocalciuric hypercalcemia type 3. Nat Genet 45(1):93–97. https://doi.org/10.1038/ng.2492
16. Gorvin CM, Rogers A, Hastoy B, Tarasov AI, Frost M, Sposini S, Inoue A, Whyte MP, Rorsman P, Hanyaloglu AC, Breitwieser GE, Thakker RV (2018) AP2sigma mutations impair calcium-sensing receptor trafficking and signaling, and show an endosomal pathway to spatially direct G-protein selectivity. Cell Rep 22(4):1054–1066. https://doi.org/10.1016/j.celrep.2017.12.089
17. Howles SA, Wiberg A, Goldsworthy M, Bayliss AL, Gluck AK, Ng M, Grout E, Tanikawa C, Kamatani Y, Terao C, Takahashi A, Kubo M, Matsuda K, Thakker RV, Turney BW, Furniss D (2019) Genetic variants of calcium and vitamin D metabolism in kidney stone disease. Nat Commun 10(1):5175. https://doi.org/10.1038/s41467-019-13145-x
18. Bouschet T, Martin S, Henley JM (2005) Receptor-activity-modifying proteins are required for forward trafficking of the calcium-sensing receptor to the plasma membrane. J Cell Sci 118(Pt 20):4709–4720. https://doi.org/10.1242/jcs.02598
19. Awata H, Huang C, Handlogten ME, Miller RT (2001) Interaction of the calcium-sensing receptor and filamin, a potential scaffolding protein. J Biol Chem 276(37):34871–34879. https://doi.org/10.1074/jbc.M100775200
20. Pi M, Spurney RF, Tu Q, Hinson T, Quarles LD (2002) Calcium-sensing receptor activation of rho involves filamin and rho-guanine nucleotide exchange factor. Endocrinology 143(10):3830–3838. https://doi.org/10.1210/en.2002-220240
21. Koh J, Dar M, Untch BR, Dixit D, Shi Y, Yang Z, Adam MA, Dressman H, Wang X, Gesty-Palmer D, Marks JR, Spurney R, Druey KM, Olson JA Jr (2011) Regulator of G protein signaling 5 is highly expressed in parathyroid tumors and inhibits signaling by the calcium-sensing receptor. Mol Endocrinol 25(5):867–876. https://doi.org/10.1210/me.2010-0277
22. Sekine-Aizawa Y, Huganir RL (2004) Imaging of receptor trafficking by using alpha-bungarotoxin-binding-site-tagged receptors. Proc Natl Acad Sci USA 101(49):17114–17119. https://doi.org/10.1073/pnas.0407563101
23. Centeno PP, Herberger A, Mun HC, Tu C, Nemeth EF, Chang W, Conigrave AD, Ward DT (2019) Phosphate acts directly on the calcium-sensing receptor to stimulate parathyroid hormone secretion. Nat Commun 10(1):4693. https://doi.org/10.1038/s41467-019-12399-9
24. Leach K, Gregory KJ, Kufareva I, Khajehali E, Cook AE, Abagyan R, Conigrave AD, Sexton PM, Christopoulos A (2016) Towards a structural understanding of allosteric drugs at the human calcium-sensing receptor. Cell Res 26(5):574–592. https://doi.org/10.1038/cr.2016.36
25. Sposini S, Jean-Alphonse FG, Ayoub MA, Oqua A, West C, Lavery S, Brosens JJ, Reiter E, Hanyaloglu AC (2017) Integration of GPCR signaling and sorting from very early endosomes via opposing APPL1 mechanisms. Cell Rep 21(10):2855–2867. https://doi.org/10.1016/j.celrep.2017.11.023
26. Konieczny V, Tovey SC, Mataragka S, Prole DL, Taylor CW (2017) Cyclic AMP recruits a discrete intracellular Ca(2+) store by unmasking hypersensitive IP(3) receptors. Cell Rep 18(3):711–722. https://doi.org/10.1016/j.celrep.2016.12.058

Part III

Calcium Measurement in Specific Organelles

Chapter 10

Measuring Calcium Signaling at the Primary Cilia

Vansh K. Goel, Ayan K. Barui, and Surya M. Nauli

Abstract

Cellular signaling is nature's ingenious way for cells to perceive their surroundings and transmit external cues to internal compartments. Due to its critical role in cellular functions, the intricate machinery of molecular signaling has been intensively studied. A diverse arsenal of techniques exists to quantify the molecules involved in these processes. Among them, calcium stands out as a ubiquitous signaling molecule with roles in countless biological pathways. To elucidate its function as a second messenger, methods for measuring intracellular calcium have steadily evolved. This chapter introduces various methods for investigating calcium signaling cascades in cells as well as in cilia (thin hairlike projections) specifically, where calcium signaling is triggered by different cilial manipulation techniques.

Key words Calcium signaling, Intracellular signaling, Mechanosensation, Primary cilia

1 Introduction

Once disregarded as vestigial remnants, primary cilia have risen to prominence as crucial sensory organelles, translating external stimuli into intricate intracellular responses [1]. Their diverse roles in mechano- and chemosensation are underscored by the devastating consequences of their dysfunction, manifested in a spectrum of human diseases known as ciliopathies. This chapter delves into the fascinating interplay between primary cilia and calcium signaling, revealing the exquisite dance of information transfer within these cellular outposts. At the heart of the cilium's sensory prowess lies a specialized protein complex: polycystin-1, an atypical G-protein-coupled receptor, and polycystin-2, a calcium-permeable channel [2]. When fluid-shear stress bends the cilium, this complex orchestrates an electrifying cascade of events. Calcium ions, the ubiquitous cellular messengers, flood into the cilium through polycystin-2 [3], igniting a symphony of signals within the cell

---

Vansh K. Goel and Ayan K. Barui have contributed equally to this work

Caroline M. Gorvin (ed.), *Calcium Signaling: Methods and Protocols*, Methods in Molecular Biology, vol. 2861,
https://doi.org/10.1007/978-1-0716-4164-4_10,

[4]. Protein modifications and membrane hyperpolarization are just a few of the intricate melodies played by this calcium influx [5].

This elegant dance relies heavily on the precise communication orchestrated by calcium. As the ultimate cellular language, calcium acts as a conductor, coordinating diverse processes like muscle contraction [6], neurotransmitter release [7], and even the spark of life itself [8]. The intricate connections between calcium and cilia extend deep into the evolutionary past, hinting at their fundamental importance in maintaining cellular and organismal health. Studies have revealed a treasure trove of signaling molecules residing within the cilium, including key players in pathways like Hedgehog, Wnt, and Notch [9–11]. These protein orchestras are transported to their designated roles by a dedicated delivery system called intraflagellar transport (IFT) [12]. Their precise localization and concerted action highlight the cilium's role as a hub for integrating diverse signaling cues from the extracellular environment [13].

But what happens when this meticulously tuned machine falters? Defects in primary cilia lead to a diverse array of debilitating conditions [14], collectively known as ciliopathies [15–17]. From cystic kidneys and obesity to blindness and mental retardation, the reach of these disorders underscores the critical role cilia play in maintaining organismal health. Understanding the precise link between cilial dysfunction and disrupted calcium signaling is crucial for developing effective therapeutic strategies.

To unravel the mysteries of ciliopathy and understand how dysfunctional cilia disrupt cellular communication, tools for studying their delicate calcium signaling are essential. However, the small size and orientation of primary cilia pose a major challenge. Traditional calcium indicators simply cannot access their confined space, necessitating innovative solutions. This chapter provides a set of protocols for cell culture to augment ciliation and describes different ciliary manipulation techniques and how to evaluate the calcium rise induced by those manipulations.

2 Materials

2.1 Cell Preparation

1. Collagen solution: Rat type I collagen (50 μg/mL) in 0.02 N acetic acid [*see* **Note 1**].
2. Phosphate-buffered saline (PBS).
3. Growth media: Dulbecco's modified Eagle medium (DMEM), 10% fetal bovine serum (FBS), 1% penicillin-streptomycin.
4. Starving media: DMEM, 2% FBS, 1% penicillin-streptomycin.
5. 6-well tissue culture plate (*see* **Note 2**).
6. 6-square glass No. 0 coverslips (*see* **Note 2**).

7. Precision tungsten microwire: Straightened and etched to a diameter of under 100 μm (wire quality #823 and surface finish #42 obtained from Luma Metall AB, Sweden, meets these criteria).
8. Imaging plate: A standard 35 mm imaging plate with a section of the bottom removed and replaced with a clear glass cover-slip. The edges are sealed with a clear silicone-based glue.

2.2 Calcium Signaling

1. Dulbecco's PBS: Special PBS solution containing free Ca^{2+} ions. Ensure it is at neutral pH.
2. Fura-2-acetoxymethyl ester (Fura-2-AM) solution: Add 50 μg Fura-2-AM to 50 μL of cell culture-grade anhydrous dimethyl sulfoxide (DMSO) to obtain a 1 mM stock solution. Mix this solution for around 5 min to ensure homogeneity. Brief centrifugation of the vial is required to ensure that no undissolved Fura-2-AM is used during the experiment (*see* **Note 3**).
3. Optional: Pluronic—0.05% nonionic, detergent Pluronic.
4. One of the three following solutions should be used as incubation/perfusion buffer:
   (a) Krebs buffer solution: 122 NaCl mM, 5.17 KCl mM, 5.56 mM dextrose, 2.49 mM $MgSO_4$, 25.6 mM $NaHCO_3$, 0.114 mM ascorbic acid, 1.60 mM $CaCl_2$, and 0.027 mM EGTA, and is continuously bubbled with 95% O_2 and 5% CO_2 to maintain the pH at 7.2.
   (b) Modified DMEM buffer: 0.81 mM MgSO4, 0.44 mM KH_2PO_4, 5.37 mM KCl, 137 mM NaCl, 1.26 mM $CaCl_2$, 0.34 mM Na_2HPO_4, 2.0 mM L-glutamine, 5.55 mM D-glucose, 1% bovine serum albumin, 1.0 mM sodium pyruvate, and 20.0 mM HEPES at pH 7.2.
   (c) HEPES buffer: 2 mM $CaCl_2$, 1 mM $MgCl_2$, 138 mM NaCl, 10 mM D-glucose, 5 mM KCl, 10.0 mM HEPES at pH 7.4.
5. Genetically encoded calcium indicator (GECI): 5HT6-mCherry-G-GECO1.0 plasmid (Addgene, Cat. 47500) reported by Su et al. [18].
6. Jetprime transfection reagent.
7. Selection media for GECI: G418 antibiotic at a concentration of 500 μg/mL in growth media (DMEM).

2.3 Ciliary Manipulation

1. Ciliary micromanipulator.
2. Magnetic nanoparticles (NPs): Fe_2O_3 NPs coated in Sunbright-40 (oleyl-O-(CH_2CH_2)nCO-CH_2CH_2-COO-N-hydroxysuccinimide; PEG MW = 4000) and subsequently covalently bonded to Alexa Fluor 594-labeled DR5 antibody

[19]. The synthesis of these particles will not be discussed in this chapter.

3. Fluid-shear stress chamber (further explained in the "Special Equipment and Technology" section of Materials).

2.4 Special Equipment and Technology

1. 100 g AlNiCo cylindrical magnet to generate a maximum field strength of 1.35 T.
2. Perfusion pump system: must be able to produce laminar flow for fluid-shear stress. Generally, this consists of a peristaltic pump (our lab used the InsTech P720 peristaltic pump) with a speed controller, a sealable flow chamber, and a silicone gasket that fits in the flow chamber. A vacuum line is also required.
3. Fluorescence microscope: requires appropriate filters to process images. Measurement of Fura-2-AM will require a microscope capable of excitation at 340 nm and 380 nm, a dichroic mirror, and a wide band 510 nm emission filter. For experiments using GECIs, filters for both green fluorescence and mCherry (red) fluorescence will be required. For example, the 5HT6-mCherry-G-GECO1.0 has enhanced green fluorescent protein (EGFP) with excitation and emission wavelengths of 495 and 515 nm, respectively. The mCherry has excitation and emission wavelengths of 587 and 610 nm, respectively. Thus, the excitation/emission filters of 495/515 nm and 587/610 nm are used to capture EGFP and mCherry, respectively. Our laboratory uses a high-speed excitation wavelength changer of the DG4/DG5 system to allow the light path to hit different filters quickly.
4. Computer system: appropriate software to communicate with the microscope. We use NIS-Elements High Content AR 4.30.02 (Nikon) to capture and analyze intracellular Ca^{2+}. The live tracking and kymograph analyses in the software allow us to separately measure the cell body and cilia. After the automated ratiometric calculation is performed, the Ca^{2+} tracking can be very effectively achieved using binary spotting tracks to quantify ciliary Ca^{2+} in bending cilium.

3 Methods

This methods section will describe cell culture techniques for optimal cilia formation, calcium signaling indicators including small molecule and genetically encoded calcium indicators (GECIs), and ciliary manipulation for calcium signaling studies. Choosing one set of instructions from each category in this methods section will lead to a successful experiment. However, the general steps are cell preparation, calcium indicator loading, cilial manipulation, and then imaging. There are some exceptions to this rule, and they will be covered below.

3.1 Cell Culture

In selecting your cell culture instructions, consider the overall goal of your experiment. Choose the situation (population vs single cell) that better suits your experimental goals.

3.1.1 Cell Population Study

This study is useful for both small molecule and GECI-based calcium measurement:

1. Incubate six glass coverslips in collagen solution for around 5 s.
2. Put a coverslip in each well of the six-well tissue culture plate.
3. Incubate coverslips for 1 h at room temperature.
4. Add 2 mL PBS to each well and rinse thoroughly.
5. Carefully aspirate the PBS.
6. Sterilize the six-well tissue culture plate under the UV light for minimum 20 min (*see* **Note 4**).
7. Seed a suitable number of cells in 2 mL of growth media in each well of the six-well plate (*see* **Note 5**).
8. After cells reach around 98% confluency, aspirate the growth media, and replace it with starving media for 2–4 days prior to imaging (*see* **Note 6**).

3.1.2 Individual Cell Study

This study is useful for only GECI-based calcium measurement:

1. Incubate a length of precision tungsten microwire in collagen solution for a minimum of 1 min.
2. Rinse the wire vigorously in PBS.
3. Sterilize the microwire for at least 30 min under UV light.
4. Mount the microwire to the imaging plate containing growth media.
5. Seed cells at low density and monitor cell density for 1–2 days.
6. Once cell confluency reaches around 95%, remove the growth media and replace it with starving media (*see* **Note 7**).

3.2 Calcium Indicator

Detection of cellular Ca^{2+} fluxes is generally detected by employing calcium indicators. The major categories of calcium indicators include the following: (i) small molecules (calcium-sensitive fluorescence dyes) and (ii) GECIs. Among small molecules, the ratiometric dye Fura-2-AM is often used for detection of calcium flux (Fig. 1). Fura-2-AM can be excited at 340 nm for calcium bound as well as 380 nm for calcium unbound. The emission wavelength of this dye is 510 nm irrespective of the excitation wavelength. The ratio of 340/380 is generally evaluated for normalizing unequal

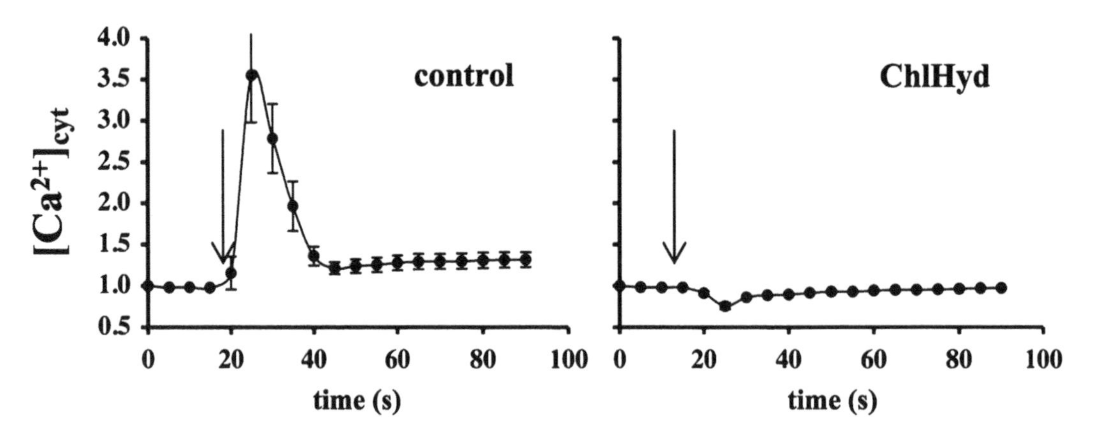

Fig. 1 Intracellular free calcium as a readout of ciliary function. Cytosolic free calcium ($[Ca^{2+}]_{cyt}$) in response to fluid-shear stress is measured with Fura-2. Wild-type embryonic endothelial cells treated without (control) and with chloral hydrate (ChlHyd; 4 m*M*, 12 h) are challenged with shear stress. Chloral hydrate reversibly blocks the formation of primary cilia or destroys cilia structure. Arrows indicate the step increase in fluid-shear stress. (Reproduced from Ref. Nauli et al. [20] with permission from Elsevier)

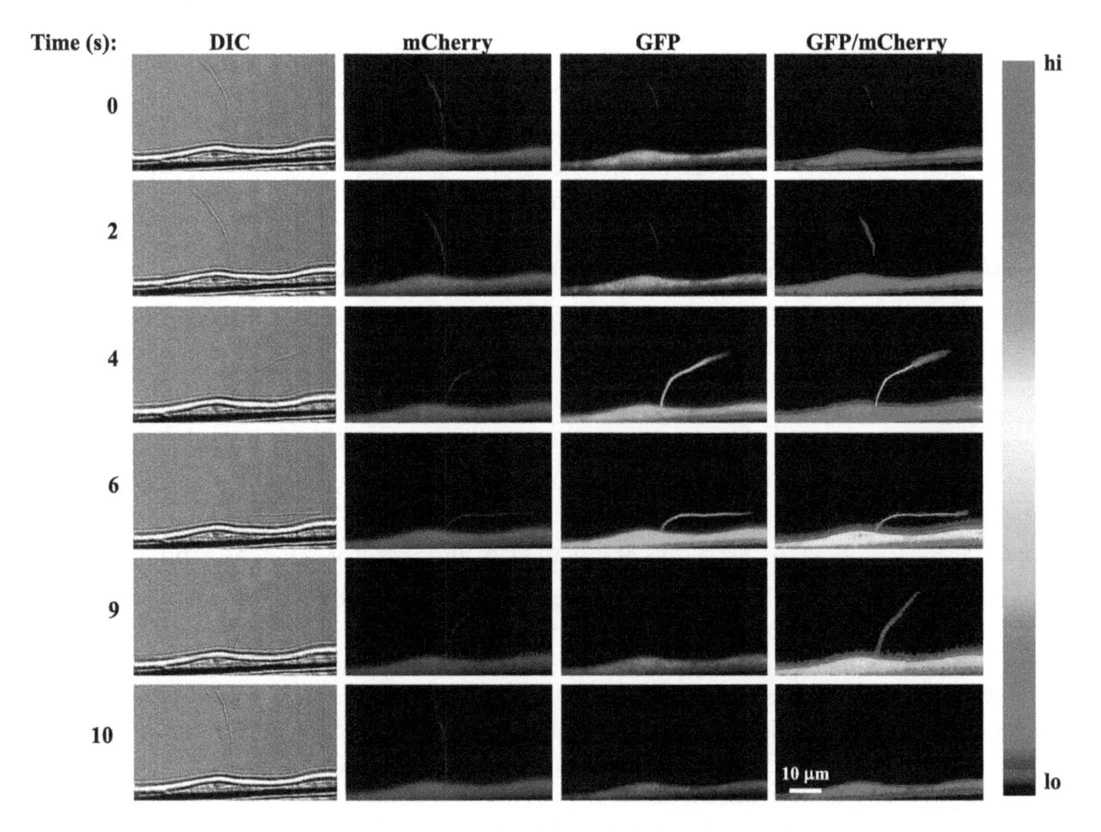

Fig. 2 Single live cell and cilia imaging with 5HT6-mCherry-G-GECO1.0. The images show (from left to right column) DIC used for tracking a cilium, mCherry fluorescent ciliary marker, the Ca^{2+}-sensitive GECO1.0 and an EGFP/mCherry ratio pseudocolored to show Ca^{2+} levels. When fluid flow is applied (time series from top to bottom), the cilium bends, inducing a Ca^{2+} increase in both the cytoplasm and cilioplasm. (Reproduced from Ref. Pala et al. [17] with permission from American Chemical Society)

loading of Fura-2-AM in the cells and potential fluorescence artifacts. On the other hand, GECIs are comprised of a Ca^{2+} binding moiety with one or two fluorophores. The free, unbound Ca2+ ions bind to the GECI binding domain, resulting in conformational changes (Fig. 2). The conformational changes then shift the fluorophores, causing its fluorescence intensity to change. GECIs need insertion of nucleic acid sequence coding for a calcium sensor into the cells. When GECIs are expressed in cells, the calcium indicator is fused in the cellular environment.

While small molecule indicators (in this method, Fura-2-AM [20]) are only able to visualize changes in cytoplasmic calcium concentration, GECIs [17] can visualize alteration of calcium concentration in both cytoplasm and cilioplasm. So, be careful in selecting your calcium indicator. Proceed with whichever set of instructions better supports one's experimental goals. The following section describes the methods of Fura-2-AM and GECIs-based calcium detection in cells.

3.2.1 Small Molecule Fura-2-AM

1. Rinse the cells twice using Dulbecco's PBS.
2. Incubate cells with ~10 μM Fura-2-AM solution.
3. Remove undissolved dye by cautiously pipetting the upper ~48 μL Fura-2-AM solution from a vial, and add it to 5 mL of your chosen incubation solution (optional: add pluronic as needed (*see* **Note 8**).
4. Incubate cells with the incubation solution bearing Fura-2-AM dye for 30 min at varying temperatures (*see* **Note 9**).
5. Replace Fura-2-AM-containing solution with fresh incubation solution, and incubate the fluid-shear stress chamber for 15 min at room temperature to equilibrate prior to any ciliary stress experiments.

3.2.2 GECI Transfection

1. GECIs are plasmids and must be transfected to the target cell line before cell culture begins. The following steps will explain how to produce seed cells that have successfully been transfected.
2. In a new tissue culture plate, grow mammalian cells at 37 °C in growth media.
3. Once 60–70% confluence is reached, transfect cells with the GECI construct using the Jetprime transfection reagent following manufacturer's instructions.
4. Replace the current growth media with selection media (i.e., media containing an antibiotic (usually geneticin or puromycin) corresponding to the mammalian selection cassette in the GECI plasmid), and continue to observe the cell growth.

5. Once cells reach around 95% confluency, use them as seed cells for either cell culture methods (individual study or cell population study) described above.

3.3 Cilial Manipulation

Cilia act as both chemo- and mechanosensors of the external cell environment. As chemosensors, cilia contain multiple G-protein-coupled receptors (GPCRs) that are sensitive to serotonin, somatostatin, melanin, and dopamine. There are three common ways for the investigation of mechanosensory function of cilia [20]. The first method is ciliary micromanipulation to bend one primary cilium using a micropipette, which is a very direct way (Fig. 3a). The second technique is the use of magnetic field employing magnetic NPs precoated with antibody specific to cilia surface (Fig. 3b). The third way is the use of fluid-shear stress across apical membrane of cells using a perfusion pump (Fig. 3c). The following section explains the experimental setups for the abovementioned ways of studying mechanosensory function of cilia.

3.3.1 Ciliary Micromanipulation

1. Use the ciliary micromanipulator to move and bend the cilia (*see* **Note 10**).

3.3.2 Magnetic NPs

1. Add 50 μL of magnetic NPs to 1 mL of growth media (*see* **Note 11**).
2. Add this growth media to cells.
3. Incubate cells at 37 °C overnight.
4. Arrange the cells under the microscope, place the electromagnet nearby, and then turn the electromagnet on. Field strength can be controlled by distance to the cells, or by increasing the amperage through the electromagnet.

3.3.3 Fluid-Shear Stress Through a Perfusion Pump System

1. Assemble the perfusion system with the cells on the bottom, then the gasket, and finally the chamber cover (Fig. 4) [21].
2. Set the cells in focus under your microscope, and then turn on the peristaltic pump (*see* **Note 12**).
3. Ensure new perfusate is drawn from a clean reservoir and used perfusate is stored as waste.

3.4 Computer Imaging

1. Set up your microscope according to the manufacturer's instructions.
2. Set up the excitation and emission filters according to the fluorophore that was selected.
3. Establish a baseline Ca^{2+} fluorescence value by taking images every 10 s for 5 min.
4. After this period, induce mechanosensation (cilia manipulation) and take images continuously until mechanosensation is

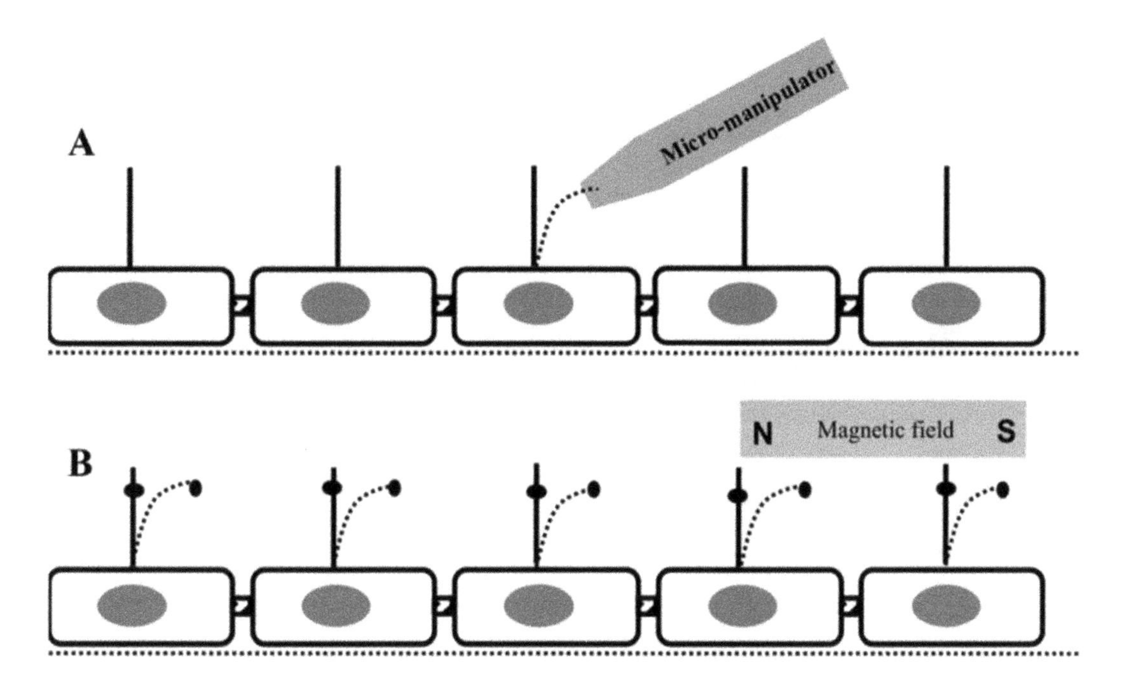

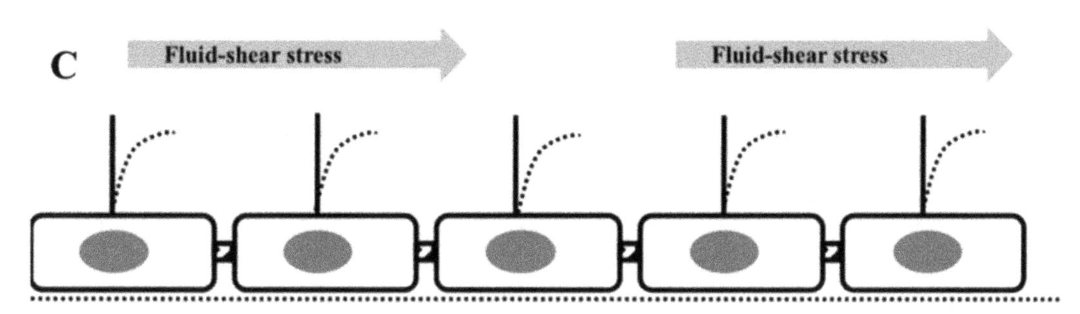

Fig. 3 Techniques to study mechanosensory function of cilia. There are at least three different ways to study mechanosensory function of cilia. (**a**) The first way is to directly bend one primary cilium by applying a negative pressure at the tip of a micropipette. (**b**) The second way to study sensory function of cilia is with nanomagnetic beads, precoated with ciliary surface-specific antibody. The nano beads can be twisted or pulled by applying a magnetic field around the cells. (**c**) The third and more physiological way to examine cilium function is to apply fluid shear on the top of the cells. This technique is the most robust and inexpensive way to analyze mechanosensory function of cilia. (Reproduced from Ref. Nauli et al. [20] with permission from Elsevier)

stopped. After mechanosensation is stopped, continue taking images for 5 more minutes.

3.5 Data Analysis

3.5.1 Fura-2-AM

1. The free calcium for Fura-2-AM based study can be calculated employing the following formula provided by the manufacturer (Invitrogen). $[Ca^{2+}] = Kd \times Q(R–Rmin)/(Rmax—R)$. Kd is dissociation constant of Fura-2-AM from the released

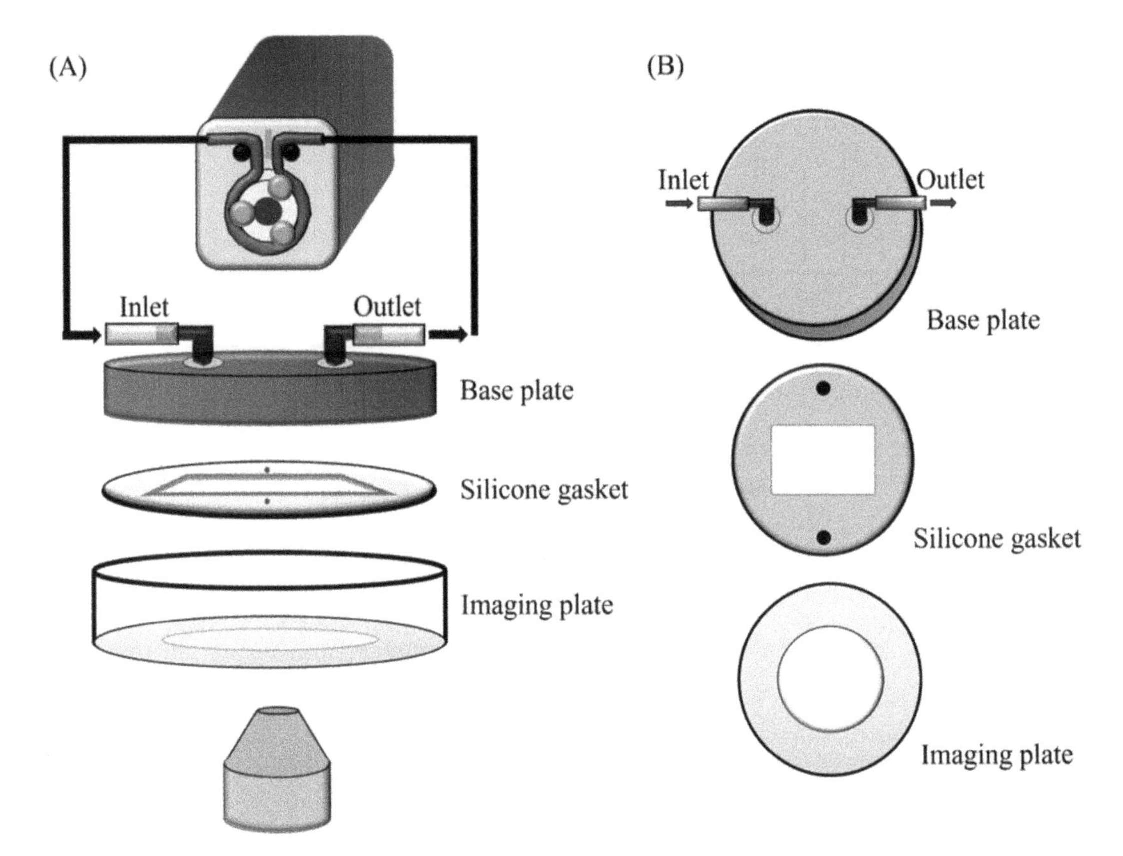

Fig. 4 Flow equipment setup for live cell and cilia imaging. (**a**) The flow setup uses a peristaltic pump in a closed loop system. The perfusion chamber consists of a top base plate with inlet and outlet connected to the pump. Then, we have a silicone gasket to form a channel for laminar flow and finally, a glass bottom imaging plate. (**b**) Top view of each perfusion chamber component. (Reproduced from Ref. Sherpa et al. [21] with permission from Elsevier)

calcium; R is the measured ratio F340/F380; Rmin and Rmax are the fluorescence intensity ratios at minimum and maximum ion concentration, respectively.

3.5.2 GECIs

1. For data analysis, construct region of interest (ROI) around the cilium employing ROI tool.
2. Make ROI adjacent to cilium for measurement of background fluorescence.
3. Also make separate ROI for cell body.
4. Modulate the ROI at different time points post-fluid flow to account for the bending of cilium.
5. Subtract the background fluorescence intensity from cilium ROI fluorescence intensity, followed by two normalization steps (i) against mCherry and (ii) basal signal fluorescence intensities applied to data.

6. Generate the EGFP/mCherry images using the ratio of fluorescence intensities of EGFP and mCherry followed by applying pseudocolor for observation purposes (Fig. 2).

4 Notes

1. Collagen is not soluble at neutral pH.
2. Ensure that a single glass coverslip can lay flat in a single well.
3. Fura-2-AM solution works best when prepared fresh—this avoids fluorescence quenching, provides good signal-to-noise ratio, and allows for optimal dye loading.
4. After sterilization, the six-well plate may be stored under sterile conditions at 4 °C for a week.
5. Aim to reach 70–90% confluency of cell density after culture overnight.
6. If a high amount of cell death occurs with 2% FBS containing starving media, serum concentration can be increased to 5% FBS to prevent cell death. If the cell differentiation does not occur properly, the serum concentration of the starving media can be reduced to 0.5% to promote ciliation.
7. 100% confluency of cells in this case would be the inability to see the surface of the microwire. If necessary, the microwire can be lightly rotated to confirm confluency.
8. Pluronic is helpful in increasing ester solubility and may decrease dye compartmentalization. Be careful of adding too much pluronic, as this will lead to cellular acidification via formaldehyde release through the hydrolysis of acetoxymethyl ester.
9. Incubation should occur at 38 °C for human cells or 33 °C for mouse cells.
10. Our lab used a micropipette. However, this technique is generally avoided as it is difficult to ensure uniformity across tests and is very time-consuming.
11. If more solution is required, maintain the same ratio of 50 μL NPs to 1 mL of growth media.
12. Ensure no pressure is building up in the perfusion chamber. This can be detected by the cells going out of focus due to the coverslip or microwire bending. If this occurs, try introducing a vacuum line to the perfusion chamber.

Acknowledgments

This work was supported by the NIH HL147311 and HL147311-S1. We are thankful to all members, including prior members of our laboratory, who have contributed so much in improving measurements of ciliary signaling in live cells.

References

1. Nauli SM, Haymour HS, AbouAlaiwi WA, Lo ST, Nauli AM (2010) Primary cilia are mechanosensory organelles in vestibular tissues. In: Mechanosensitivity and Mechanotransduction. Springer Nature, London
2. Nauli SM, Alenghat FJ, Luo Y, Williams E, Vassilev P, Li X et al (2003) Poly-cystins 1 and 2 mediate mechanosensation in the primary cilium of kidney cells. Nat Genet 33:129–137
3. Jin X, Mohieldin AM, Muntean BS, Green JA, Shah JV, Mykytyn K et al (2014) Cilioplasm is a cellular compartment for calcium signaling in response to mechanical and chemical stimuli. Cell Mol Life Sci 71:2165–2178
4. Lee KL, Guevarra MD, Nguyen AM, Chua MC, Wang Y, Jacobs CR (2015) The primary cilium functions as a mechanical and calcium signaling nexus. Cilia 4:7
5. Praetorius HA, Spring KR (2001) Bending the MDCK cell primary cilium increases intracellular calcium. J Membr Biol 184:71–79
6. Ebashi S, Endo M (1968) Calcium ion and muscle contraction. Prog Biophys Mol Biol 18:123–183
7. Kerr R, Lev-Ram V, Baird G, Vincent P, Tsien RY, Schafer WR (2000) Optical imaging of calcium transients in neurons and pharyngeal muscle of C. elegans. Neuron 26:583–594
8. Wier W, Cannell M, Berlin M, Marban E, Lederer W (1987) Cellular and subcellular heterogeneity of [Ca^{2+}]i in single heart cells revealed by fura-2. Science 235:325–328
9. Huang P, Schier AF (2009) Dampened hedgehog signaling but normal Wnt signaling in zebrafish without cilia. Development 136: 3089–3098
10. Ishikawa H, Thompson J, Yates JR, Marshall WF, Marshall WF (2012) Proteomic analysis of mammalian primary cilia. Curr Biol 22:414–419
11. Pazour GJ, Agrin N, Leszyk J, Witman GB (2005) Proteomic analysis of a eukaryotic cilium. J Cell Biol 170:103–113
12. Liem KF, Ashe A, He M, Satir P, Moran J, Beier D et al (2012) The IFT-A complex regulates Shh signaling through cilia structure and membrane protein trafficking. J Cell Biol 197:789–800
13. Mukhopadhyay S, Wen X, Chih B, Nelson CD, Lane WS, Scales SJ et al (2010) TULP3 bridges the IFT-A complex and membrane phosphoinositides to promote trafficking of G protein-coupled receptors into primary cilia. Genes Dev 24:2180–2193
14. Christensen ST, Clement CA, Satir P, Pedersen LB (2012) Primary cilia and coordination of receptor tyrosine kinase (RTK) signalling. J Pathol 226:172–184
15. Kathem SH, Mohieldin AM, Abdul-Majeed S, Ismail SH, Altaei QH, Alshimmari IK et al (2014) Ciliotherapy: a novel intervention in polycystic kidney disease. J Geriatr Cardiol 11: 63–73
16. Nauli SM, Sherpa RT, Reese CJ, Nauli AM (2016) Mechanosensory and chemosensory primary cilia in ciliopathy and ciliotherapy. In: Mechanobiology: exploitation for medical benefit. Wiley, New York
17. Pala R, Mohieldin AM, Shamloo K, Sherpa RT, Kathem SH, Zhou J et al (2018) Personalized nanotherapy by specifically targeting cell organelles to improve vascular hypertension. Nano Lett 19:904–914
18. Su S, Phua SC, Derose R, Chiba S, Narita K, Kalugin PN et al (2013) Genetically encoded calcium indicator illuminates calcium dynamics in primary cilia. Nat Methods 10:1105–1107
19. Pala R, Mohieldin AM, Sherpa RT, Kathem SH, Shamloo K, Luan Z et al (2019) Ciliotherapy: remote control of primary cilia movement and function by magnetic nanoparticles. ACS Nano 13:3555–3572
20. Nauli SM, Jin X, AbouAlaiwi WA, El-Jouni W, Su X, Zhou J (2013) Non-motile primary cilia as fluid shear stress mechanosensors. In: Methods in enzymology. Elsevier Inc, Amsterdam
21. Sherpa RT, Pala R, Mohieldin AM, Nauli SM (2019) Measurement of cytoplasmic and cilioplasmic calcium in a single living cell. In: Methods in cell biology. Elsevier Inc, Amsterdam

Chapter 11

Monitoring ER Ca^{2+} by Luminescence with Low Affinity GFP-Aequorin Protein (GAP)

Macarena Rodriguez-Prados, Jonathan Rojo-Ruiz, Belen Calvo, Javier Garcia-Sancho, and Maria Teresa Alonso

Abstract

The endoplasmic reticulum (ER) is the main cellular reservoir of Ca^{2+}, able to accumulate high amounts of calcium close to the millimolar range and to release it upon cell activation. Monitoring of Ca^{2+} dynamics within the ER lumen is best achieved using genetically encoded and targeted reporters. Luminescent probes based on the photoprotein aequorin have provided significant insight to measure subcellular Ca^{2+}. Here we describe a robust and quantitative method based on the Ca^{2+} indicator of the GFP-Aequorin Protein (GAP) family, targeted to the ER lumen. A low Ca^{2+} affinity version of GAP, GAP1, carrying mutations in two EF-hands of aequorin, reconstituted with coelenterazine *n* has a reduced affinity for Ca^{2+} such that it conforms with the $[Ca^{2+}]$ values found in the ER and it slows the consumption of the probe by Ca^{2+}. This feature is advantageous because it avoids fast aequorin consumption allowing long-term (longer than 1 h) ER Ca^{2+} measurements. GAP1 targeted to the ER allows monitoring of resting $[Ca^{2+}]_{ER}$ and Ca^{2+} dynamics in intact cells stimulated with IP_3-produced agonists. In addition, GAP1 can record Ca^{2+} mobilization in permeabilized cells challenged with IP_3. We also provide a detailed calibration procedure which allows to accurately convert the luminescence signal into $[Ca^{2+}]_{ER}$.

Key words Luminescence, Aequorin, GAP, Ca^{2+} calibration, ER

1 Introduction

The endoplasmic reticulum (ER) is the largest organelle in animal cells and the location of important cellular functions such as synthesis and folding of proteins, lipid metabolism, or Ca^{2+} storage. The ER and its muscle counterpart, the sarcoplasmic reticulum (SR), are the major stores of intracellular Ca^{2+}, which can reach up to millimolar concentrations. This function is achieved by the combination of three distinct elements: (a) Ca^{2+} ATPases of the SERCA family located in the ER membrane that efficiently accumulates Ca^{2+} against its concentration gradient; (2) luminal Ca^{2+}-binding proteins, like calreticulin or calsequestrin, which buffer the stored Ca^{2+}; and (3) Ca^{2+} channels that release the stored Ca^{2+} into

Caroline M. Gorvin (ed.), *Calcium Signaling: Methods and Protocols*, Methods in Molecular Biology, vol. 2861,
https://doi.org/10.1007/978-1-0716-4164-4_11,

the cytosol along its concentration gradient. The two main Ca^{2+} channels located in the ER membrane are the inositol 1,4,5-trisphosphate receptors (IP_3Rs) [1] and the ryanodine receptors (RyRs) [2]. Ca^{2+} is typically released upon cellular stimulation via PLC activation through G-protein-coupled receptors and cleavage of phosphatidylinositol 4,5 bisphosphate (PIP_2) into IP_3 and diacylglycerol. Ryanodine receptors act through Ca^{2+}-induced Ca^{2+} release (CICR). Proper regulation of intraluminal Ca^{2+} concentration via these elements is crucial for organelle functions since increase or reduction of $[Ca^{2+}]_{ER}$ may trigger processes such as ER stress and apoptosis [3].

The release of Ca^{2+} from the ER/SR controls many cellular functions such as muscle contraction, secretion, and synaptic activity [4]. It also controls the Ca^{2+} entry through the plasma membrane via the so-called store-operated Ca^{2+} entry (SOCE) mechanism [5]. This comprises the Ca^{2+} channel STIM, present in the ER membrane, and the Orai channel, present in the plasma membrane. Upon cell activation, Ca^{2+} released from the ER triggers STIM translocation to the plasma membrane where it interacts with Orai, which opens and activates SOCE. This ubiquitous mechanism is especially relevant for immune function [6]. In addition, the sustained depletion of the ER Ca^{2+} store can trigger unfolded protein response (UPR) activation (ER stress) and, eventually, apoptosis. In fact, SERCA inhibitors have been extensively used as pharmacological tools to trigger UPR and, therefore, have been considered as therapeutic tools against cancer [7].

Aequorin is a small protein of 22-kDa from *Aequorea victoria* that contains three binding sites for Ca^{2+} [8]. Apoaequorin binds to the prosthetic group coelenterazine that gets oxidized upon Ca^{2+} binding, and this active form triggers the emission of one photon of blue light. Coelenterazine is highly hydrophobic, and when added to cells, it freely diffuses through the plasma membrane and binds to apoaequorin to generate its active form in a process called reconstitution. The emission of light is practically irreversible in terms of the time duration of the experiment, and this implies that in environments of high Ca^{2+}, aequorin gets consumed across the duration of the experiment [9].

We have recently developed a new family of genetic Ca^{2+} indicators dubbed GAPs (*GFP-Aequorin Proteins*) that consists of a mutated version of GFP fused to the N-terminus of apoaequorin (AEQ) via a 16-residue linker (Fig. 1a) [10]. One of the unique features of the GAP family is its dual mode of Ca^{2+} measuring, either fluorescent or luminescent. In the first mode, only the apoaequorin (without its cofactor) is required, whereas in the second mode, aequorin needs to be reconstituted with its cofactor coelenterazine. GAP1 is a low Ca^{2+} affinity GAP variant due to the three substitutions in the aequorin moiety in the second and third EF hands of aequorin (D117A, D119A, and D163A). This affinity

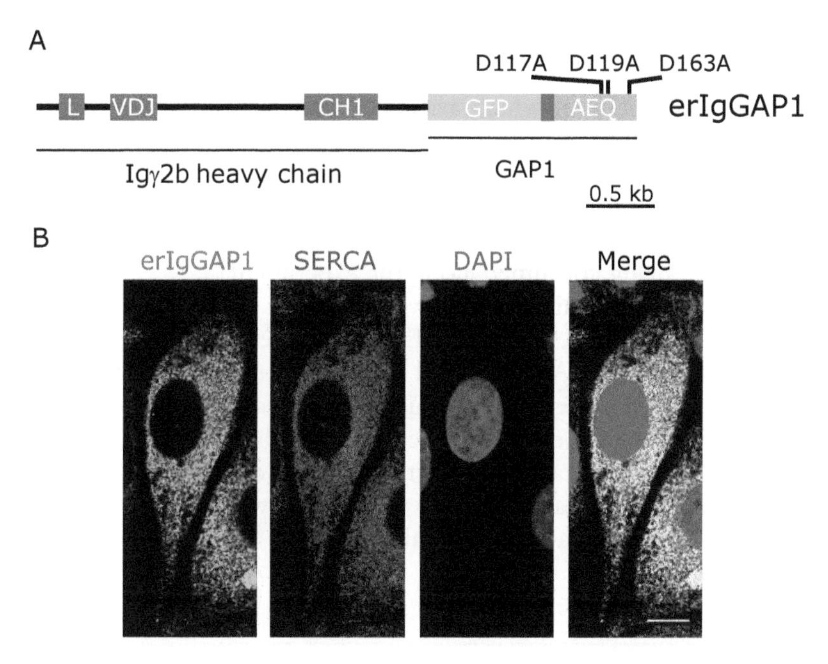

Fig. 1 Design and localization of GAP1 targeted to the ER. (**a**). Domain structure of erIgGAP1 construct drawn approximately to scale. A segment of the Igγ2b heavy chain gene containing the leader sequence (L), and the exons VDJ and CH1 (purple) fused with a GFP variant (dubbed uvGFP or C3) followed by a linker sequence (blue) and apoaequorin (AEQ, cyan), carrying the indicated mutations. (**b**). Localization of erIgGAP1 in the ER. Confocal fluorescence images of erIgGAP1 HeLa cells. GAP1 endogenous green fluorescence colocalizes with SERCA revealed with a specific anti-SERCA antibody (red fluorescence). Nuclei were stained with DAPI (blue fluorescence). Scale bar, 10 μm

for Ca^{2+} conforms to the high $[Ca^{2+}]$ reported in the ER, close to the millimolar range. Moreover, its lower affinity reduces the light emission and, in turn, its consumption upon Ca^{2+} binding, allowing long-term (longer than 1 h) measurements in high $[Ca^{2+}]$ environments such as the ER [11]. GAP1 expressed in the ER displays a strong fluorescent signal, and its distribution pattern is characteristic for the ER, being excluded from the nucleus and colocalized with the SERCA pump (Fig. 1b). In this chapter we describe the luminescent mode of operation of GAP1, when it is reconstituted with coelenterazine *n* to monitor resting and Ca^{2+} dynamics within the ER lumen.

Monitoring of Ca^{2+} through luminescence has a number of advantages: firstly, a high signal-to-noise ratio (SNR), mainly due to the low background of mammalian cells that do not contain endogenous chemiluminescent proteins. As a consequence, reliable Ca^{2+} measurements can be performed with relatively low levels of indicator expression. Secondly, it has a broad dynamic range due to the combination of aequorin mutations and coelenterazine derivatives. Thirdly, it has a low Ca^{2+} buffering effect, in comparison with fluorescent synthetic probes. For example, cytosolic Ca^{2+} peaks evoked by caffeine recorded by aequorin can be greatly dampened by loading cells with Fura-2 [12]. Fourthly, it does not require excitation, which avoids problems associated with phototoxicity. Fifthly, it provides a reliable quantitative method based on the calibration of the luminescence signal at the end of each experiment. Calibration is relatively straightforward, although the interpretation of the results has to take into account that aequorin is being consumed along the experiment.

In this chapter, we provide a method for monitoring Ca^{2+} dynamics within the ER by luminescence using the genetic Ca^{2+} sensor GAP1 targeted to the ER lumen in stably expressing erIg-GAP1 HeLa cells. This protocol can be applied for studying ER Ca^{2+} responses in intact cells stimulated with IP_3-forming agonists (e.g., histamine) or with inhibitors of the SERCA pump, as well as in permeabilized cells challenged with IP_3. In addition, we also provide a detailed calibration of the luminescence signal into $[Ca^{2+}]_{ER}$.

2 Materials

2.1 Equipment

1. Four-well plates.
2. A luminometer (also dubbed aequorinometer) is schematized in Fig. 2. This apparatus is custom-designed and built by Cairn Research Ltd. (UK). It is composed of a dark light-tight chamber with a holder for a four-well plate containing the adherent cells. It has a shutter that is opened by the experimenter from outside. Cells are continuously perfused by a gravity-based perfusion system and connected to a vacuum. The perfusion is controlled by a six-valve controller, and perfusion flow is set at a speed of 5 mL/min (*see* **Note 1**). Emitted light is collected by a photomultiplier tube placed underneath the cell plate.

2.2 Cell Culture

1. HeLa cells stably expressing erIgGAP1 (*see* **Note 2**). GAP1 consists of GFP fused to the N-terminus of apoaequorin via a 16-residue linker (Fig. 1a). The GFP variant exhibits four substitutions (Q80R, F99S, M153T, and V163A), with respect to the wild-type GFP, and the apoaequorin mutant carries three mutations (D117A, D119A, and D163A; named after the aequorin protein). To target GAP1 to the ER, the GAP1 N-terminus is fused to a segment of the Igγ2b heavy chain that includes the leader (L) sequence, the VDJ, and the CH1 domains. The latter domain is responsible for protein retention within the ER lumen due to its interaction with the ER resident protein BIP [13]. In cells lacking the immunoglobulin light chain, GAP will be retained in the ER.

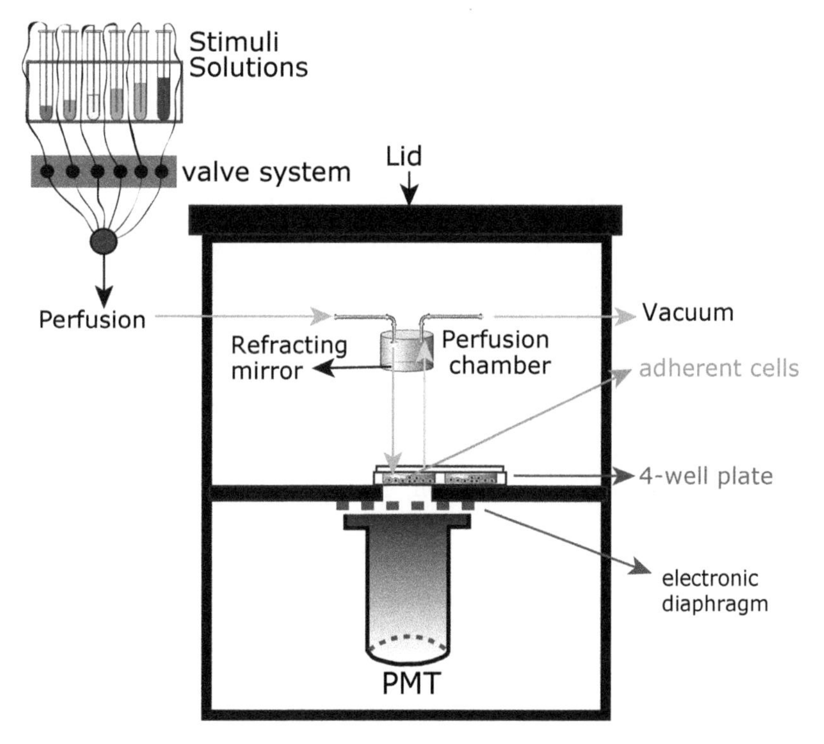

Fig. 2 Scheme of the luminescence apparatus (aequorimeter) used for measuring $[Ca^{2+}]_{ER}$ in cellular batches. The flow of the perfusion system is indicated by the blue arrows. *PMT* photomultiplier tube. (More details in Rodriguez-Prados [17])

2. Culture media: Dulbecco's modified Eagle medium (DMEM) supplemented with 10% heat-inactivated fetal bovine serum, 2 mM L-glutamine, 100 μg/mL streptomycin, 100 U/ml penicillin, and 0.2 mg/mL G-418.
3. Trypsinization media: 0.05% Trypsin-EDTA.

2.3 Stock and Working Solutions

The total volumes of each solution to be perfused are tentative, and the user must adjust it according to the flow speed, the number of wells, and the actual protocol. Use Milli-QR distilled water for all recipes in which it is included:

1. *ATP-Mg stock solution, 0.1 M*: Dissolve 5.51 g adenosine 5′--triphosphate disodium salt hydrate in 10 mL of a 1 M $MgCl_2$ solution (see recipe). Adjust pH to 7.0 with NaOH. Bring to 0.1 L final volume with distilled water. Store frozen at −20 °C for 24 months in 10 mL aliquots.
2. *BHQ stock solution, 10 mM*: Dissolve 44.46 mg BHQ in 20 mL DMSO (10 mM final concentration). Store frozen at −20 °C in 0.5 mL aliquots.

3. *BHQ 10 μM in EM0Ca solution*: Add 50 μL of 10 mM BHQ stock solution to 50 mL EM0Ca solution. Prepare fresh immediately before use.
4. *$CaCl_2$ solution, 1 M*: Dissolve 73.51 g $CaCl_2$ dihydrate in 0.5 L distilled water to obtain a 1 M stock solution. Store frozen at −20 °C for 24 months in 10 mL aliquots.
5. *Coelenterazine n, 200 μM*: Prepare a 200 μM stock in methanol and aliquot it in 30 μL on ice. Gas briefly with nitrogen before closing the tubes. Wrap with aluminum foil and store at −80 °C for up to 6 months. The working solution can be stored at −20 °C for a few weeks. During the working day, each aliquot should be kept on ice and sealed with parafilm to avoid methanol evaporation.
6. *D-glucose solution, 1 M*: Dissolve 198.17 g D-glucose in 1 L distilled water to obtain a 1 M stock solution. Store frozen at −20 °C for 12 months in 50 mL aliquots.
7. *EGTA solution, 1 M*: Dissolve 190.75 g EGTA in 0.5 L distilled water to make a 1 M stock solution. Adjust to pH 8 with NaOH for complete solubilization. Store frozen at −20 °C for 12 months in 50 mL aliquots.
8. *Extracellular medium (EM), 10x*: To prepare 1 L, add the following: 84.74 g of NaCl (1.450 M final concentration), 3.73 g of KCl (50 mM final concentration), 23.83 g of Na-HEPES (100 mM final concentration), 2.03 g of $MgCl_2 \times 6H_2O$ (10 mM final concentration). Dissolve in 0.99 L distilled water. Adjust pH to 7.6 with NaOH. Bring to a final volume of 1 L with distilled water. This solution can be stored at 4 °C or frozen for 12 months.
9. *Extracellular medium (EM), 1x*: Add 10 mL of a 1 M D-glucose stock solution (10 mM final concentration) to 0.1 L 10x × EM solution. Bring to a final volume of 1 L with distilled water. The pH of this solution should be 7.4 and it is not necessary to confirm it. This solution can be stored for 1 month at 4 °C.
10. *EM with 1 mM $CaCl_2$ (EMC)*: To 1 L EM (1x), add 1 ml of 1 M $CaCl_2$. Store up to 6 months at 4 °C.
11. *EM without $CaCl_2$ (0.5 mM EGTA; EM0Ca)*: To 1 L EM, add 0.5 mL of 1 M EGTA (0.5 mM final concentration). The final volume of EMC and EM0Ca depends on the specific protocol. Calculate volumes beforehand considering the concrete protocol (e.g., number of repeated pulses added) and the perfusion flow speed. Store up to 6 months at 4 °C.
12. *G 418 solution*: Prepare 100 mg/mL stock in H_2O. Solutions should be filter-sterilized prior to storage at 2 °C to 8 °C for 24 months. Solutions can also be stored at −20 °C.

13. *$MgCl_2$ stock solution, 1 M*: Dissolve 10.10 g $MgCl_2$ dihydrate in 0.05 L distilled water to obtain a 1 M stock solution. Store frozen at −20 °C for 24 months in 10 mL aliquots.
14. *Histamine working solution*: Make 25 mL of 100 μM histamine solution by taking 25 μL of a 0.1 M stock solution (prepared in distilled water) and diluting it in 25 mL of EMC medium. Prepare fresh immediately before use.
15. *Intracellular medium (IM)*: Prepare a solution containing 10 mM NaCl, 140 mM KCl, 1 mM KH_2PO_4, 1 mM $MgCl_2$ (use 1 M stock solution), 1 mM ATP-Mg (use 0.1 M ATP-Mg stock solution), 2 mM sodium succinate, 1 mM sodium-pyruvate, 20 mM Na-HEPES.
16. *IM with 100 nM $CaCl_2$ (IMC)*: To one volume of IM, add 1.21 mM EGTA and 0.73 mM EGTA-Ca^{2+} (values calculated using the program *MaxChelator*). Adjust to 7.2 pH with 2 M Tris Base the same day of the measurements (*see* **Note 3**).
17. *IM without $CaCl_2$ (0.5 mM EGTA; IM0Ca)*: To one volume of IM, add 0.5 mM EGTA. Adjust to 7.2 pH with 2 M Tris Base.
18. *IP_3 solution*: Make a solution of 5 μM IP_3 into IMC medium. Prepare fresh immediately before use.
19. *Lysis solution*: Prepare a solution of 0.1 mM digitonin and 10 mM $CaCl_2$ in 50 mL distilled water.
20. *Permeabilization solution*: Prepare a solution of 50 μM digitonin in IM0Ca medium. Prepare fresh immediately before use.

3 Methods

3.1 $[Ca^{2+}]_{ER}$ in Intact Cells

1. A day prior to measurements, seed 1 × 10^5 erIgGAP1-HeLa cells in each of the 4-well plates (*see* **Note 4**).
2. Remove cell media and wash cells once with 0.2 mL EM0Ca.
3. Remove medium and add 0.2 mL EM0Ca containing 10 μM BHQ. Incubate 10 min (*see* **Note 5**).
4. Add 1 μL coelenterazine *n* of a 200 μM stock to each well, and incubate 1 h protected from light (**Notes 6** and 7).
5. During this incubation time, prepare all the solutions, purge the perfusion system, and adjust the flow speed at 4.5–5 mL/min. Transfer an empty four-well plate to the aequorimeter, and connect the perfusion and vacuum system (*see* **Note 8**).
6. Open program. Start perfusion with EM0Ca. Note down the luminescence background value. This will be subtracted in the analysis. Stop perfusion.
7. Replace the emptied plate with a four-well plate with cells. Start perfusion with EM0Ca. Start recording. Perfuse for (at least)

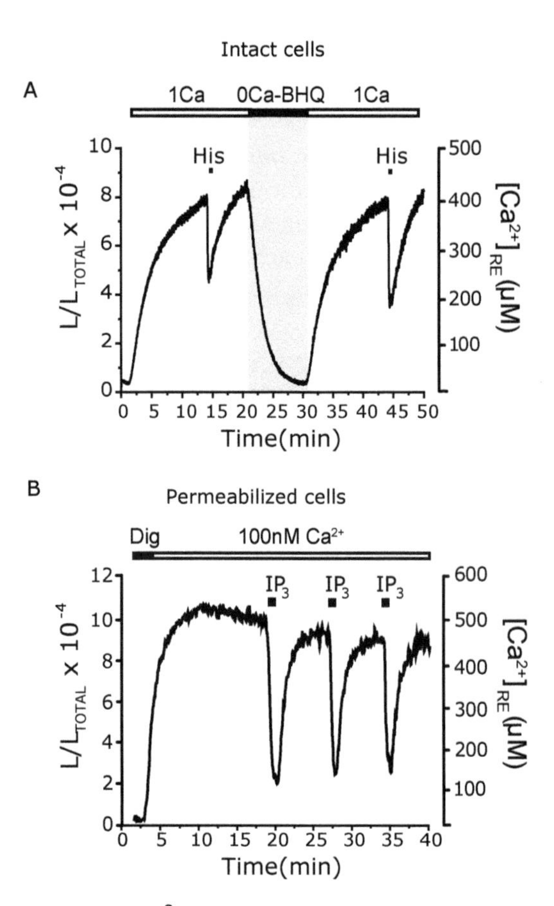

Fig. 3 Representative ER Ca^{2+} dynamics in (**a**) intact and (**b**) permeabilized cells monitored by luminescence with erIgGAP1. (**a**). Stably expressing erIgGAP1 HeLa cells stimulated with histamine (His, 100 μM; 30 s) in an extracellular medium containing 1 mM Ca^{2+} or perfused with the membrane-permeable SERCA inhibitor 2,5-di-tert-butylhydroquinone (BHQ, 10 μM) in Ca^{2+}-free (containing 0.5 mM EGTA) extracellular medium (EM0Ca) which produces a complete passive emptying of the ER Ca^{2+} store. (**b**). HeLa cells were permeabilized in an intracellular-like Ca^{2+}-free medium with digitonin (Dig, 50 μM) and perfused with an intracellular-like medium containing 100 nM Ca^{2+} (buffered with EGTA) to refill the ER. Then, they were repeatedly stimulated with IP_3 (5 μM). Figure modified from Rodriguez-Prados et al. [11]

1 min to obtain a basal signal. Make sure that the signal is stable.

8. At min 1, switch to EMC solution and perfuse for 13 min. This time period is enough to reach a stable value of the luminescence signal. This indicates that the ER Ca^{2+} store is refilled and it reached a steady-state $[Ca^{2+}]_{ER}$ value (Fig. 3a).
9. At min 14, perfuse histamine solution (100 μM) for 30 s. This provokes a quick drop in the signal, indicating a Ca^{2+} release from the ER.

10. At min 14.30 perfuse EMC.
11. At min 20.30 perfuse EM0Ca/BHQ. BHQ is a reversible SERCA inhibitor, and its application provokes a passive emptying of the ER Ca^{2+} store through the ER Ca^{2+} leak.
12. At min 30 change to EMC solution.
13. At min 44 perfuse histamine (100 μM) for 30 s.
14. At min 44.30 change to EMC solution.

3.2 $[Ca^{2+}]_{ER}$ in Permeabilized Cells

Perform **steps 1–7** as described in the Protocol 3.1 for intact cells:

1. At min 1 perfuse permeabilization solution (Fig. 3b).
2. At min 2 perfuse IMC solution.
3. At min 18.30 switch to IP3 (5 μM) and perfuse for 30 s. A fast and notable drop in the signal is observed.
4. At min 19 replace IMC solution.
5. Repeat **steps 3** and **4** at min 25.30 and 32.30.

3.3 Calibration

The calibration of luminescence signal into $[Ca^{2+}]$ requires calculating the L/L_{TOTAL} ratio, where L is the luminescence emission (in cps) at a given time and L_{TOTAL} is the sum of all the counts remaining at that time. Quantification of L_{TOTAL} requires the luminescence discharge of the residual unconsumed aequorin (in cps) at the end of each run, and this is experimentally performed by lysing the cells with digitonin in medium containing saturating (10 mM) Ca^{2+}:

1. Perfuse lysis solution for 1 min. This step releases all counts of remaining aequorin. A peak in luminescence (in counts per second (cps)) is recorded (Fig. 4a).
2. At min 51, change to solution containing 10 mM $CaCl_2$ in distilled water and continue perfusing for ~7 min (*see* **Note 9**).
3. Pause perfusion. Save data and analyze.

3.4 Data Analysis

To accurately determine the total luminescence (L_{TOTAL}), it is important to compute all the aequorin luminescence; that implies to keep on recording till the cps value returns to the basal level recorded at the beginning of the experiment. For GAP1 this can take longer than 30 min, which would lengthen markedly the duration of each experiment. In order to shorten this time period, we estimate the remaining luminescence by assuming that the consumption decay function follows the equation:

$$L_t = L \cdot e^{-kt}$$

where k is the first-order rate constant for the consumption decay. To estimate $T_{1/2}$, a box with a height equivalent to two units of increase on the cps axis (in logarithmic scale) is depicted over the

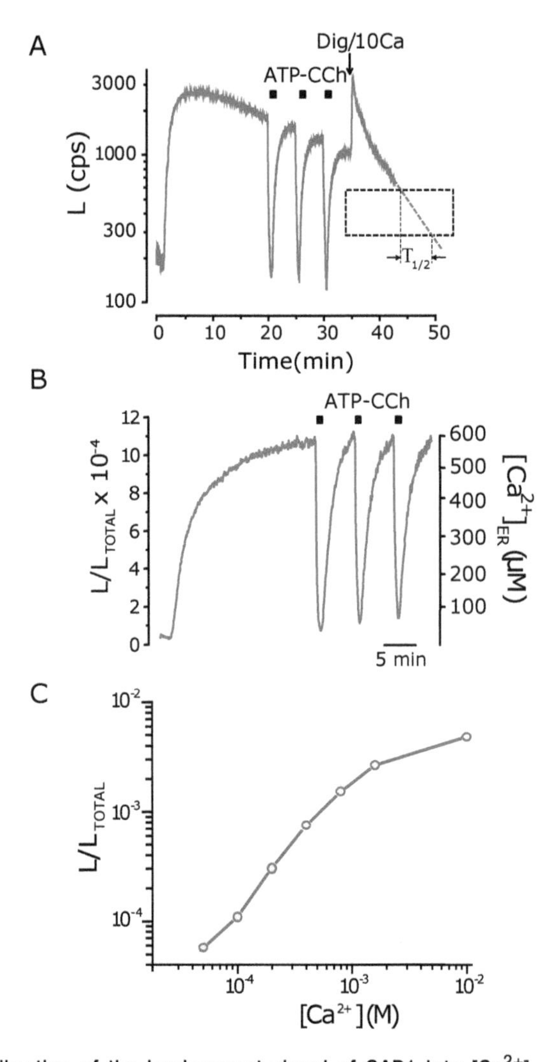

Fig. 4 Calibration of the luminescent signal of GAP1 into $[Ca^{2+}]_{ER}$ values. (**a**). erlgGAP1-stably expressing HeLa cells were handled as in Fig. 3a, and repeatedly stimulated with pulses of ATP plus carbachol (ATP-CCh, 100 μM each; 30 s) as indicated. At 35 min the remaining aequorin counts were released by lysis with digitonin in saturating Ca^{2+} (Dig/10 Ca). Luminescence emission (cps) is plotted in log scale. The figure also illustrates the procedure for $T_{1/2}$ estimation. The box, whose height is equivalent to a twofold increase, is displayed over the graph, and $T_{1/2}$ is calculated from the extrapolation of the luminescence consumption decay (dashed) as indicated; in this case, the value is 5.60 min (336 s). (**b**). Calibrated data from (A) are shown as L/L_{TOTAL} (left axis). Calibration scale for $[Ca^{2+}]_{ER}$ (in μM) is shown at the right. Note that the $[Ca^{2+}]_{ER}$ at the steady state always returns to the same value after washing out each addition. (**c**). Calibration curve of GAP1 reconstituted with coelenterazine *n*. Briefly, erlgGAP1-stably expressing HeLa cells reconstituted with coelenterazine *n* were lysed with 0.01% Triton X-100 in a solution containing 140 mM KCl, 10 mM NaCl, 1 mM $MgCl_2$, 10 mM Na-MOPS, pH 7.2, and increasing calcium concentrations (50, 100, 200, 400, 800, and 1600 μM) were added to trigger luminescence emission. The total luminescence (L_{TOTAL}) was obtained by applying saturating Ca^{2+} concentrations (10 mM). Data are shown as the ratio between luminescence/total luminescence at each time point ($L/L_{TOTAL}{\cdot}s^{-1}$). Figures modified from Rodriguez-Prados et al. [11]

graph (Fig. 4a). $T_{1/2}$ is calculated from the extrapolation of the luminescence consumption decay (dashed line), as indicated. In the experiment shown, the value is 5.60 min (336 s). Then, the last cps readout is multiplied by the $T_{1/2}$ value (in seconds) and by 1.49. The resulting value replaced the last value in the sum of all the obtained values along the experiment in order to calculate L_{TOTAL}. The background light value (without cells, typically ~70–100 cps) recorded at the beginning of the experiment must be subtracted from L_{TOTAL}. We typically obtain 1–2 × 10^7 cps per 1 × 10^5 erIgGAP1-HeLa cells. The calibrated data in L/L_{TOTAL} can be computed and plotted as shown in Fig. 4b. The $[Ca^{2+}]_{ER}$ is estimated by interpolation of the calibration curve (Fig. 4c) as previously reported [11].

4 Notes

1. The routine experiments are performed at 22–25 °C, but if required, the experiments can also be performed at 37 °C. In that case, solutions can be placed in a water bath, and an *Inline Solution Heater* can be installed at the entry of the dark chamber.
2. This gene is termed erIgGAP1 to distinguish it from erGAP1, previously described as a fluorescent indicator, where GAP1 is targeted to the ER by fusing it to the calreticulin signal peptide and the ER retention signal KDEL, on its N- and C-terminal domains, respectively [10]. This targeting approach cannot be utilized in the luminescence mode, because the free C-terminal proline residue of aequorin is required for light emission [14].
3. It is important to adjust the pH of this solution the same day of the measurements. If there are some solutions remaining at the end of the day, they can be stored frozen until the next working day. The solutions should not be repeatedly frozen.
4. HeLa cells attach well to the plate bottom, but other cells such as HEK293 are less adherent, and they might detach from the plate especially if they are very confluent when applying the perfusion flow. To avoid this, we routinely coat the plates with sterile poly-L-lysine (0.1% in distilled water (w/v).
5. This treatment guarantees that the ER Ca^{2+} store is empty, and it avoids premature consumption of aequorin (of GAP1) upon its reconstitution with coelenterazine.
6. In addition to native coelenterazine, other derivatives (dubbed as *cp*, *f*, *h*, *hcp*, *n*, or *i*) have been synthesized [15]. They reconstitute aequorin and change the properties of luminescence quantum yield and the emission speed upon Ca^{2+} binding. Coelenterazine *n* displays a reduced light emission and a

slower response to Ca^{2+} binding. The combination of the mutated D119A aequorin with coelenterazine *n* reduces the aequorin affinity for Ca^{2+} and has been instrumental for allowing monitoring of Ca^{2+} in the ER lumen [16].

7. Unbound coelenterazine is very sensitive to light. Stocks should be wrapped with aluminum foil, and cells should be protected from light during the reconstitution step. In general, experiments should be performed in a room with dimmed light.
8. It is important to set the flow speed at the start of the experiment and to check that it is not modified during the course of the experiment. If the speed is changed, the kinetics can be severely modified.
9. In our experience, the 6–7-min period is enough to obtain a reliable decay plot that allows for making the measurements.

Acknowledgements

This study was supported by grants from the *Ministerio de Economía y Competitividad* (BFU2017-83066-P and PID2020-116086RB-I00), from the *Consejería de Educación de la Junta de Castilla y León* (GR175), and from the *Programa Estratégico Instituto de Biología y Genética Molecular (IBGM), Escalera de Excelencia, Junta de Castilla y León* (CLU-2019-02). We thank Miriam Fernández, Jesús Fernández, Carla Rodríguez, and Iris Fernández for expert technical help. We also thank former members of our laboratory and Dr. Schimmang for helpful comments on the manuscript.

References

1. Berridge MJ (2016) The inositol trisphosphate/calcium signaling pathway in health and disease. Physiol Rev 96(4):1261–1296. https://doi.org/10.1152/physrev.00006.2016
2. Zalk R, Lehnart SE, Marks AR (2007) Modulation of the ryanodine receptor and intracellular calcium. Annu Rev Biochem 76:367–385
3. Carreras-Sureda A, Pihán P, Hetz C (2018) Calcium signaling at the endoplasmic reticulum: fine-tuning stress responses. Cell Calcium 70:24–31
4. Clapham DE (2007) Calcium signaling. Cell 131(6):1047–1058
5. Putney JW (2017) Store-operated calcium entry: an historical overview. Adv Exp Med Biol 981:205–214. https://doi.org/10.1007/978-3-319-55858-5_9
6. Vaeth M, Kahlfuss S, Feske S (2020) CRAC channels and calcium signaling in T cell-mediated immunity. Trends Immunol 41(10):878–901. https://doi.org/10.1016/j.it.2020.06.012
7. Mahalingam D, Wilding G, Denmeade S, Sarantopoulas J, Cosgrove D, Cetnar J et al (2016) Mipsagargin, a novel thapsigargin-based PSMA-activated prodrug: results of a first-in-man phase I clinical trial in patients with refractory, advanced or metastatic solid tumours. Br J Cancer 114(9):986–994. https://doi.org/10.1038/bjc.2016.72
8. Shimomura O, Johnson FH, Saiga Y (1962) Extraction, purification and properties of aequorin, a bioluminescent protein from the

luminous hydromedusan. Aequorea J Cell Comp Physiol 59:223–239

9. Alonso MT, Rodriguez-Prados M, Navas-Navarro P, Rojo-Ruiz J, Garcia-Sancho J (2017) Using aequorin probes to measure Ca ($^{2+}$) in intracellular organelles. Cell Calcium 64:3–11. https://doi.org/10.1016/j.ceca.2017.01.006
10. Rodriguez-Garcia A, Rojo-Ruiz J, Navas-Navarro P, Aulestia FJ, Gallego-Sandin S, Garcia-Sancho J et al (2014) GAP, an aequorin-based fluorescent indicator for imaging Ca^{2+} in organelles. Proc Natl Acad Sci USA 111(7):2584–2589. https://doi.org/10.1073/pnas.1316539111
11. Rodriguez-Prados M, Rojo-Ruiz J, Aulestia FJ, Garcia-Sancho J, Alonso MT (2015) A new low-Ca(2)(+) affinity GAP indicator to monitor high Ca(2)(+) in organelles by luminescence. Cell Calcium 58(6):558–564. https://doi.org/10.1016/j.ceca.2015.09.002
12. Alonso MT, Chamero P, Villalobos C, Garcia-Sancho J (2003) Fura-2 antagonises calcium-induced calcium release. Cell Calcium 33(1):27–35
13. Montero M, Brini M, Marsault R, Alvarez J, Sitia R, Pozzan T et al (1995) Monitoring dynamic changes in free Ca^{2+} concentration in the endoplasmic reticulum of intact cells. EMBO J 14(22):5467–5475
14. Watkins NJ, Campbell AK (1993) Requirement of the C-terminal proline residue for stability of the Ca($^{2+}$)-activated photoprotein aequorin. Biochem J 293(Pt 1):181–185
15. Shimomura O, Musicki B, Kishi Y, Inouye S (1993) Light-emitting properties of recombinant semi-synthetic aequorins and recombinant fluorescein-conjugated aequorin for measuring cellular calcium. Cell Calcium 14(5):373–378
16. Barrero MJ, Montero M, Alvarez J (1997) Dynamics of [Ca^{2+}] in the endoplasmic reticulum and cytoplasm of intact HeLa cells. A comparative study. J Biol Chem 272(44):27694–27699
17. Rodriguez-Prados M (2017) Dinámica del calcio reticular en astrocitos corticales. Tesis Doctoral

Chapter 12

Analyzing Mitochondrial Calcium Influx in Isolated Mitochondria

Nasab Ghazal and Jennifer Q. Kwong

Abstract

Mitochondria play a crucial role in Ca^{2+} signaling and homeostasis and can contribute to shaping the cytosolic Ca^{2+} landscape as well as regulate a variety of pathways including energy production and cell death. Dysregulation of mitochondrial Ca^{2+} homeostasis promotes pathologies including neurodegenerative diseases, cardiovascular disorders, and metabolic syndromes. The significance of mitochondria to Ca^{2+} signaling and regulation underscores the value of methods to assess mitochondrial Ca^{2+} import. Here we present a plate reader-based method using the Ca^{2+}-sensitive fluorescent probe calcium green-5 N to measure mitochondrial Ca^{2+} import in isolated cardiac mitochondria. This technique can be expanded to measure Ca^{2+} uptake in mitochondria isolated from other tissue types and from cultured cells.

Key words Calcium, Heart, Mitochondria, Mitochondrial permeability transition pore, Signaling, Uniporter

1 Introduction

Ca^{2+} is a critical intracellular second messenger that governs cellular processes ranging from metabolism to synaptic transmission, muscle contraction, and even cell death [1–3]. This ion serves as a dynamic signal that encodes information through temporal and spatial changes in its concentration. Under resting conditions, cytosolic Ca^{2+} concentrations are kept low and free Ca^{2+} levels are tightly regulated. This regulation can involve plasma membrane transporters like plasma membrane Ca^{2+} transport ATPase and Na^+/ Ca^{2+} exchanger, Ca^{2+} buffering proteins, and organelles that can store Ca^{2+} like the endoplasmic/sarcoplasmic reticulum and, importantly, the mitochondria [4, 5].

Indeed, mitochondria are central sites of intracellular Ca^{2+} storage and signaling. These organelles import and buffer Ca^{2+}, thereby contributing to the creation of Ca^{2+} signaling microdomains, shaping the cytosolic Ca^{2+} landscape, and modulating

Caroline M. Gorvin (ed.), *Calcium Signaling: Methods and Protocols*, Methods in Molecular Biology, vol. 2861,
https://doi.org/10.1007/978-1-0716-4164-4_12,

intercellular Ca^{2+} signals [6–10]. Additionally, mitochondrial Ca^{2} is intimately linked to cellular energetics; Ca^{2+} can modulate the activity of key matrix dehydrogenases of the tricarboxylic acid cycle and upregulate the activity of the mitochondrial ATP synthase [11–14]. In contrast, dysregulation of cytosolic Ca^{2+} homeostasis causes mitochondrial Ca^{2+} overload, triggering opening of the mitochondrial permeability transition pore (MPTP), mitochondrial inner membrane permeability, loss of the electrochemical gradient required to drive ATP synthesis, and, ultimately, cell death [15]. Thus, mitochondrial calcium signaling exerts roles in both physiological and pathological processes, with established implications for neurodegenerative diseases, muscular dystrophy, ischemia reperfusion injury, and heart failure [16–19].

Despite the long recognized importance of mitochondrial Ca^{2+} signaling to physiology and disease, the identity of the mitochondrial Ca^{2+} import machinery has only recently been defined [20]. At the molecular level, mitochondrial Ca^{2+} uptake is mediated by the mitochondrial calcium uniporter (MCU) complex. Originally identified as a 480 kDa holocomplex with the MCU subunit serving as the pore-forming subunit of the channel [20, 21], studies on mitochondrial Ca^{2+} regulation have revealed that the complex also comprises the regulatory subunits MCU-dominant negative β-subunit (MCUb) [9, 22, 23], essential MCU regulator (EMRE) [24], MCU regulator 1 (MCURl) [25], mitochondrial calcium uptake (MICU) 1 [26], MICU2, MICU3 [27], and SLC25A23 [28]. However, given the importance of mitochondrial Ca^{2+} signaling, and its roles in physiology and pathology, there is considerable interest in identifying pathways and therapeutics that can modulate mitochondrial Ca^{2+} entry.

In this chapter, we detail an experimental protocol to analyze mitochondrial Ca^{2+} influx (Fig. 1). While there are a number of different methods to assess mitochondrial Ca^{2+} influx, as well as different types of Ca^{2+}-sensitive probes, many methods require specialized equipment like confocal microscopes or fluorimeters [29–32]. Here, we present a plate reader-based platform for measuring mitochondrial Ca^{2+} uptake using calcium green-5 N, a Ca^{2+}-sensitive fluorescent probe that is well-suited for use in isolated mitochondria [13]. Moreover, while our example uses mitochondria isolated from cardiac tissue, this protocol can be adapted for use in mitochondria isolated from any tissue or even from cultured cells. This protocol also can measure mitochondrial Ca^{2+} retention capacity, or the maximum amount of Ca^{2+} that can be sequestered by mitochondria before the induction of permeability transition, to capture mitochondrial function and sensitivity to Ca^{2+} overload-induced dysfunction.

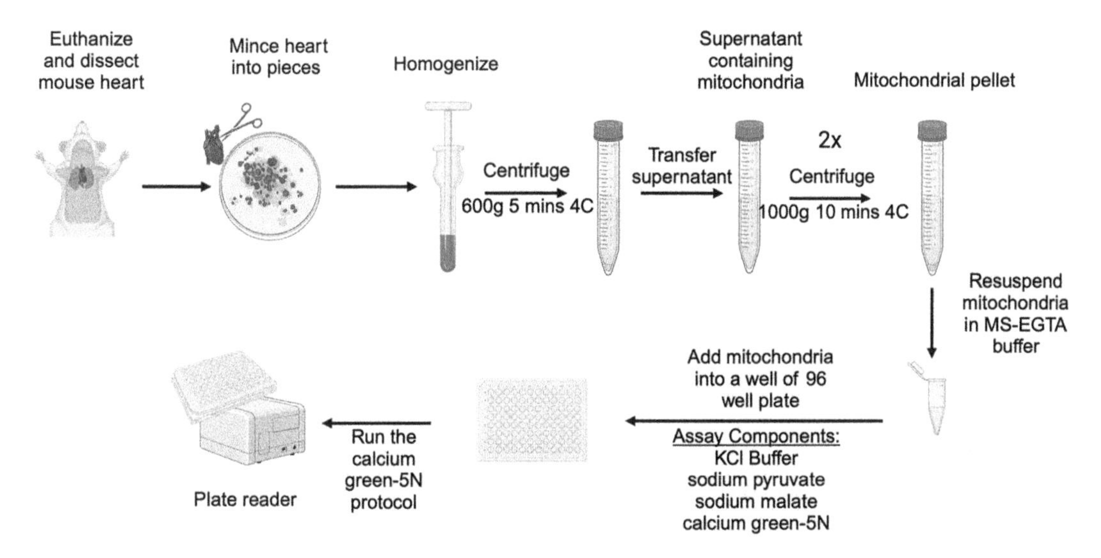

Fig. 1 Schematic of the workflow for cardiac mitochondrial isolation and the mitochondrial Ca^{2+} uptake assay. Mouse heart is collected and homogenized in MS-EGTA buffer. Cardiac mitochondria are isolated by differential centrifugation. Components for the mitochondrial calcium uptake assay (mitochondria, substrates, calcium green-5 N, and KCl buffer) are mixed in a single well of a 96-well plate. A plate reader is used to read calcium green-5 N fluorescence following $CaCl_2$ injections into the well. Figure created using BioRender

2 Materials

2.1 Animal Necropsy

1. 70% ethanol.
2. 1× phosphate-buffered saline (PBS).
3. Isoflurane.

2.2 Mitochondrial Isolation

1. MS-EGTA buffer: 225 mM mannitol, 75 mM sucrose, 5 mM 4-(2-hydroxyethyl)-1-piperazineethanesulfonic acid (HEPES), 1 mM ethylene glycol-bis(β-aminoethyl ether)-N,N,N′,N′-tetraacetic acid (EGTA), pH adjusted to 7.4 with KOH. The MS-EGTA buffer should be sterile-filtered using a 0.22 μm filter. Sterile-filtered MS-EGTA buffer can be prepared in advance and stored at 4 °C for a few weeks.
2. 0.22 μm filter.

2.3 Mitochondrial Calcium Uptake Assay

1. KCl buffer: 125 mM KCl, 20 mM HEPES, 1 mM KH_2PO_4, 2 mM $MgCl_2$, 40 μM EGTA, and pH adjusted to 7.2 with KOH. KCl buffer can be prepared in advance and stored at room temperature.
2. 1 M sodium pyruvate, pH 7.4. 1 M sodium pyruvate can be prepared in advance, aliquoted, and stored at −20 °C. This can be thawed and reused multiple times.

3. 500 mM sodium malate, pH 7.4. 500 mM sodium malate can be prepared in advance, aliquoted, and stored at −20 °C. This can be thawed and reused multiple times.
4. 1 mM calcium green-5 N prepared in DMSO. This calcium green-5 N stock can be aliquoted and stored at −20 °C. Frozen aliquots can be thawed and reused multiple times.
5. 0.7 M $CaCl_2$ (*see* **Note 1**).
6. Anhydrous dimethyl sulfoxide (DMSO).
7. Ru360.
8. 96-well clear, flat-bottom microplates.

2.4 Protein Quantification

1. BioRad Protein Assay Dye Reagent Concentrate.
2. BioRad Quick Start Bovine Serum Albumin (BSA) standard set.
3. Ultrapure water.
4. 96-well clear, flat-bottom microplates.

2.5 Equipment

1. Surgical instruments (scissors and forceps).
2. Kontes Potter-Elvehjem size 22 tissue grinder with PTFE pestle and an 8 mL glass tube (Thermo Scientific).
3. Eppendorf 5810R refrigerated centrifuge with an F-34-6-38 fixed-angle rotor and adaptors for 15 mL conical tubes.
4. Biotek Synergy Neo2 plate reader with reagent injectors (*see* **Note 2**).

3 Methods

3.1 Assay Preparation

1. Program the plate reader for kinetic measurements of calcium green-5 N fluorescence (excitation/emission of 506/532 nm) and reagent injections from the reagent dispensers (see Subheading 3.5 Mitochondrial calcium uptake assay **step 1**).
2. Ensure that the MS-EGTA buffer and 1× PBS are pre-chilled to 4 °C before use.
3. Pre-chill the tissue grinder on ice.
4. Pre-chill the centrifuge and adaptors to 4 °C before use.

3.2 Harvesting Hearts from Mice

Euthanize the mouse, and perform tissue collection according to the Institutional Animal Care and Use Committee standards for your institution:

1. Anesthetize the mouse by isofluorane overdose.
2. Euthanize the mouse by cervical dislocation.
3. Sterilize the chest of the mouse with 70% ethanol.

4. Using surgical scissors and forceps, and remove the skin above the chest.
5. Open the thoracic cavity by cutting the diaphragm along the bottom of the rib cage, and then make two vertical incisions on the lateral sides of the rib cage to expose the heart.
6. Excise out the heart and place the tissue into a dish containing 10 mL ice-cold 1× PBS.
7. Rinse the heart thoroughly in the 1× PBS, ensuring that all blood is squeezed out of the ventricles.
8. Transfer the rinsed heart into 5 mL of fresh ice cold 1× PBS.
9. The heart is ready for mitochondrial isolation (*see* **Note 3**).

3.3 Mitochondrial Isolation

1. Using sharp surgical scissors, and mince the heart into small pieces in 1× PBS.
2. Discard the PBS and transfer the tissue to a pre-chilled 8 mL glass tissue homogenizer tube.
3. Add 5 mL of ice-cold MS-EGTA buffer, and homogenize heart tissue until large pieces are no longer visible (*see* **Note 4**).
4. Transfer the tissue homogenate into a pre-chilled 15 mL conical tube, and centrifuge at 600 *g* for 5 min at 4 °C to pellet nuclei and unbroken cells.
5. Transfer the supernatant containing mitochondria into a new pre-chilled 15 mL conical tube, and discard the pellet.
6. Centrifuge the supernatant at 10,000 *g* for 10 min at 4 °C.
7. Discard the supernatant and keep the pellet containing mitochondria.
8. Wash the mitochondria by resuspending the mitochondrial pellet in 5 mL MS-EGTA buffer, and then centrifuge at 10,000 *g* for 10 min at 4 °C.
9. Discard the supernatant and repeat **step** 7 to perform a second wash of the mitochondrial pellet.
10. Discard the supernatant.
11. Resuspend the mitochondrial pellet in 100 μL MS-EGTA buffer and transfer to a 1.5 mL Eppendorf tube.
12. Keep isolated mitochondria on ice (*see* **Note 5**).
13. Proceed to protein quantification.

3.4 Protein Quantification

1. Dilute the BioRad Protein Assay Dye Reagent Concentrate into 1× according to manufacturer's instructions.
2. Prepare an aliquot of mitochondria diluted in ultrapure water, ensuring that the dilution factor allows the protein

concentration to fall within the range of the BSA standards (0.125–2 mg/mL).

3. Pipette 2 μL of all samples (blank, BSA standards, and diluted mitochondria) into a 96-well plate in duplicates.
4. Pipette 298 μL of the 1× Protein Assay Dye Reagent into each well of the 96-well plate and mix by pipetting up and down.
5. Incubate the reaction for 5 min at room temperature.
6. Measure the absorbance at 595 nm using a plate reader.
7. Calculate the average absorbance values for each sample (average of the duplicates).
8. Subtract the average absorbance of the blank control samples from average absorbances of the standards and the mitochondrial sample.
9. Plot a standard curve with the protein concentration (μg/μL) of each BSA standard on the x-axis and the absorbance on the y-axis.
10. Calculate the protein concentration of the mitochondrial sample (adjusted for the dilution factor used).

3.5 Mitochondrial Calcium Uptake Assay

1. Program the plate reader protocol for the kinetic measurements of calcium green-5 N fluorescence with automated reagent injections. Configure the plate reader for the excitation and emission wavelengths for calcium green-5 N fluorescence (excitation/emission of 506/532 nm) (*see* **Notes 6** and **7**):
   - Set the total assay time to 1000 s.
   - Set the measurement interval to 1 s.
   - Set up the reagent injectors to dispense 5 μL $CaCl_2$ solution (*see* **Note 1**) at specified time points (30, 150, 300, 480, and 690 s).
2. Prime reagent injectors with the 0.7 M $CaCl_2$ solution.
3. Pipette 100 μg mitochondria into a well of a 96-well plate.
4. Add KCl buffer to a final volume of 197 μL.
5. Add mitochondrial substrates: 1 μL of 1 M sodium pyruvate and 1 μL of 500 mM sodium malate to the reaction (*see* **Note 8**).
6. Mix gently by pipetting, and incubate the reaction for 2 min at room temperature to allow the mitochondria to become energized.
7. Add 1 μL of the 1 mM calcium green-5 N to reaction and pipette gently to mix (*see* **Note 9**).
8. Start the calcium green-5 N fluorescence measurement protocol.

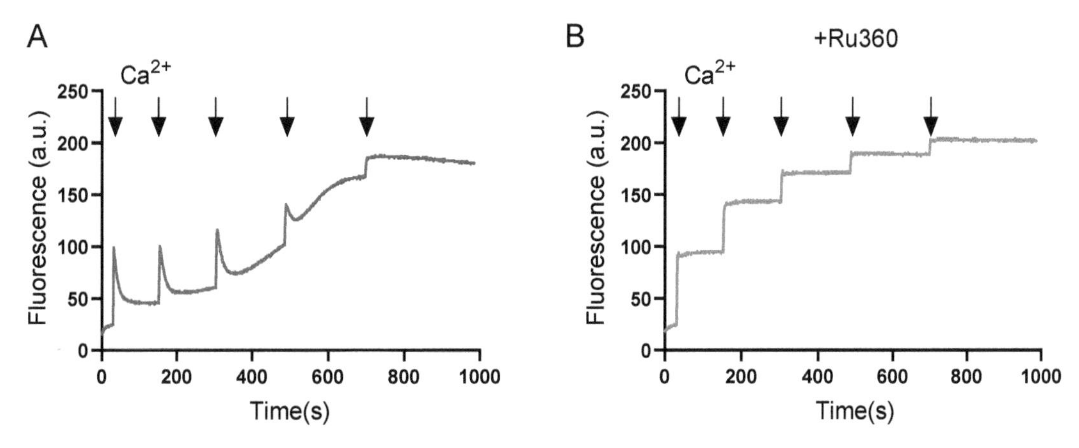

Fig. 2 Mitochondrial Ca^{2+} uptake measured by calcium green-5 N fluorescence. (**a**) Representative graph of mitochondrial Ca^{2+} uptake in isolated cardiac mitochondria. Heart mitochondria were challenged with 5 μL injections of 0.7 mM $CaCl_2$ indicated by black arrows. Mitochondrial Ca^{2+} uptake is indicated by the sharp decrease in calcium green-5 N fluorescence following $CaCl_2$ addition (observed following the first three additions). MPTP opening is induced following the third injection of $CaCl_2$ as evidenced by the progressive increase in calcium green-5 N fluorescence. (**b**) Calcium green-5 N fluorescence in isolated heart mitochondria challenged with 5 μL of 0.7 mM $CaCl_2$ following incubation with the uniporter complex inhibitor Ru360 (5 μM for 10 min). Ru360-treated mitochondria are unable to uptake Ca^{2+} as indicated by the stepwise increases in calcium green-5 N fluorescence following $CaCl_2$ addition

9. Plot calcium green-5 N fluorescence as a function of time (Fig. 2a).

3.6 Control Experiment for the Mitochondrial Calcium Uptake Assay

The mitochondrial calcium uniporter complex and mitochondrial calcium uptake can be inhibited pharmacologically by the compound Ru360. Thus, an important control experiment is to conduct the mitochondrial calcium uptake assay in the presence of Ru360:

1. Prepare a 1 mM stock solution of Ru360 in ultrapure water and place on ice (*see* **Note 10**).
2. Assemble the mitochondrial calcium uptake assay reaction as described in Subheading 3.5, **steps 3** and **4**.
3. Add 1 μL of 1 mM Ru360 to the reaction.
4. Pipette gently to mix.
5. Incubate the reaction at room temperature for 10 min.
6. Add 1 μl of the 1 mM calcium green-5 N to reaction and pipette gently to mix.
7. Start the calcium green-5 N fluorescence measurement protocol.
8. Plot calcium green-5 N fluorescence as a function of time (Fig. 2b).

3.7 Data Analysis and Interpretation

1. Mitochondrial Ca^{2+} retention capacity (μmol of Ca^{2+} per μg of protein) can be calculated by determining the minimal amount of Ca^{2+} that can be taken up by mitochondria before induction of permeability transition (identified by a sharp increase in calcium green-5 N fluorescence whereby mitochondria no longer take up Ca^{2+} and subsequent Ca^{2+} additions result in a stepwise increase in fluorescence) (Fig. 2a) (*see* **Note 11**).
2. For the control experiment, treatment of mitochondria with Ru360 should inhibit mitochondrial calcium uptake, resulting in a stepwise increase in calcium green-5 N fluorescence with each addition of $CaCl_2$ solution (Fig. 2b).

4 Notes

1. The $CaCl_2$ concentration is user-defined depending on the amount of mitochondria used in the assay and can be adjusted in subsequent runs to titrate the amount of Ca^{2+} necessary to induce MPTP opening.
2. This protocol describes the use of a plate reader in a 96-well plate format with automated reagent injectors, which presents advantages as plate readers are often readily accessible and the 96-well plate format allows for a reduced reaction volume requiring fewer isolated mitochondria per assay. However, fluorimeters capable of detecting calcium green-5 N with the sample stirring and injection ports can also be used.
3. For this assay, functional mitochondria must be used. Proceed to the mitochondrial isolation step immediately following tissue collection.
4. Do not overhomogenize the tissue to ensure that mitochondria remain intact.
5. Mitochondria must be functional for this assay and should be used for experimentation within 1 h.
6. The total assay time, as well as the timing and frequency of $CaCl_2$ additions, is user-defined and should be experimentally optimized.
7. Consult the user manual for your specific plate reader for detailed instructions on how to program the instrument and control the reagent injectors.
8. The combination of pyruvate and malate can be substituted with different substrates that support mitochondrial function.
9. Keep the reaction protected from light.
10. Ru360 is unstable in solution. Ru360 must be prepared fresh each day, should be reconstituted immediately before use, and should be discarded following completion of the experiment.

11. The minimal amount of Ca^{2+} required to induce permeability transition can be experimentally determined by adjusting the concentration of the $CaCl_2$ solution added, the volume of the $CaCl_2$ added, and the number of $CaCl_2$ additions used in the assay.

Funding

This study was supported by T32GM135060-04 to Nasab Ghazal and by R01GM144729 to Jennifer Q. Kwong.

References

1. Denton RM, McCormack JG (1980) The role of calcium in the regulation of mitochondrial metabolism. Biochem Soc Trans 8(3):266–268
2. Berridge MJ, Bootman MD, Lipp P (1998) Calcium – a life and death signal. Nature 395(6703):645–648
3. Szent-Gyorgyi AG (1975) Calcium regulation of muscle contraction. Biophys J 15(7): 707–723
4. Carafoli E (2002) Calcium signaling: a tale for all seasons. Proc Natl Acad Sci USA 99(3): 1115–1122
5. Brini M, Carafoli E (2011) The plasma membrane Ca(2)+ ATPase and the plasma membrane sodium calcium exchanger cooperate in the regulation of cell calcium. Cold Spring Harb Perspect Biol 3(2):a004168
6. Rizzuto R, Brini M, Murgia M, Pozzan T (1993) Microdomains with high Ca2+ close to IP3-sensitive channels that are sensed by neighboring mitochondria. Science 262(5134):744–747
7. Deluca HF, Engstrom GW (1961) Calcium uptake by rat kidney mitochondria. Proc Natl Acad Sci USA 47(11):1744–1750
8. D'Angelo D, Vecellio Reane D, Raffaello A (2023) Neither too much nor too little: mitochondrial calcium concentration as a balance between physiological and pathological conditions. Front Mol Biosci 10:1336416
9. Kwong JQ (2017) The mitochondrial calcium uniporter in the heart: energetics and beyond. J Physiol 595(12):3743–3751
10. Kwong JQ, Molkentin JD (2015) Physiological and pathological roles of the mitochondrial permeability transition pore in the heart. Cell Metab 21(2):206–214
11. McCormack JG, Denton RM (1980) Role of calcium ions in the regulation of intramitochondrial metabolism. Properties of the Ca2+-sensitive dehydrogenases within intact uncoupled mitochondria from the white and brown adipose tissue of the rat. Biochem J 190(1):95–105
12. Jouaville LS, Pinton P, Bastianutto C, Rutter GA, Rizzuto R (1999) Regulation of mitochondrial ATP synthesis by calcium: evidence for a long-term metabolic priming. Proc Natl Acad Sci USA 96(24):13807–13812
13. Kwong JQ, Lu X, Correll RN, Schwanekamp JA, Vagnozzi RJ, Sargent MA et al (2015) The mitochondrial calcium uniporter selectively matches metabolic output to acute contractile stress in the heart. Cell Rep 12(1):15–22
14. Kwong JQ, Huo J, Bround MJ, Boyer JG, Schwanekamp JA, Ghazal N et al (2018) The mitochondrial calcium uniporter underlies metabolic fuel preference in skeletal muscle. JCI Insight 3(22):e121689
15. Duchen MR (2000) Mitochondria and calcium: from cell signalling to cell death. J Physiol 529(Pt 1):57–68
16. Kwong JQ, Beal MF, Manfredi G (2006) The role of mitochondria in inherited neurodegenerative diseases. J Neurochem 97(6): 1659–1675
17. Zablocka B, Gorecki DC, Zablocki K (2021) Disrupted calcium homeostasis in Duchenne muscular dystrophy: a common mechanism behind diverse consequences. Int J Mol Sci 22(20):11040
18. Murphy E, Steenbergen C (2008) Mechanisms underlying acute protection from cardiac ischemia-reperfusion injury. Physiol Rev 88(2):581–609
19. Popoiu TA, Dudek J, Maack C, Bertero E (2023) Cardiac involvement in mitochondrial disorders. Curr Heart Fail Rep 20(1):76–87
20. Baughman JM, Perocchi F, Girgis HS, Plovanich M, Belcher-Timme CA, Sancak Y

et al (2011) Integrative genomics identifies MCU as an essential component of the mitochondrial calcium uniporter. Nature 476(7360):341–345

21. De Stefani D, Raffaello A, Teardo E, Szabo I, Rizzuto R (2011) A forty-kilodalton protein of the inner membrane is the mitochondrial calcium uniporter. Nature 476(7360):336–340
22. Raffaello A, De Stefani D, Sabbadin D, Teardo E, Merli G, Picard A et al (2013) The mitochondrial calcium uniporter is a multimer that can include a dominant-negative pore-forming subunit. EMBO J 32(17):2362–2376
23. Huo J, Lu S, Kwong JQ, Bround MJ, Grimes KM, Sargent MA et al (2020) MCUb induction protects the heart from Postischemic remodeling. Circ Res 127(3):379–390
24. Sancak Y, Markhard AL, Kitami T, Kovacs-Bogdan E, Kamer KJ, Udeshi ND et al (2013) EMRE is an essential component of the mitochondrial calcium uniporter complex. Science 342(6164):1379–1382
25. Mallilankaraman K, Cardenas C, Doonan PJ, Chandramoorthy HC, Irrinki KM, Golenar T et al (2012) MCUR1 is an essential component of mitochondrial Ca2+ uptake that regulates cellular metabolism. Nat Cell Biol 14(12): 1336–1343
26. Perocchi F, Gohil VM, Girgis HS, Bao XR, McCombs JE, Palmer AE et al (2010) MICU1 encodes a mitochondrial EF hand protein required for Ca(2+) uptake. Nature 467(7313):291–296
27. Plovanich M, Bogorad RL, Sancak Y, Kamer KJ, Strittmatter L, Li AA et al (2013) MICU2, a paralog of MICU1, resides within the mitochondrial uniporter complex to regulate calcium handling. PLoS One 8(2):e55785
28. Hoffman NE, Chandramoorthy HC, Shanmughapriya S, Zhang XQ, Vallem S, Doonan PJ et al (2014) SLC25A23 augments mitochondrial Ca(2)(+) uptake, interacts with MCU, and induces oxidative stress-mediated cell death. Mol Biol Cell 25(6):936–947
29. Wallace DJ, zum Alten Borgloh SM, Astori S, Yang Y, Bausen M, Kügler S et al (2008) Single-spike detection in vitro and in vivo with a genetic Ca2+ sensor. Nat Methods 5(9):797–804
30. Eberhard M, Erne P (1991) Calcium binding to fluorescent calcium indicators: calcium green, calcium orange and calcium crimson. Biochem Biophys Res Commun 180(1): 209–215
31. Tinning PW, Franssen A, Hridi SU, Bushell TJ, McConnell G (2018) A 340/380 nm light-emitting diode illuminator for Fura-2 AM ratiometric Ca(2+) imaging of live cells with better than 5 nM precision. J Microsc 269(3): 212–220
32. Rizzuto R, Simpson AW, Brini M, Pozzan T (1992) Rapid changes of mitochondrial Ca2+ revealed by specifically targeted recombinant aequorin. Nature 358(6384):325–327

Part IV

Tissue-Specific Assessment of Calcium Signaling

Chapter 13

Measuring Calcium Levels in Bone-Resorbing Osteoclasts and Bone-Forming Osteoblasts

Morten S. Hansen

Abstract

Bone remodeling is a crucial, dynamic process that renews bone and maintains mineral homeostasis. It consists of several steps, including osteoclastic bone resorption and osteoblastic bone formation and mineralization. Intracellular calcium signaling is essential for osteoclast and osteoblast differentiation and activity. Here, we describe the differentiation of human osteoclasts and osteoblasts in vitro and provide common methods to determine cell differentiation and activity. We then describe protocols for measuring intracellular calcium in these cells using Fura2-AM.

Key words Bone resorption, Bone formation, Bone mineralization, Fura-2, Osteoblasts, Osteoclasts

1 Introduction

The skeleton is a highly dynamic and metabolically active tissue that undergoes remodeling throughout life. The process of bone remodeling is mainly carried out by bone-resorbing multinucleated osteoclasts and bone-forming osteoblasts. Osteoclasts are derived from hematopoietic stem cells [1], and precursors of osteoclasts include monocytes, macrophages, and (mononucleated) preosteoclasts. Osteoblasts are derived from mesenchymal stem cells [1], and osteoprogenitors and pre-osteoblasts represent the initial stages of osteoblastogenesis [2]. Osteoclasts and osteoblasts use intracellular calcium as a key signaling molecule to initiate gene transcription, induce differentiation, and maintain cell survival and apoptosis [3–6]. In the early steps of osteoclastogenesis, monocytes differentiate into macrophages following macrophage colony-stimulating factor (M-CSF) signaling [7], which induces expression of the cell surface protein, receptor activator of nuclear factor-κβ (RANK) [8]. The interaction between RANK and its ligand (RANKL), which is secreted by osteoblasts, is a crucial step in osteoclast differentiation [9]. RANKL evokes intracellular calcium

Caroline M. Gorvin (ed.), *Calcium Signaling: Methods and Protocols*, Methods in Molecular Biology, vol. 2861,
https://doi.org/10.1007/978-1-0716-4164-4_13,

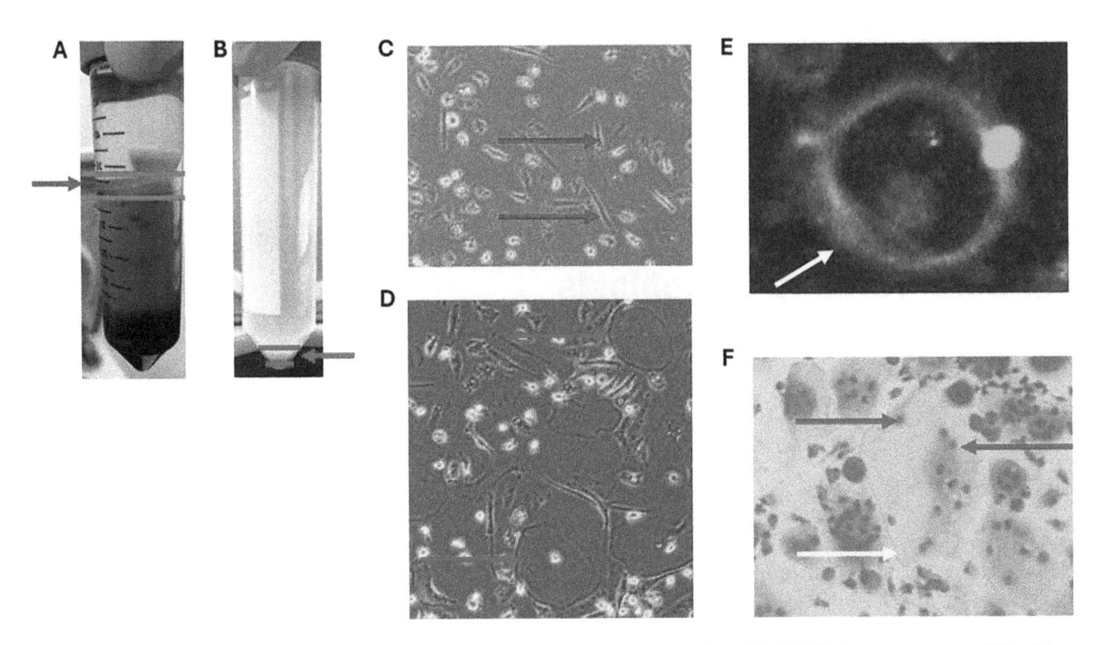

Fig. 1 Density gradient centrifuged with PBMC layer (**a**) without break, with PBMC layer shown within blue bars and marked by a blue arrow, and (**b**) with break with the PBMC layer shown within red bars and marked by a red arrow. (**c**) Microscopy images of preosteoclasts marked by a blue arrow and (**d**) mature, multinucleated osteoclasts marked by a red arrow. (**e**) Actin ring structure of a mature osteoclast marked by a white arrow. (Adapted from Hansen et al. [13]). (**f**) May-Grünwald and Giemsa staining of mature osteoclasts showing nuclei (marked by red arrows) and osteoclastic cell membrane (marked by yellow arrow). (Adapted from Hansen et al. [13])

($Ca^{2+}{}_i$) oscillations which upregulate NFATc1, a key regulator of osteoclast differentiation and activity, thereby promoting osteoclastogenesis [3, 4]. Via secretion of osteoprotegerin (OPG), a RANKL decoy receptor [10], osteoblasts may also inhibit osteoclast differentiation and activity, highlighting the important coupling between osteoclasts and osteoblasts. Human osteoclasts and osteoblasts can be grown in culture, and examination of intracellular calcium signaling can be used to assess their activity.

Human osteoclasts can be grown in culture by isolating peripheral blood mononuclear cells (PBMCs) from blood samples using density gradient centrifugation. Blood samples are layered on Ficoll–Paque; then, centrifugation separates the sample into its constituent parts (e.g., plasma, PBMCs, granulocytes). CD14+ monocytes can then be isolated from the PBMC layer using magnetic beads or enrichment cocktails (e.g., RosetteSep) (Fig. 1a, b). Following monocyte isolation, cells can be differentiated to macrophages by culturing cells in M-CSF (Fig. 1c). The addition of RANKL to macrophage cultures is then used to induce differentiation to human osteoclasts. Nine days after monocyte isolation, cells should be fully mature osteoclasts that are as follows: (i) multinucleated (Fig. 1d); (ii) structurally characterized by actin

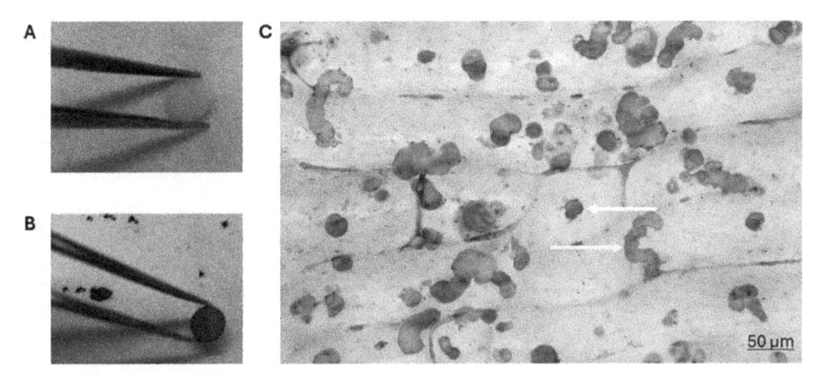

Fig. 2 Tweezers holding bone chips used for resorption analysis (**a**) before and (**b**) after toluidine blue staining. (**c**) Microscopical pictures of resorption areas on toluidine blue-stained bone chips. White arrow, pit formation; yellow arrow, trench formation

rings (Fig. 1e), which are necessary for bone resorption [11]; and (iii) express specific molecular and genetic markers including those for the protease cathepsin K and the enzyme tartrate-resistant acid phosphatase (TRAcP). The differentiation and activity of osteoclasts can be assessed by examining each of these features. May--Grünwald Giemsa staining followed by microscopy analysis can be used to quantify the number of nuclei per cell (Fig. 1f), growth of osteoclasts on bone chips followed by staining with fluorescent actin-binding protein, and toluidine blue stain (Fig. 2a, b) can be used to assess actin ring formation and bone resorption areas (Fig. 2c), respectively, Western blot analysis can be used to assess protein expression of cathepsin K, and staining of cell culture media can be used to examine secretion of TRAcP. Osteoblast differentiation and the cellular activity of mature osteoblasts are commonly analyzed using osteogenic markers, e.g., alkaline phosphatase (ALP) and runt-related transcription factor 2 (RUNX2), or by mineralization assay, such as Alizarin Red (see Fig. 3).

Here, we first describe how to isolate and culture human osteoclasts and osteoblasts and how to assess their differentiation and activity. While other methods to differentiate osteoclasts and osteoblasts exist, we present protocols for osteoclasts differentiated from $CD14^{+}$-monocytes and osteoblasts differentiated from human mesenchymal stem cells (hMSCs). These protocols have been used extensively to assess the effects of several compounds, including cathepsin K inhibitors, glucocorticoids, and the gut hormone glucose-dependent insulinotropic polypeptide (GIP) on osteoclast differentiation and activity [12–15]. In addition, these protocols have been used to examine the impact of different biological stages, such as obesity and aging [16, 17], and the role of several factors, including legumain [18] and delta-like 1/pre-adipocyte factor-1 [19] on osteoblast function. Besides this, investigations aimed to characterize genetic changes during osteoclast

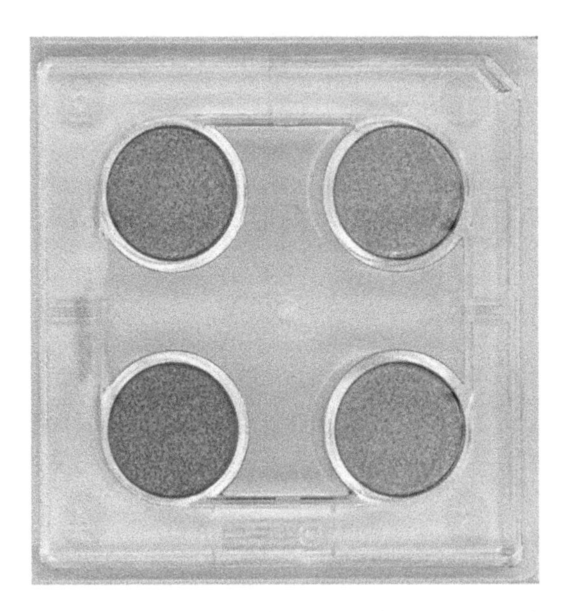

Fig. 3 Three intensities of Alizarin Red S staining of mature osteoblasts reflecting mineralization ability

[20] and osteoblast [21] differentiation in vitro have included methods presented in these protocols. We then describe imaging using Fura-2 AM to assess intracellular calcium signaling in these cells and how to analyze the data. Fura-2 AM is ideal for labeling primary osteoclasts and osteoblasts as these cells are difficult to transfect. We have used these protocols to assess the effect of GIP on differentiated human osteoclasts and osteoblasts [13].

2 Materials

Handle human samples with care and transport them in sterile containers or medium. Ethical approval for obtaining, analyzing, and storing human samples must be sought prior to commencement of any of these studies, and local health regulations (e.g., hepatitis B vaccination) adhered to before handling blood samples. Conditions for differentiating cells or assessment of cell function include incubation at 37 °C with 5% CO_2. Be aware that some reagents may be harmful, and the use of these may be restricted. Prepare and store all reagents at room temperature unless indicated otherwise. Please follow all local disposal regulations when discarding waste materials.

1. Peripheral human blood (in buffy coats) (*see* **Note 1**).
2. PBS.
3. Ficoll–Paque®.

2.1 Differentiation of Osteoclasts from Peripheral Human Blood Mononuclear Cells (PBMCs)

4. Biotin-free bovine serum albumin (BSA): Make a working solution of 0.1% BSA by mixing 0.1 g BSA with 100 mL PBS. Aliquot to smaller vials (e.g., 2 mL microcentrifuge tubes) and store at −18 °C until use.
5. Cell culture media: Use α-Minimal Essential Media (MEM) containing 10% heat-inactivated fetal bovine serum (FBS) and 1% penicillin/streptomycin.
6. Anti-Human CD14 Magnetic Particles.
7. Cell strainer with pore size of 40 μm for filtration.
8. Cell separation magnet that can hold two 17 × 100 mm round-bottom test tubes in an upright position.
9. Human M-CSF stock solution: Dissolve to 25 μg/mL in PBS containing 0.1% BSA. Aliquot to 0.5 mL microcentrifuge tubes and store at −80 °C (*see* **Note 2**).
10. Human RANKL stock solution: Dissolve to 25 μg/mL in PBS containing 0.1% BSA. Aliquot to 0.5 mL microcentrifuge tubes and store at −80 °C (*see* **Note 2**).
11. Incubation buffer: Weigh 1 g biotin-free BSA. Dissolve the BSA in 200 mL PBS. Filter the dilution through a sterile filter. Add 0.8 mL of 0.5 M sterile EDTA. Store at 4 °C.

2.2 Staining Osteoclast Nuclei

1. Giemsa solution: For 25 mL, mix 2.3 mL phosphate buffer, pH 7.0, with 1.4 mL Giemsa. Add demineralized water to a total volume at 25 mL.
2. May-Grünwald solution.
3. PBS.
4. 3.7% formaldehyde.
5. Methanol.
6. Phase-contrast microscope with 20× objective.

2.3 Assessment of Osteoclast Bone-Resorptive Activity

1. Toluidine blue: Add 0.5 g toluidine blue and 0.5 g sodium borate to 50 mL milliQ water, pH 7.0. Store at 4 °C.
2. Cortical bone slices stored in ethanol at 4 °C (*see* **Note 3**).
3. Accutase.
4. Phase-contrast microscope with ×10 objective.

2.4 TRAcP Activity of Human Osteoclasts

1. Reaction buffer: 1 M acetate, 0.5% Triton X-100, 1 M NaCl, 10 mM EDTA, pH 5.5. Store at 4 °C.
2. L-Ascorbic acid: 8.8 mg/mL in milliQ water. Prepare solution just before use.
3. DiSodium tartrate: 46 mg/mL in milliQ water. Prepare solution just before use.

4. 4-Nitrophenylphosphate: 18 mg/mL in milliQ water. Prepare solution just before use.
5. TRAcP solution buffer: Make fresh. Use the dilutions as described in **steps 1–4**. For 8 mL solution, use 1 mL L-ascorbic acid, 1 mL DiSodium tartrate, 1 mL 4-nitrophenylphosphate, 2 mL reaction buffer and add 3 mL milliQ water.
6. Stop buffer: 0.3 M NaOH stored at 4 °C.
7. Plate reader capable of reading absorbance.

2.5 Differentiation of Osteoblasts from Human Mesenchymal Stem Cells

1. Heparinized MEM: Sterile-filter 10 mL MEM mixed with 1 mL heparin in a 50 mL tube. Keep at 4 °C.
2. Cell culture media for stem cell cultivation: MEM with 10% FBS and 1% penicillin/streptomycin.
3. Cell culture media for osteoblast differentiation: Use MEM with 10% FBS and 1% penicillin/streptomycin.
4. For induction of osteoblast differentiation, supplement the media with 10 mM β-glycerolphosphate, 10 nM dexamethasone, 0.2 mM ascorbic acid, and 10 nM vitamin D3.
5. Heparin.
6. Lymphoprep.

2.6 Alizarin Red Staining and Quantification in Osteoblasts

1. Alizarin Red solution: Add 10 mL demineralized water to 137 mg Alizarin Red, pH 4.2. Adjust pH using 1 M NaOH (sodium hydroxide).
2. PBS.
3. 70% ethanol.
4. Deionized water.
5. *Quantification solution*: 20% methanol, 10% acetic acid, and 70% water.
6. Plate reader capable of reading absorbance.
7. Phase-contrast microscope with ×10 objective.

2.7 ALP Staining and Quantification of Activity in Osteoblasts

1. Staining solution: Resuspend 2 mg Naphtol-AS-TR-phosphate in 100 μL N,N-dimethylpormamide, and add 10 mL demineralized water. Mix with 10 mg Fast Red TR that has been resuspended in 12 mL 0.1 M Tris buffer.
2. Fixation buffer: Mix acetone with 10 mM citrate buffer 1.5:1, pH 4.2 (adjust with citrate acid).
3. PBS.
4. Reaction buffer for quantification of activity: 1 mg/mL nitrophenylphosphate in 50 mM $NaHCO_3$, pH 9.6 (adjust with 1 mM $MgCl_2$).

5. Stop buffer: 3 M NaOH.
6. Fixation solution: 3.7% formaldehyde and 90% ethanol.
7. Tris-buffered saline (TBS): 20 mM Trizma base, 150 mM NaCl, pH 7.5.
8. Deionized water.
9. Plate reader capable of reading absorbance.
10. Phase-contrast microscope with ×10 objective.

2.8 Intracellular Calcium Imaging Using Fura-2-AM

1. Imaging media: make 50 mL Hanks balanced salt solution (HBSS), Ca^{2+}- and Mg^{2+}-free, 10 mM HEPES. Keep in the fridge for long-term storage.
2. Fura-2-acetoxymethylester (Fura-2-AM): Add 50 μL DMSO to 50 μg Fura-1-AM. Protect solution from the light and vortex briefly. Desiccated form should be stored at −20 °C. Once reconstituted, Fura-2 should be stored in the fridge for 3–4 weeks (*see* **Note 4**). A working solution of 4 μM in imaging media (Fura-2 loading solution) should be made immediately before the experiment and leftovers discarded at the end of the day.
3. 100 μM stock of the positive control ionomycin: dissolve in DMSO to make a stock solution, and then add 10 μL to samples (therefore cells are exposed to a final concentration of 1 μM due to the 1:100 dilution).
4. Microscope capable of measuring excitation at 340 nm and 380 nm and emission at 510 nm. We used a Crest X-light spinning disk system coupled to a Ti-E base and 10×/0.4 air objective. Excitation was delivered at 340 nm and 385 nm using a FuraLED system, with emitted signals detected at 470–550 nm.
5. 35 mm imaging dishes or 6-well plate.
6. Phosphate-buffered saline: dissolve one tablet (available from Gibco) in 500 mL sterile water. Autoclave and store at room temperature.
7. 24 mm glass coverslips.
8. Immersion oil-type LDF.
9. ImageJ: available as a free download at https://imagej.net/ij/

3 Methods

All procedures should be carried out at room temperature in a tissue culture laminar flow cabinet unless stated otherwise.

3.1 Establishment of Osteoclast Cultures

3.1.1 CD14+-Monocyte Isolation

1. Mix the blood from the buffy coat with PBS (1:1) in a T75 cell culture flask. For 50 mL blood, add 50 mL PBS to a final volume of 100 mL.
2. For every 100 mL dilution of blood and PBS, take four 50 mL sterile tubes. Add 25 mL Ficoll-Paque® to each of the tubes just before use (*see* **Note 5**).
3. Layer 25 mL of blood/PBS dilution on top of the Ficoll–Paque to a final volume of 50 mL (*see* **Note 6**). Repeat until all the blood has been added.
4. Centrifuge at 2000 rpm for 20 min without break. Meanwhile, prepare two new 50 mL tubes.
5. Gently take the tubes from the centrifuge and place them in a rack. Take the lid off the tubes, and transfer the interphase (the white cell layer) using a disposable pipette (*see* **Note** 7). Combine the interphase layer from two tubes into a new 50 mL centrifuge tube. Discard the remaining liquid in the original tubes.
6. Add PBS up to 50 mL and dissolve the pellet.
7. Centrifuge at 2000 rpm for 12 min with brake.
8. Discard the supernatant and add PBS up to 50 mL (*see* **Note 8**). Dissolve the pellet.
9. Centrifuge at 2000 rpm for 12 min with brake. Meanwhile, prepare one 50 mL tube by placing it in a rack, remove the lid and put the cell strainer on the tube.
10. Discard the supernatant, and dissolve the cell pellet in 2 mL incubation buffer per 50 mL tube.
11. Collect the cells using a Pasteur pipette and add them on top of the filter.
12. When the cell pellets from both tubes have been added, gently add 1 mL incubation buffer on the filter (*see* **Note 9**).
13. Discard the 40 μm filter.
14. Add 600 μL Anti-Human CD 14 Magnetic Particles to the cell suspension, and gently mix it using a pipette (*see* **Note 10**). You do not need to preheat the CD14 Magnetic Particles. Place the lid on the tube without closing it and leave the tube for 30 min in a rack.
15. Transfer the cell suspension to a 14 mL falcon tube, mix gently and place the tube on a cell separation magnet. Leave for 8–10 min. Meanwhile, prepare cell culture medium (*see* **Note 11**).
16. Discard the supernatant while the tube is still placed on the magnet (*see* **Note 12**).
17. Remove the tube from the magnet.

18. Add 5 mL incubation buffer and gently resuspend the cells (*see* **Note 13**).
19. Place the tube on the magnet for 4–5 min.
20. Discard the supernatant while the tube is still on the magnet.
21. Remove the tube from the magnet.
22. Add 5 mL incubation buffer and gently resuspend the cells.
23. Place the tube on the magnet for 4–5 min.
24. Discard the supernatant while the tube is still on the magnet.
25. Remove the tube from the magnet and place it in a rack.
26. Add 5 mL cell culture media and dissolve the cell pellet. Count the number of cells (*see* **Note 14**).
27. For every T75 cell culture flask, add cell suspension corresponding to 5×10^6 cells to a total volume of 20 mL cell culture media. For T25 cell culture flasks, add cell suspension corresponding to 1.67×10^6 cells to a total volume of 6.7 mL. Add 25 ng/mL M-CSF to each cell culture flask (*see* **Note 15**). Do not add RANKL at this point. Incubate the cells at 37 °C with 5% CO_2 for 2 days.

3.1.2 Osteoclast Differentiation

1. After 2 days, remove the cell culture media from the flask, and centrifuge at 1500 rpm for 5 min (*see* **Note 16**). Remove the supernatant.
2. Dissolve the pellet in fresh media containing both 25 ng/mL M-CSF and 25 ng/mL RANKL. The total volume of cell culture media is 20 mL per T75 flask and 6.7 mL per T25 flask.
3. After another 3 days, change media. At this point, all living cells should have attached to the cell culture flask. Remove all media and replace with fresh media containing M-CSF and RANKL.
4. After another 2 days, remove all media and replace with fresh media containing M-CSF and RANKL.
5. After another 2 days, the cells should be ready to use as mature osteoclasts. Always check for multinucleation using a phase-contrast microscope before use to ensure the success of differentiation (*see* **Note 17**).

3.2 Preparation of Mature Osteoclasts for Cell Differentiation and Activity Assays

1. Remove the cell media from mature cells.
2. Wash the cell culture flask in PBS twice.
3. Add Accutase (1 mL per T25 flask and 3 mL per T75 flask) (*see* **Note 18**).
4. Incubate for 8 min at 37 °C with 5% CO_2.
5. Gently detach cells with a cell scraper. Avoid making bubbles.

6. Collect the cells by adding cell culture media (2 mL if T25 flask, 7 mL if T75 flask). Pipette up and down and collect the 3 or 10 mL, respectively. Transfer to a sterile tube.
7. Add cell culture media (3 mL if T25 flask, 10 mL I T75 flask) to wash the flask to collect residual cells. Collect the suspension and transfer to the sterile tube from **step 6**.
8. Centrifuge at 1500 rpm for 5 min. Remove the supernatant. Resuspend the cell pellet (0.5 mL per T25 flask used or 1 mL per T75 flask used). Count the cells. Cells are now ready for plating for assays.
9. Dilute the cells ($2.5*10^5$ cells/mL) in cell culture medium containing M-CSF and RANKL (25 ng/mL each).
10. Cells are now ready for use in assays.

3.2.1 Assessing Osteoclast Nuclei and Number with May-Grünwald Giemsa Staining

1. Plate cells at 25,000 cells per well in 100 μL cell suspension (*see* **Note 19**).
2. Incubate at 37 °C with 5% CO_2 for 40–60 min.
3. Add 100 μL of any compound/treatment diluted in cell culture medium containing M-CSF and RANKL (25 ng/mL each). For vehicle group, use the substance the compound was diluted in, e.g., saline.
4. Incubate at 37 °C, 5% CO_2 for 72 h.

   Following 72 h, remove the cell culture media and store at −80 °C for later analyses.
5. Wash the wells twice with PBS.
6. Fix with 150 μL 3.7% formalin and leave for 10 min. Then discard fixative.
7. Gently add 100 μL methanol and leave for 15 min. Then discard methanol.
8. Let the wells dry in a fume hood for 10–20 min. Then add 100 μL May-Grünwald solution. Leave for 5 min. Then discard the solution.
9. Add 100 μL Giemsa solution. Leave for 15 min. Then discard solution.
10. Wash with 200 μL demineralized water twice, and let the wells dry overnight in the fume hood.
11. Visualize the stained wells under a microscope with a 20× objective. Count the number of osteoclasts and the number of nuclei per osteoclast (*see* **Note 20**).

3.2.2 Assessment of Osteoclast-Resorptive Activity

1. Remove bone slices with tweezers from their storage in ethanol (*see* **Note 21**).
2. Place the required number into a 50 mL tube and wash in cell culture media on a rocker.

3. Place a single bone slice into each well of a 96-well plate with sterile tweezers (only fill the number of wells required for your study).
4. Add 200 μL cell suspension corresponding to 50,000 cells per well. Incubate for 40–60 min (*see* **Note 22**).
5. Prepare compound/treatment at 25 times higher than the final concentration.
6. Add 10 μL of 25× compound/treatment and add another 40 μL cell suspension. Gently mix without flipping the bone slices.
7. Incubate at 37 °C with 5% CO_2 for 72 h.
8. Remove and save cell culture media at −80 °C.
9. Add 150 μL sterile water to each bone slices to stop the experiment.
10. Scrape the cells off the bone slice using a cotton stick and rinse the bone slice in water. Dry the bone slice on paper. Preferably use the cotton stick on both sides of the bone slice to optimize the staining procedure.
11. While holding the bone slice with a pair of tweezers, put the bone slice in toluidine blue solution for 15 s (see Fig. 2a, b). Dry on paper. The bone slice can be stored at room temperature.
12. Analyse resorption areas using a microscope (*see* **Note 23**). We determine pits as round excavations and trenches as elongated excavations (see Fig. 2c). Resorption areas have sharp borders, and varying depth sizes that can be visualized microscopically. Be sure to distinguish resorption areas from artifacts, such as diffuse staining, osteons, or blood vessels, which usually do not have sharp borders and only superficial staining.

3.3 Measurement of TRAcP Enzymatic Activity

This protocol uses the cell media collected and stored at −80 °C in Subheadings 3.2.1 and 3.2.2:

1. Transfer 20 μL cell culture medium into a new 96-well plate.
2. Add 80 μL TRAcP solution buffer.
3. Cover with adhesive tape and protect from light, e.g., by packing the plate in foil. Incubate at 37 °C for 20 min.
4. Add 100 μL stop buffer.
5. Read absorbance 400 and 645 nm on a plate reader.

1. Harvest 10 mL bone marrow in 10 mL heparinized MEM. Bone marrow is usually harvested from the hip, e.g., the iliac crest (*see* **Note 24**).
2. Dilute 1:1 in PBS and mix gently.

3.4 Establishment of Osteoblast Cultures

3.4.1 Isolation of Mesenchymal Stem Cells for Osteoblast Culture

3. Equally distribute the dilution to two 50 mL tubes.
4. Gently add 10 mL Lymphoprep in the bottom of each tube.
5. Centrifuge at 2500 rpm for 25 min without break.
6. Transfer the interphase (*see* **Note 6**) from both tubes to a new 50 mL tube containing 20 mL MEM with 10% FBS and 1% penicillin/streptomycin.
7. Centrifuge at 2000 rpm for 10 min with break.
8. Discard the supernatant.
9. Dissolve the cell pellet in 20 mL MEM with 10% FBS and 1% penicillin/streptomycin.
10. Count the cells. For T80 flasks, add 1×10^7 cells and incubate at 37 °C with 5% CO_2 for 5–7 days.
11. When the cells have attached, remove the cell medium, and either replace it with MEM containing 10% FBS, 1% penicillin/streptomycin, 1% GlutaMAX, 1% sodium pyruvate, and 1% nonessential amino acids or freeze the cells (*see* **Notes 25** and **26**).
12. Change the medium twice weekly until cells are 80% confluent. Trypsinize and count the cells.

3.4.2 Osteoblast Differentiation

1. Immediately add cells at a density of 20,000 cells/cm^2 in α-MEM containing 10% FBS and 1% penicillin/streptomycin. Incubate at 37 °C with 5% CO_2 overnight.
2. Discard the cell culture medium.
3. Induce osteoblast differentiation by adding osteoblast base media with supplements.
4. Change the cell medium every third day (*see* **Note 27**). The duration of differentiation is 12–14 days (*see* **Note 28**).

3.5 Quantification of Alkaline Phosphatase (ALP) Activity

1. Discard the cell media.
2. Wash once with TBS and discard.
3. Add fixation solution for 30 s.
4. Discard the fixation solution.
5. Wash once with PBS and discard. At this point, it needs to be taken into consideration if the ALP activity should be normalized to the number of cells (*see* **Note 29**).
6. Add reaction buffer (100 μL per well in a 96-well plate).
7. Incubate at 37 °C for 20 min.
8. Add stop buffer (50 μL per well in a 96-well plate).
9. Transfer 80 μL of the media to a transparent 96-well plate.
10. Measure the absorbance at 405 nm.

3.6 ALP Staining

1. Follow Subheading 3.7, **steps 1–4**.
2. Discard the cell media.
3. Wash once with PBS.
4. Add fixation buffer and leave for 5 min.
5. Discard fixation buffer.
6. Add ALP solution and leave for 1 h.
7. Remove the ALP solution and wash with distilled water.
8. Take images on a phase-contrast microscope with 10× objective.

3.7 Alizarin Red S Quantification

1. Add solution (e.g., 400 μL solution per well in a 24-well plate).
2. Leave for 10 min.
3. Transfer 100 μL to a 96 well plate.
4. Measure absorbance at 450 and 570 nm.

3.8 Alizarin Red S Staining

Alizarin Red staining should be done when cells have differentiated to mature osteoblasts:

1. Discard the cell media.
2. Wash once with PBS.
3. Fix with 70% ethanol for 1 h at −20 °C.
4. Discard the ethanol.
5. Wash with distilled water.
6. Stain with 40 mM Alizarin Red S, pH 4.2, for 10 min with rotation.
7. Wash once with water to remove excess dye.
8. Wash once with PBS for 2–3 min with rotation.
9. Take images on a phase-contrast microscope with 10× objective.

3.9 Intracellular Calcium Imaging Using Fura-2-AM

3.9.1 Preparation of Coverslips for Imaging

1. Lift coverslips with tweezers and place in a 35 mm imaging dish or 6-well plate.
2. Add 70% ethanol to the coverslip (150–200 μL should cover), and incubate at room temperature for 10 mins.
3. Remove ethanol. Wash coverslips in ~1 mL PBS 2–3 times.
4. Coverslips are now ready to use for cell seeding or can be stored in PBS at 4 °C for 2–3 weeks.

3.9.2 Preparation of Cells for Imaging

1. Plate mature osteoclasts or osteoblasts in imaging chambers with coverslips at 100,000 cells per well.
2. Leave cells to settle for at least 4 h (*see* **Note 30**).
3. Remove media from coverslip.

4. Add 100 μL 4 μM Fura-2 loading solution directly to the coverslip, and incubate at 37 °C for 40 min.
5. Remove loading solution and wash coverslip 2× with imaging solution.
6. Place coverslip in imaging chamber and add 1 mL of imaging buffer.

3.9.3 Intracellular Calcium Imaging on the Microscope

1. Switch on the microscope system and humidity chamber to set the temperature to 37 °C 1 h before imaging.
2. Identify a region with multiple cells within frame.
3. Sequentially image cells at 340 and 380 nm for 3 min.
4. After 3 min, pause the imaging software and add 10 μL of compound (at 100× the final desired concentration) carefully without disturbing the cells or chamber.
5. Image for another 30 min.
6. Pause the imaging software, and add 10 μL of 1 μM ionomycin carefully without disturbing the cells or chamber.
7. Image for 3 min (*see* **Note 31**).

3.9.4 Analysis

1. Convert images from your software to .tif if this was not the default file type.
2. Open .tif in ImageJ and generate two stacks of images, one for 340 nm and the other for 380 nm.
3. Draw a region of interest (ROI) around each cell (Analyze> Tools> ROI manager) using the circle selection tool, and quantify the fluorescence intensity for the whole stack.
4. Generate 340/380 ratios.
5. Normalize each value to the baseline value.
6. Plot in GraphPad Prism or your chosen software. From these traces several subanalyses can be performed, e.g., total number of oscillations, amplitude of responses, and maximal responses [13].

4 Notes

1. PBMCs can be isolated from whole blood samples, buffy coats, or leukocyte cones. Researchers should investigate which of these will be the easiest to obtain from their location. Often, buffy coats and leukocyte cones can be purchased from local blood centres (e.g., the NHS blood and transplant service in the UK).

2. Cytokines should be purchased from reputable sources (e.g., Biotechne and Cambridge Biosciences).
3. We use bovine cortical bone slices of 6 mm diameter and 0.4 mm depth. These fit into a well of a 96-well plate.
4. Addition of 0.03–0.04% Pluronic F-127 can increase the aqueous solubility of Fura-2 AM.
5. Ficoll-Paque® is sensitive to light and should be stored accordingly. Please be aware when adding blood on top of Ficoll–Paque® that the tube is immediately put into a centrifuge.
6. Be careful when adding blood to Ficoll–Paque® as you may risk mixing them if you add the blood too fast. Think of a bartender pouring a beer; tip the tube approximately 45 degrees and then slowly add the blood along the inside of the tube. When 2/3 of the blood has been added, carefully and slowly tip the tube back to an upright position when adding the final 5–7 mL. Carefully close the tube and add it to a rack before repeating this with the other tubes.
7. When pipetting the interphase, use a disposable pipette that you can manage with your fingers. We usually use a disposable 1 mL Pasteur pipette. We do not recommend you use a pipette pump for this step.
8. At this point, the cell pellet may not be attached to the tube so be careful when removing the supernatant not to remove any cells. When adding PBS to the pellet, start by adding only 10–20 mL to resuspend the pellet. Then add up to a final volume of 50 mL and dissolve again.
9. Although only a total of 5 mL incubation buffer is added, the final volume of the cell suspension is approximately 6 mL.
10. If there is more or less than 6 mL after filtration, consider adjusting the volume of CD14 magnetic particles so the ratio between cell suspension and CD14 magnetic particles is 10:1.
11. When the tube is on the magnet, prepare and/or heat the cell culture medium.
12. When discarding the supernatant from the falcon tube on the magnet, be sure to gently push the tube as close to the magnet as possible using your fingers. You should discard the supernatant by aspirating on the inside of the tube that is opposite the magnet. This will reveal the cell suspension as an orange- or brown-colored suspension that is attached to the inside of the tube closest to the magnet.
13. Be sure to dissolve the entire cell suspension that is attached to the inside of the tube.
14. We use an automated cell counter, but recommend you refer to local guidelines or procedures. When we calculate cell number,

we mix 90 μL trypan blue with 10 μL cell suspension in a new microcentrifuge tube and leave it for 1–2 min. Then, 10 μL cell suspension is placed in a counting slide that is placed in the cell counter. The cell number is then calculated by multiplying the number on the automated cell counter by 5.

15. M-CSF and RANKL are diluted to 25 μg/mL before storage to make calculations easier (i.e., simple 1:1000 dilution) when adding them to cell culture media during osteoclast differentiation and osteoclast activity experiments.
16. At the time of first media change during osteoclast differentiation, not all cells have attached to the cell culture flask yet. Therefore, collect the old cell medium and put it in a 50 mL tube. Centrifuge at 1500 rpm for 5 min. Remove the supernatant. Dissolve the cell pellet (1 mL cell culture media per T75 flask).
17. We advise, if possible, to take images of the mature osteoclasts on a phase-contrast microscope (×10 objective) before use (see Fig. 1d). Take at least 6–9 pictures of each cell culture flask at different regions of the flask. This can be used to clarify how well the cells have differentiated and can be used to support findings in publications.
18. For removal of osteoclasts, use Accutase (1 mL if T25 cell culture flask and 3 mL if T75 cell culture flask) or a similar mild cell detachment buffer. Upon incubation with Accutase, consider the osteoclasts as small bubbles that should not erupt when scraping the cell culture flasks with a cell scraper. Preferably use repeated gentle movements than harsh scrapes. Always check if the cells have loosened from the cell culture flask using a microscope.
19. When doing experiments with preosteoclasts or osteoclasts, we recommend at least four technical replicates.
20. For counting, we use a 10 × 10 counting grid that fits in the ocular of the microscope to navigate and use a systematic counting system. Use 20× objective. This will create 20 possible visual fields (see Fig. 4a). We count the number of nuclei and the number of osteoclasts within the borders of the counting grid on 7 visual fields for each 96-well plate by starting at either visual field 1 or 2. Count one visual field, then skip two, then count one and so on until you reach a total of seven counted visual fields. The first visual field should be North on the middle of the well so each of the two corners of the upper bar of the counting grid touch the edge of the bone slice. Then move South so the lower bar of the counting grid is "replaced" by the upper bar of counting grid and go West until the top left corner of the upper bar of the counting grid touches the edge of the bone slice for visual field 2. Then move East so the left

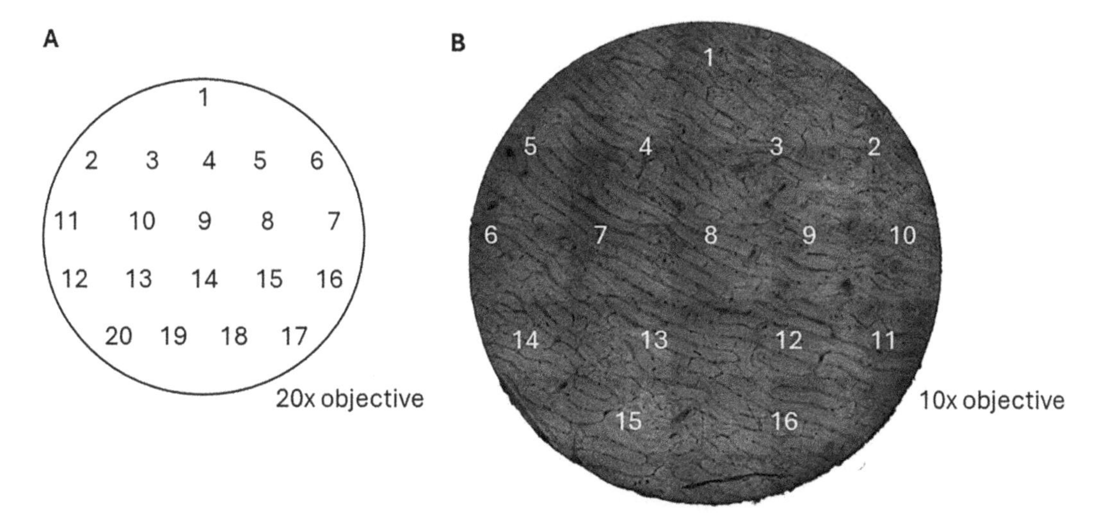

Fig. 4 (**a**) Schematic overview of a 96-well plate used for analysis of osteoclast differentiation with an approach to evaluate visual fields (1–20) within a 10 × 10 grid at 20× objective. (**b**) Microscope images showing an overview of bone chips used for resorption analysis with an approach to systematically evaluate resorption areas with numbers 1–16 representing visual fields within a 10 × 10 grid at 10× objective

bar of the counting grid is "replaced" by the right bar of the counting grid for visual field 3 and so on.

21. For osteoclast activity analyses, we use cortical bovine bone slices designed to fit in a 96-well plate. The bone slices are stored in the fridge in ethanol and needs to be washed in cell culture medium, α-MEM, before use. Take the number of bone slices required for the experiment and place in a 50 mL tube. Add 25 mL cell culture media and shake for 30–45 min. Use sterile tweezers to place the bone slices in the wells.
22. When seeding osteoclasts on bone slice, incubate for at least 40 min before adding any treatment to make sure that osteoclasts have attached to the bone surface and restored morphology after Accutase treatment.
23. For counting resorption areas, we use a 10 × 10 counting grid that fits in the ocular of the microscope to navigate and use a systematic counting system. Use 10× objective. This will create 16 possible visual fields (see Fig. 4b). Start North on the middle of the bone slice so each of the two corners of the upper bar of the counting grid touch the edge of the bone slice. Then move South so the lower bar of the counting grid is "replaced" by the upper bar of counting grid and go East until the top right corner of the upper bar of the counting grid touches the edge of the bone slice. Then move West so the left bar of the counting grid is "replaced" by the right bar of the counting grid and so on. Preferably analyze resorption areas in pits, i.e., round cavities, and trenches, i.e., prolonged cavities.

24. Researchers should investigate ethical approvals from their location.
25. For cryopreservation of cells, remove the cell culture medium, and wash twice with PBS without calcium and magnesium. Detach the cells using trypsin-EDTA (3 mL if T75 flask) and incubate at 37 °C with 5% CO_2 for 3 min. Take the cells out of the incubator, and check under the microscope that all cells have detached. If not, you may tap the flask to loosen the cells or incubate for another 1–2 min. When the cells have detached, add 7 mL MEM with 10% FBS and 1% penicillin/streptomycin and transfer to a 50 mL tube. Centrifuge at 1000 rpm for 5 min. Discard the supernatant and dissolve the cell pellet in 10 mL MEM. Count the cells. Centrifuge at 1000 rpm for 5 min. Discard the supernatant and dissolve the cell pellet in cryopreservation medium (for 10 mL solution, use 7 mL MEM, 2 mL FBS, 100 μL penicillin/streptomycin, and 1 mL DMSO). The final volume depends on the number of cells. Add 1 mL cells (1×10^6 cells/mL) to a freezing tube and leave it at −80 °C overnight. Then transfer the cells to a liquid nitrogen freezer.
26. To thaw cells, preheat cell culture medium (MEM with 10% FBS and 1% penicillin/streptomycin). Obtain the vial of cells from the liquid nitrogen and put it in a 37 °C water bath. Be sure the lid is closed completely to avoid contamination with the water. When cells have thawed, transfer the cells to a 15 mL tube and gently add 10 mL cell culture medium. Centrifuge 10 min with break. Discard supernatant. Dissolve the cell pellet and transfer the cells to a cell culture flask (1×10^6 cells/ 80 cm^2). Change media and check the cells the following day, and start differentiation when cells are approximately 80% confluent.
27. It is recommended to prepare the osteoblast differentiation medium fresh before every media change as some of the compounds have a short half-life at room temperature.
28. While the duration of osteoblast differentiation is approximately 2 weeks, it should be considered to include a plate that can be analyzed for alkaline phosphatase activity after 7 days. The reason for this is to check the differentiation process, but it can also add important knowledge regarding whether a given treatment has effects on early, late, or both stages of osteoblast differentiation.
29. For normalization of ALP activity, add DAPI working solution (for stock, mix 1 μL DAPI in 1 mL PBS and dilute this 1:50 for the working solution, e.g., 1 μL DAPI stock solution in 49 μL PBS) for 30 min protected from light, and count the cells. Discard DAPI solution and gently wash twice with TBS.

30. Cells can be plated on day 9, then left overnight to settle and imaged on day 10.
31. Cells should have a large response to ionomycin. If cells did not respond to ionomycin, we discarded data for that cell.

Acknowledgments

MSH chapter is funded by the Region of Southern Denmark (grant number 18/17553) and Novo Nordisk Foundation (grant number NNF18OC0052699).

References

1. Hadjidakis DJ, Androulakis II (2006) Bone remodeling. Ann N Y Acad Sci 1092:385–396. https://doi.org/10.1196/annals.1365.035
2. Kim J-M, Lin C, Stavre Z, Greenblatt MB, Shim J-H (2020) Osteoblast-osteoclast communication and bone homeostasis. Cells 9(9): 2073
3. Takayanagi H, Kim S, Koga T, Nishina H, Isshiki M, Yoshida H, Saiura A, Isobe M, Yokochi T, Inoue J, Wagner EF, Mak TW, Kodama T, Taniguchi T (2002) Induction and activation of the transcription factor NFATc1 (NFAT2) integrate RANKL signaling in terminal differentiation of osteoclasts. Dev Cell 3(6):889–901. https://doi.org/10.1016/s1534-5807(02)00369-6
4. Hwang SY, Putney JW Jr (2011) Calcium signaling in osteoclasts. Biochim Biophys Acta 1813(5):979–983. https://doi.org/10.1016/j.bbamcr.2010.11.002
5. Danciu TE, Adam RM, Naruse K, Freeman MR, Hauschka PV (2003) Calcium regulates the PI3K-Akt pathway in stretched osteoblasts. FEBS Lett 536(1–3):193–197. https://doi.org/10.1016/s0014-5793(03)00055-3
6. Romanello M, Padoan M, Franco L, Veronesi V, Moro L, D'Andrea P (2001) Extracellular NAD(+) induces calcium signaling and apoptosis in human osteoblastic cells. Biochem Biophys Res Commun 285(5):1226–1231. https://doi.org/10.1006/bbrc.2001.5325
7. Martinez FO, Gordon S, Locati M, Mantovani A (2006) Transcriptional profiling of the human monocyte-to-macrophage differentiation and polarization: new molecules and patterns of gene expression. J Immunol 177(10): 7303–7311. https://doi.org/10.4049/jimmunol.177.10.7303
8. Mun SH, Park PSU, Park-Min K-H (2020) The M-CSF receptor in osteoclasts and beyond. Exp Mol Med 52(8):1239–1254. https://doi.org/10.1038/s12276-020-0484-z
9. Boyce BF, Xing L (2008) Functions of RANKL/RANK/OPG in bone modeling and remodeling. Arch Biochem Biophys 473(2): 139–146. https://doi.org/10.1016/j.abb.2008.03.018
10. Boyle WJ, Simonet WS, Lacey DL (2003) Osteoclast differentiation and activation. Nature 423(6937):337–342. https://doi.org/10.1038/nature01658
11. Matsubara T, Kinbara M, Maeda T, Yoshizawa M, Kokabu S, Takano Yamamoto T (2017) Regulation of osteoclast differentiation and actin ring formation by the cytolinker protein plectin. Biochem Biophys Res Commun 489(4):472–476. https://doi.org/10.1016/j.bbrc.2017.05.174
12. Soe K, Delaisse JM (2010) Glucocorticoids maintain human osteoclasts in the active mode of their resorption cycle. J Bone Miner Res 25(10):2184–2192. https://doi.org/10.1002/jbmr.113
13. Hansen MS, Soe K, Christensen LL, Fernandez-Guerra P, Hansen NW, Wyatt RA, Martin C, Hardy RS, Andersen TL, Olesen JB, Hartmann B, Rosenkilde MM, Kassem M, Rauch A, Gorvin CM, Frost M (2023) GIP reduces osteoclast activity and improves osteoblast survival in primary human bone cells. Eur J Endocrinol 188(1):144. https://doi.org/10.1093/ejendo/lvac004
14. Panwar P, Soe K, Guido RV, Bueno RV, Delaisse JM, Bromme D (2016) A novel approach to inhibit bone resorption: exosite inhibitors against cathepsin K. Br J Pharmacol

173(2):396–410. https://doi.org/10.1111/bph.13383

15. Vanderoost J, Soe K, Merrild DM, Delaisse JM, van Lenthe GH (2013) Glucocorticoid-induced changes in the geometry of osteoclast resorption cavities affect trabecular bone stiffness. Calcif Tissue Int 92(3):240–250. https://doi.org/10.1007/s00223-012-9674-6
16. Abdallah BM, Haack-Sorensen M, Fink T, Kassem M (2006) Inhibition of osteoblast differentiation but not adipocyte differentiation of mesenchymal stem cells by sera obtained from aged females. Bone 39(1):181–188. https://doi.org/10.1016/j.bone.2005.12.082
17. Tencerova M, Frost M, Figeac F, Nielsen TK, Ali D, Lauterlein JL, Andersen TL, Haakonsson AK, Rauch A, Madsen JS, Ejersted C, Hojlund K, Kassem M (2019) Obesity-associated hypermetabolism and accelerated senescence of bone marrow stromal stem cells suggest a potential mechanism for bone fragility. Cell Rep 27(7):2050–2062. e2056. https://doi.org/10.1016/j.celrep.2019.04.066
18. Jafari A, Qanie D, Andersen TL, Zhang Y, Chen L, Postert B, Parsons S, Ditzel N, Khosla S, Johansen HT, Kjaersgaard-Andersen P, Delaisse JM, Abdallah BM, Hesselson D, Solberg R, Kassem M (2017) Legumain regulates differentiation fate of human bone marrow stromal cells and is altered in postmenopausal osteoporosis. Stem Cell Rep 8(2):373–386. https://doi.org/10.1016/j.stemcr.2017.01.003
19. Abdallah BM, Jensen CH, Gutierrez G, Leslie RG, Jensen TG, Kassem M (2004) Regulation of human skeletal stem cells differentiation by Dlk1/Pref-1. J Bone Miner Res 19(5):841–852. https://doi.org/10.1359/JBMR.040118
20. Hansen MS, Madsen K, Price M, Soe K, Omata Y, Zaiss MM, Gorvin CM, Frost M, Rauch A (2024) Transcriptional reprogramming during human osteoclast differentiation identifies regulators of osteoclast activity. Bone Res 12(1):5. https://doi.org/10.1038/s41413-023-00312-6
21. Rauch A, Haakonsson AK, Madsen JGS, Larsen M, Forss I, Madsen MR, Van Hauwaert EL, Wiwie C, Jespersen NZ, Tencerova M, Nielsen R, Larsen BD, Rottger R, Baumbach J, Scheele C, Kassem M, Mandrup S (2019) Osteogenesis depends on commissioning of a network of stem cell transcription factors that act as repressors of adipogenesis. Nat Genet 51(4):716–727. https://doi.org/10.1038/s41588-019-0359-1

Chapter 14

Quantitative Intravital Calcium Imaging in Mouse Kidney

Andreja Figurek, Nevena Jankovic, and Andrew M. Hall

Abstract

Intracellular calcium is an important regulator of solute transport in renal epithelial cells, and disordered calcium signaling may underlie the pathogenesis of certain kidney diseases. Intravital multiphoton imaging of the kidney in transgenic mice expressing highly sensitive fluorescent reporters allows detailed study of calcium signals within different specialized segments of the renal tubule and how these are integrated with other cellular processes. Moreover, changes in activity can be observed in real time in response to physiological interventions or disease-causing insults. In this chapter, we will provide a detailed protocol for performing this powerful research technique.

Key words Calcium, Intravital imaging, Kidney, Multiphoton microscopy, Renal tubule

1 Introduction

The kidneys are vital organs that regulate body homeostasis. The filtering unit of the kidney is the nephron, which is divided into specialized segments, each of which is adapted to specific tasks. Calcium is an intracellular second messenger that translates external stimuli into signaling responses, and studies suggest that calcium can regulate renal tubular solute transport [1–3]. Moreover, disordered calcium signaling has been implicated in the pathogenesis of kidney diseases [4, 5].

Intravital imaging with multiphoton microscopy in the externalized kidney provides the possibility to study dynamic cellular processes in unprecedented detail [6]. Moreover, responses to physiological perturbations or pathological insults can be recorded in real time. Renal epithelial cells in living animals are resistant to either loading of calcium dyes or transfection with viral vectors [7]. Uptake of dyes can be achieved in ex vivo models—including isolated perfused organs and tissue slices [8]—but here the critical in vivo context is lost.

An important breakthrough in the field was the generation of genetically modified mice stably expressing calcium-sensitive

Caroline M. Gorvin (ed.), *Calcium Signaling: Methods and Protocols*, Methods in Molecular Biology, vol. 2861,
https://doi.org/10.1007/978-1-0716-4164-4_14,

fluorescent reporters in the kidney. Initial studies used GCaMP2 and GCaMP3 [9, 10], but these were somewhat limited by the low sensitivity of the probes. The subsequent development of more sensitive reporters such as GCaMP6s has been a game changer [11]. Crucially, these can be excited at lower laser powers, enabling repetitive imaging of signals over time without inducing phototoxicity (Fig. 1a–c). Moreover, the enhanced signal resolution allows detailed visualization of intracellular patterns of calcium signals [12]. Current limitations include depth of imaging, which is restricted to the outer cortex. However, medullary structures can be visualized using the tissue slice technique [13].

Once data are acquired, choosing the most appropriate analysis approach is essential. At the most simplistic level, regions of interest (ROIs) can be drawn around single cells or tubules to derive signal changes over time within the enclosed structures (Fig. 1d). Limitation of tissue movements due to breathing, circulation, and temperature changes is critical to obtaining usable traces. Image stabilization approaches can correct for any remaining movement artifacts [14]. Techniques such as temporal color coding can be a very effective method to visually depict differences in calcium activity between different cell types (Fig. 1d).

Quantitative analysis of calcium imaging can be partially or fully automated. Semiautomated analysis does not require advanced programming skills, but combines the utilization of widely used software, such as Fiji/ImageJ, together with basic knowledge of programming languages, like Python. Fully automated analysis is complex to perform and necessitates expertise for using machine learning, but once established can be applied to analyze large data sets in a high-throughput manner. For example, in a previous study, we used a calcium-independent marker (the nuclear dye Hoechst) and supervised machine learning to segment >600 cells per field of view and extract calcium traces in an unbiased manner [12]. By projecting data back onto the structure of the kidney, we could construct large-scale maps of dynamic calcium activity and discovered major differences between nephron segments.

This chapter will describe how to generate transgenic mice expressing fluorescent calcium sensors and how to use these to perform detailed imaging of calcium signals in kidney cells using intravital multiphoton microscopy.

2 Materials

2.1 Generation of Transgenic Mice Expressing a Calcium-Sensitive Probe

GCaMP6s-expressing mice were generated by breeding Ai96 (RCL-GCaMP6s) (JAX stock no. 024106) with CMV-Cre mice (JAX stock no. 006054).

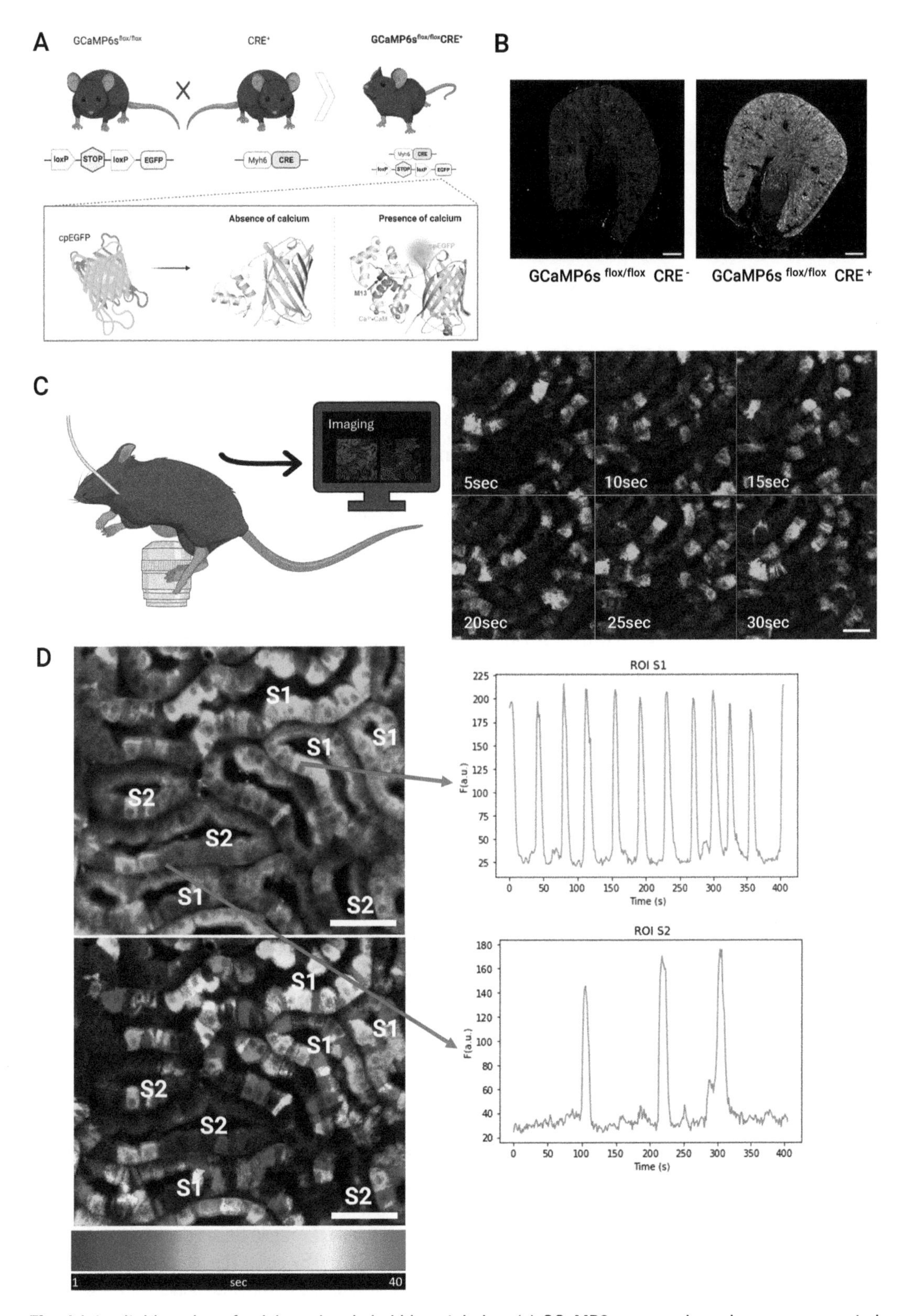

Fig. 1 Intravital imaging of calcium signals in kidney tubules. (**a**) GCaMP6s-expressing mice were generated by breeding Ai96(RCL-GCaMP6s) (JAX stock no. 024106) with CMV-Cre mice (JAX stock no. 006054). (**b**)

2.2 Agarose

Prepare 4.5% Agarose in 0.9% NaCl solution: dissolve 0.9 g of NaCl in 100 mL of water. When NaCl is dissolved, add 4.5 g of agarose. Heat up and stir thoroughly until agarose is boiled and dissolved. Store in aliquots at −20 °C (*see* **Note 1**).

2.3 Analgesia

Buprenorphine 0.1 mg/kg subcutaneously 30 min before anesthesia.

2.4 Anesthesia

Isoflurane connected to oxygen bottle source—VetFlo. A cocktail containing ketamine (0.065 g/kg) and xylazine (0.01 g/kg) given intraperitoneally can be used instead of isoflurane (*see* **Note 2**).

2.5 Surgery

1. Surgical tools (forceps, clamps, fine scissors, spring scissors), needle, and thread.
2. Tissue glue—Histoacryl Flexible.
3. Polyethylene tubing (ID 0.28 mm) for cannulating the jugular vein
4. Feedback loop warming system—Complete Homeothermic Blanket System with Flexible Probe
5. Pump 11 Elite—Standard Infuse/Withdraw Programmable Syringe Pumps.
6. Multiphoton microscope with 25× water-immersion inverted objective and capability of emitting light at 950 nm and collecting signal at 475–575 nm.

3 Methods

Cannulation of the jugular vein allows injection of dyes and reagents, and externalization of the kidney enables intravital imaging. Monitoring of vital signs during experiments is critical to ensure adequate depth of anesthesia and to preserve normal kidney function. These include respiration rate, absence of withdrawal reflexes after leg or tail pinch, and body temperature. To keep the latter stable, the animal needs to be on a temperature-controlled stage during the experiment.

Fig. 1 (continued) Immunofluorescence staining of whole kidney sections verifying the expression of GCaMP6s. Scale bar = 500 μm. (**c**) Schematic of intravital multiphoton kidney imaging in an anesthetized mouse. An image sequence is depicted showing spontaneous calcium changes in tubular epithelial cells. GCaMP6s was excited at 950 nm and signal collected at 475–575 nm. Scale bar = 50 μm. (**d**) Sum image of calcium signals in early (S1) and late (S2) regions of the proximal tubule. Example traces depict fluorescence signal changes in regions of interest (ROIs) drawn around individual S1 and S2 cells. The temporal color-coded image underneath denotes timing of appearance of calcium signals in tubules over time. Scale bars = 50 μm. Schematic images were created using BioRender

3.1 Jugular Vein Cannulation

1. Besides analgesia, animals need to be deeply anesthetized with either isoflurane or ketamine/xylazine before starting surgery. Therefore, a withdrawal reflex should be checked before cutting the skin. The right jugular vein needs to be dissected and free from fat and connective tissue, cut, and cannulated with a polyethylene catheter. After the vein has been cannulated, make sure it is securely sutured with thread and cover it with histoacryl-flexible adhesive to stabilize the catheter.

3.2 Externalization of the Kidney

1. A small abdominal incision on the flank is used to expose the left kidney (*see* **Note 3**). In order to provide the kidney with support and steady conditions for image acquisition, after externalization and resting on the microscopic stage cover slide, heated agarose needs to be applied.

3.3 Imaging

1. To identify the region of interest excite at 850 nm to visualize the capsule and gather emission light in the blue and green channels (330–550 nm), then zoom in to the underlying tubules. Different tubular segments can be identified by their morphology and autofluorescence patterns [12].
2. For imaging of calcium activity with GCaMP6s, excite at 950 nm and collect at 475–575 nm. Use the lowest laser power possible to prevent phototoxicity (*see* **Note 4**).
3. To label intracellular structures, inject suitable dyes intravenously. For example, TMRM (0.4 mg/kg body weight) can be injected to label mitochondria in tubules.
4. Quantitative image analysis:
   (a) If lateral movement during imaging has occurred, the time series can be stabilized using the MultiStackReg (v.1.46.2) plug-in for FIJI [15] or NoRMCorre rigid transformation [14].
   (b) Start the analysis of acquired videos by summing the calcium activity. In addition, the temporal evolution of calcium events can be displayed by using color-coded macros. This will indicate a potential difference in calcium activity between cells.
   (c) For more detailed analysis of calcium events in specific cells or tubules, draw ROIs around structures as appropriate.
   (d) Data acquired from ROIs can be further analyzed and presented as mean fluorescence, or, if necessary, amplitude, duration, and frequency of events can be calculated (*see* **Note 5**).

4 Notes

1. It is recommended to let agarose dissolve for about 1 h. It is crucial to check that dissolved agarose is not too hot during kidney stabilization, as it can result in collapsed tubules.
2. Avoid anesthesia overdose since it can lead to a blood pressure drop, a decrease in renal blood flow, and collapsed tubules.
3. Before the kidney is externalized, make a suture around the abdominal incision and close it after externalization of the kidney. Pay attention that the suture is not too tight, as it can cause ischemia.
4. Using higher laser powers during long imaging might induce calcium rises in tubules due to phototoxicity.
5. To analyze larger data sets, it is recommended to use programming languages to speed up the process.

Acknowledgments

This work was supported by the Swiss National Centre for Competence in Research (NCCR) Kidney Control of Homeostasis and by a project grant from the Swiss National Science Foundation (310030_184688). The authors also acknowledge support from the Zurich Centre for Microscopy and Image Analysis and the Zurich Centre for Integrative Human Physiology.

References

1. Praetorius HA, Leipziger J (2010) Intrarenal purinergic signaling in the control of renal tubular transport. Annu Rev Physiol 72:377–393. https://doi.org/10.1146/annurev-physiol-021909-135825
2. Du Z, Weinbaum S, Weinstein AM, Wang T (2015) Regulation of glomerulotubular balance. III. Implication of cytosolic calcium in flow-dependent proximal tubule transport. Am J Physiol Ren Physiol 308(8):F839–F847. https://doi.org/10.1152/ajprenal.00601.2014
3. Raghavan V, Rbaibi Y, Pastor-Soler NM, Carattino MD, Weisz OA (2014) Shear stress-dependent regulation of apical endocytosis in renal proximal tubule cells mediated by primary cilia. Proc Natl Acad Sci USA 111(23):8506–8511. https://doi.org/10.1073/pnas.1402195111
4. Siroky BJ et al (2006) Loss of primary cilia results in deregulated and unabated apical calcium entry in ARPKD collecting duct cells. Am J Physiol Ren Physiol 290(6):F1320–F1328. https://doi.org/10.1152/ajprenal.00463.2005
5. Ilatovskaya DV, Staruschenko A (2015) TRPC6 channel as an emerging determinant of the podocyte injury susceptibility in kidney diseases. Am J Physiol Ren Physiol 309(5):F393–F397. https://doi.org/10.1152/ajprenal.00186.2015
6. Martins JR, Haenni D, Bugarski M, Polesel M, Schuh C, Hall AM (2021) Intravital kidney microscopy: entering a new era. Kidney Int 100(3):527–535., Elsevier B.V. https://doi.org/10.1016/j.kint.2021.02.042
7. Jobsis PD, Rotstein EC, Balaban RS (2007) Limited utility of acetoxymethyl (AM)-based intracellular delivery systems, in vivo: interference by extracellular esterases. J Microsc 226(1):74–81. https://doi.org/10.1111/j.1365-2818.2007.01755.x

8. Peppiatt-Wildman CM, Crawford C, Hall AM (2012) Fluorescence imaging of intracellular calcium signals in intact kidney tissue. Nephron Exp Nephrol 121(1–2):e49–e58. https://doi.org/10.1159/000342812
9. Burford JL et al (2014) Intravital imaging of podocyte calcium in glomerular injury and disease. J Clin Invest 124(5):2050–2058. https://doi.org/10.1172/JCI71702
10. Szebényi K et al (2015) Visualization of calcium dynamics in kidney proximal tubules. J Am Soc Nephrol 26:11. [Online]. Available: https://journals.lww.com/jasn/fulltext/2015/11000/visualization_of_calcium_dynamics_in_kidney.18.aspx
11. Chen T-W et al (2013) Ultrasensitive fluorescent proteins for imaging neuronal activity. Nature 499(7458):295–300. https://doi.org/10.1038/nature12354
12. Martins JR, Haenni D, Bugarski M, Figurek A, Hall AM (2020) Quantitative intravital Ca2+ imaging maps single cell behavior to kidney tubular structure. Am J Physiol Ren Physiol 319(2):F245–F255. https://doi.org/10.1152/ajprenal.00052.2020
13. Ghazi S et al (2020) Multiparametric imaging reveals that mitochondria-rich intercalated cells in the kidney collecting duct have a very high glycolytic capacity. FASEB J 34(6):8510–8525. https://doi.org/10.1096/fj.202000273R
14. Pnevmatikakis EA, Giovannucci A (2017) NoRMCorre: an online algorithm for piecewise rigid motion correction of calcium imaging data. J Neurosci Methods 291:83–94. https://doi.org/10.1016/j.jneumeth.2017.07.031
15. Thevenaz P, Ruttimann UE, Unser M (1998) A pyramid approach to subpixel registration based on intensity. IEEE Trans Image Process 7(1):27–41. https://doi.org/10.1109/83.650848

Chapter 15

Methods for Imaging Intracellular Calcium Signals in the Mouse Mammary Epithelium in Two and Three Dimensions

Mathilde Folacci, Silke B. Chalmers, and Felicity M. Davis

Abstract

The mammary gland has a central role in optimal mammalian development and survival. Contractions of smooth muscle-like basal (or myoepithelial) cells in the functionally mature mammary gland in response to oxytocin are essential for milk ejection and are tightly regulated by intracellular calcium (Ca^{2+}). Using mice expressing a genetically encoded Ca^{2+} indicator (GCaMP6f), we present in this chapter a method to visualize at high spatiotemporal resolution changes in intracellular Ca^{2+} in mammary epithelial cells, both in vitro (2D) and ex vivo (3D). The procedure to optimally prepare mammary tissue and primary cells is presented in detail.

Key words Calcium signaling, Calcium imaging, Genetically encoded $ca2^{+}$ indicator, Gcamp6f, Mammary gland, Mammary epithelial cells, Lactation, Oxytocin

1 Introduction

Essential and eponymous, the mammary gland fosters optimal development and survival at the most vulnerable stage of mammalian life [1–3]. In mice, the postpubertal mammary gland consists of a network of branching ducts that are situated within a rich, vascularized stromal matrix containing fibroblasts, adipocytes, and immune cells [4, 5]. During gestation and early lactation, the mammary epithelium undergoes alveologenesis, a process that involves the rapid expansion and reorganization of the epithelium, enabling it to synthesize, store, and secrete milk [6]. Alveolar structures created during this process consist of an inner layer of milk synthesizing luminal cells encaged by star-shaped, contractile basal cells [6, 7]. Upon weaning, the gland undergoes a highly

The authors Mathilde Folacci and Silke B. Chalmers have equally contributed to this chapter.

Caroline M. Gorvin (ed.), *Calcium Signaling: Methods and Protocols*, Methods in Molecular Biology, vol. 2861, https://doi.org/10.1007/978-1-0716-4164-4_15,

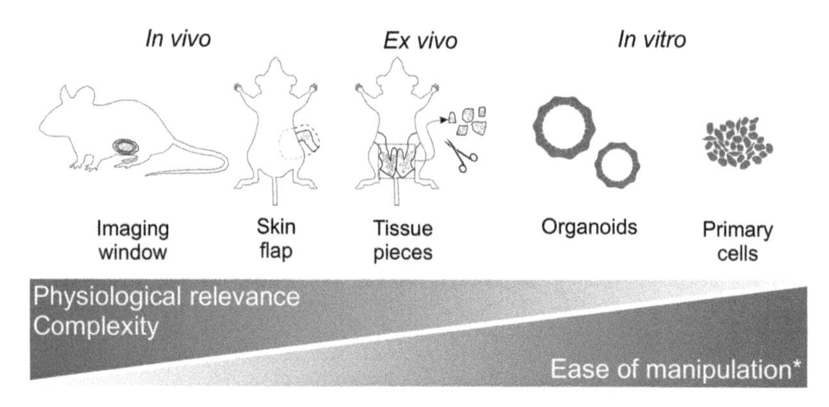

Fig. 1 Methods to image the mouse mammary gland range from relatively simple 2D cell preparations to serial in vivo manipulations. Complexity (time, expertise, equipment, and analysis pipelines) tends to increase with 3D/4D and in vivo methods; however, these methods offer the advantage of improved physiological relevance. Asterisk signifies that not all methods are currently available or readily reproducible for all stages of mammary gland development, e.g., while it is routine to isolate healthy primary cells from nonpregnant mammary tissue, it is more difficult to isolate healthy cells from lactating mouse mammary tissue. Similarly, methods for implanting imaging windows before or during gestation remain a major challenge for intravital imaging during alveologenesis and lactation

coordinated phase of regression—known as post-lactational involution—enabling it to return to a near pre-pregnant state, ready for future recurrent cycles of expansion and milk production with subsequent pregnancies [8].

The role and function of the mammary gland is inextricably linked to Ca^{2+} [9, 10]. Historically considered simply a nutritional constituent of milk, Ca^{2+} is now understood to be a key regulator of various mammary functions through its role as a ubiquitous second messenger [9, 11–14]. Ca^{2+} influx via ORAI1 is required for oxytocin-induced contractions of basal cells [15, 16]. Moreover, the onset of involution involves a rapid increase in luminal cytosolic Ca^{2+} concentrations, likely due to the early downregulation of apically expressed plasma membrane Ca^{2+} ATPase (PMCA)2 [17].

Recent improvements in our understanding of Ca^{2+}-mediated control of mammary biology largely come from advances in live imaging and Ca^{2+} signaling fields. Improved methods for visualizing the mammary epithelium have enabled researchers to explore mammary morphogenesis and function, from in vitro to in vivo scales (Fig. 1) [15, 16, 18–22]. Concomitantly, transgenic mouse models enabling targeted expression of fast and sensitive genetically encoded Ca^{2+} indicators (GECIs) have become widely available [4, 16, 23]. Together, these advances enable researchers to probe mammary Ca^{2+} signaling with a hitherto unknown clarity. Herein, we describe methods routinely used within our laboratory to visualize mammary Ca^{2+} signaling across 2D and 3D scales while offering insights and suggestions regarding choice of experimental technique and mouse model.

1.1 Model Considerations

Model selection, including choice of GECI and the method of cell type-specific targeting, influences suitability for flow-on applications. Of the range of GECIs that have been expressed in mouse models, only GCaMP6f has been assessed in mammary epithelia to date [16, 18, 23, 24]. GCaMP6f has the advantages of high sensitivity and fast kinetics; however, as a green, non-ratiometric sensor that exhibits low fluorescence in a Ca^{2+}-unbound state, it poses challenges for imaging Ca^{2+} signals in thick, contractile, and autofluorescent mammary tissue [23, 25]. Previous studies have addressed some of these issues by employing more complex breeding strategies to generate mice expressing both GCaMP6f and a bright, Ca^{2+}-insensitive red fluorescent protein (TdTomato) for ratioing [16, 24]. Recently, mouse models with dual GCaMP6f-TdTomato transgene insertion (known as "Salsa6f") have become available through worldwide distributors, ameliorating the time and cost required to obtain both GCaMP6f and TdTomato on the Rosa26 locus purely by breeding [24]. In any case, fluctuations in intracellular Ca^{2+} in GCaMP6f-expressing models should be interpreted as response-over-baseline (i.e., f/f_0 or $\Delta f/f_0$).

The Cre-driver used to direct expression of the GECI to the mammary cell type of interest is another important consideration in the experimental design [6, 15, 26]. Commonly utilized mammary Cre models include K5-CreERT2, K8-CreERT2, whey acidic protein (WAP) Cre, and beta-lactoglobulin (BLG or LGB) Cre [4, 26–28]. However, these models are not without pitfalls. Use of tamoxifen in creERT2 models can affect gland development and may also increase the likelihood of breeding complications (sterility, dystocia, cannibalization) in breeding studies. Moreover, WAP-Cre and BLG-Cre models require a cycle of pregnancy and lactation for optimal transgene expression, making them time intensive and unsuitable for assessment of Ca^{2+} signaling in the nulliparous state [4, 29]. Whether cell type-specific expression is required, however, may depend on the downstream application. The method of primary cell isolation discussed in this paper produces a mixed population of cells; thus, nonspecific expression of a GECI would make it difficult to discern cell lineages during live Ca^{2+} imaging experiments [18]. Conversely, the differing morphologies of luminal and basal cells in gestation and lactation may make in vivo or ex vivo imaging of tissue with nonspecific GECI expression viable. Interpretation and analysis of results from these experiments, however, could be challenging [30–32].

Of the methodical approaches depicted in Fig. 1, all but serial intravital imaging have been employed for mammary Ca^{2+} signaling studies [16, 18]. Each method has its advantages and difficulties, but perhaps more notably, they are also distinguished by which developmental stage they are best suited to (Fig. 1). This paper will focus on the two most commonly used methods within our laboratory: ex vivo imaging of live, thick tissue during gestation or lactation and in vitro imaging of isolated primary mammary epithelial cells from nonpregnant/non-lactating mice. These two methods provide scope for imaging different stages of the mature mammary gland. Of note, many published mammary dissection protocols describe methods that ensure optimal dissection of the abdominal (fourth) mammary glands, but not the inguinal (fifth) glands [33–35]. In our experience, the pair of inguinal mammary glands provide valuable tissue for both ex vivo imaging and primary cell isolation due to the high ratio of mammary epithelial cells in comparison with surrounding stroma. We therefore refer readers to our previous technical paper for a detailed protocol of mammary gland dissection [4].

1.2 Advantages and Disadvantages of Ex Vivo Mammary Imaging

Tissues isolated during gestation and in lactation are well suited to ex vivo imaging (Fig. 2). At late gestation, mammary alveoli have not yet undergone secretory activation; their smaller diameter means that entire lobuloalveolar units and the subtending duct can be visualized in a single field of view with a 20–30× confocal objective [16]. Lactating tissue, in comparison, contains larger, more mature structures [16]. In nonpregnant mice, most of the ductal epithelium is buried within an adipocyte-rich stroma, making it difficult to visualize mammary ducts and end buds using conventional confocal microscopes.

One major limitation of ex vivo imaging is the limited diffusion of pharmacological agents through thick tissue to the cell type of interest. As will be described below, this issue can be, in part, circumvented through delicate dissection of mammary tissue into small pieces, and selection of imaging regions near the edge of the tissue piece. Where relevant, in vitro or in vivo applications may overcome this challenge and could be considered.

1.3 Advantages and Disadvantages of In Vitro Imaging of Primary Mammary Epithelial Cells

Isolated primary mammary epithelial cells are a valuable resource for visualizing Ca^{2+} signaling in luminal and basal cell populations (Fig. 3) [18]. In our experience, isolation from nonpregnant/non-lactating mammary tissue provides the highest yield of viable cells, in comparison with tissue isolated during gestation or lactation. Due to the propensity of primary cells to dedifferentiate when cultured on stiff surfaces, we consider isolated mammary cells most suitable for Ca^{2+} imaging up to 24 h post-dissection [18, 36, 37].

The cell isolation method described in this chapter utilizes the proprietary gentleMACS™ Dissociator. The dissociation step can, however, be performed using other published protocols [38–40]. While outside the scope of this chapter, following isolation, primary epithelial cells can additionally be utilized for the development of mammary organoids, which our group has shown are also suitable for Ca^{2+} imaging studies. For further information on this, we recommend previous descriptions of this process [18, 35].

Features of *ex vivo* mammary gland calcium imaging

- Time required for tissue extraction: ~30 min
- Duration for optimal imaging: 4 to 5 h
- Offers relatively intact microenvironment (tissue architecture)
- Enables 4D (*x*, *y*, *z*, *t*) assessment of intracellular Ca^{2+} responses

Lactation ***Late gestation***

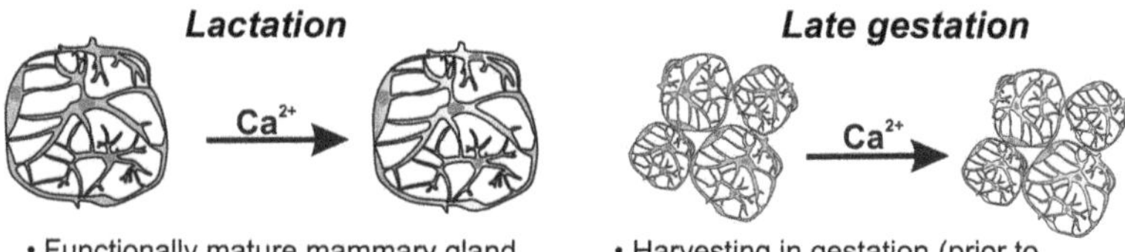

- Functionally mature mammary gland
- Enables visualisation of contraction leading to alveolar milk expulsion

- Harvesting in gestation (prior to secretory activation) allows more alveoli be observed in the imaging field of view

Method overview

Generate cell type specific GECI model — Harvest mammary tissue — Prepare tissue for live imaging — Perform imaging

Anticipated calcium imaging results during lactation

00:00 | 01:59 | 02:12 | 02:30 | 02:48
Oxytocin addition
05:49 | 07:18 | 10:23 | 15:15 | 21:05
50 μm
CellTracker™ Lo Hi $[Ca^{2+}]_i$

Fig. 2 Overview of ex vivo Ca^{2+} imaging of basal cells in the mammary gland. Figure shows features of ex vivo imaging during lactation and late gestation; schematic representation of the method; and anticipated results (lactation). For the live tissue imaging, time is indicated at the top left (min:sec). Luminal cells are stained with CellTracker (magenta), and changes in cytosolic Ca^{2+} in basal cells expressing GCaMP6f are shown in blue/green/yellow intensity LUT

2 Materials

Use ultrapure deionized water for all solutions. Storage temperatures are at room temperature unless otherwise specified. Safety measures, ethical approvals, and waste disposal should be followed according to local and national regulations. Materials related to the mouse mating procedure and mammary gland harvest are described in Stewart and Davis [4].

Features of *in vitro* mammary gland calcium imaging

- Time required for cell isolation: ~9 h
- Duration for optimal imaging: ~24 h
- Highest yield of viable cells obtained from nulliparous tissue
- Ca^{2+} responses generally representative of *ex vivo* 4D responses
- Well suited for pharmacological studies

Method overview

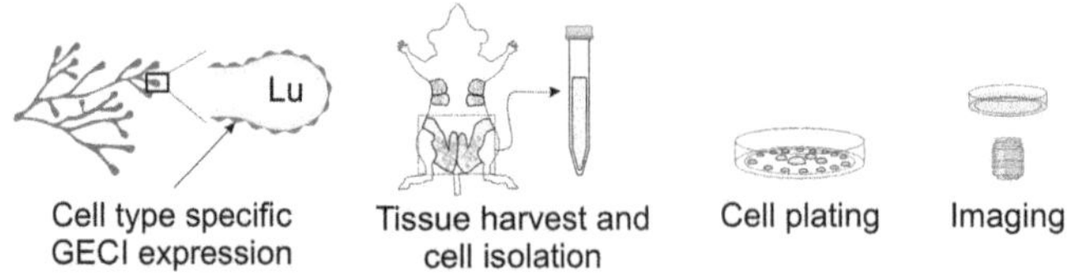

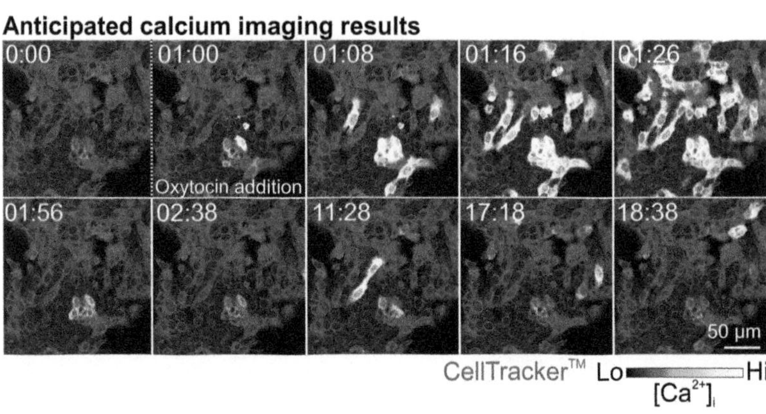

Fig. 3 Overview of in vitro Ca^{2+} imaging of primary mammary epithelial cells. Isolation of mammary epithelial cells from nulliparous tissue produces monolayers of cells that remain viable for Ca^{2+} imaging studies up to 24 h post-dissection. The Ca^{2+} signaling responses of these cells are generally representative of responses in ex vivo tissue. For the live cell imaging experiment, time is indicated at the top left (min:sec). Cells have been stained with CellTracker (magenta), and changes in cytosolic Ca^{2+} in basal cells expressing GCaMP6f are shown in blue/green/yellow intensity LUT. Lu: lumen

2.1 Mammary Gland Ex Vivo Ca^{2+} Imaging

1. Transgenic mouse expressing a GECI in mammary tissue (*see* **Note 1**).
2. Sterile disposable 0.2 μm filtration unit (500 mL) and vacuum/vacuum pump.
3. Magnetic stirrer and sterile laboratory glassware.
4. 70% ethanol.
5. Humidified cell culture incubator (37 °C, 5% CO_2).
6. Biosafety cabinet.
7. Disposable scalpel blades (#20).
8. 140 mm petri dish.
9. 50 mL centrifuge tube.
10. Sterile live tissue media: Dulbecco's modified Eagle medium/nutrient mixture F-12 (DMEM/F-12) with HEPES and

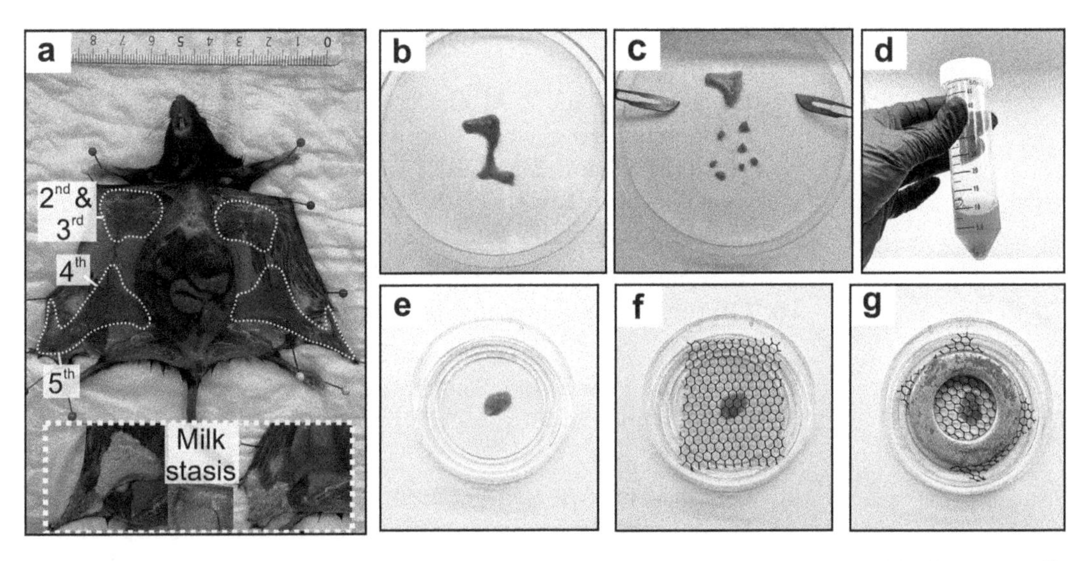

Fig. 4 Mammary tissue preparation for ex vivo imaging. (**a**) Mouse thoracic (second and third, or "upper"), abdominal (fourth), and inguinal (fifth) mammary glands during lactation. Inset (yellow dotted box) shows mammary glands presenting with milk stasis that are thus generally considered not suitable for live imaging. (**b**) Fourth and fifth mammary glands, (**c**) cut into pieces and (**d**) incubated in live imaging media. (**e**) Preparation of a mammary gland piece for live imaging using (**f**) a net and (**g**) a washer to minimize tissue displacement upon liquid addition

L-glutamine supplemented with 10%, fetal bovine serum (FBS). Aliquot and store at 4 °C until use.

11. Optical grade (0.17 mm, #1.5) glass bottom culture dishes.
12. Net and stainless-steel washer as shown in Fig. 4g, or equivalent.
13. *Optional*, CellTracker Red CMTPX (#C34552 ThermoFisher) (*see* **Note 2**). For the 6 mM stock solution, reconstitute in dimethyl sulfoxide (DMSO). Store at −20 °C. Avoid excessive freeze-thawing. For the 1.5 μM working solution, dilute the 6 mM stock in sterile live tissue media (Subheading 2.1, **item 10**), and mix well by vortexing. Prepare working solution on the day of the experiment and warm to 37 °C prior to use.
14. 2 M $CaCl_2$ solution: Weigh $CaCl_2$ and dissolve in H_2O. Store at 4 °C.
15. 1 M NaH_2PO_4 solution: Weigh NaH_2PO_4 and dissolve in H_2O. Store at 4 °C.
16. 10× HEPES stock solution: 2.199 g KCl, 7 mL of 1 M $MgCl_2$ solution, 11.915 g HEPES, 6 mL of 1 M NaH_2PO_4 solution, 2.1003 g $NaHCO_3$. Weigh and add dry components to a clean glass bottle. Dissolve in H_2O. Add $MgCl_2$ and NaH_2PO_4 solutions. Adjust the volume to 500 mL with H_2O. Pass through a 0.2 μm filtration unit. Store at 4 °C. Maintain sterility and discard after 6 months.

17. Physiological salt solution with 1.8 mM Ca^{2+} (PSS-Ca^{2+}): 50 mL 10× HEPES stock, 4.0908 g NaCl, 1.0359 g D-(+)-Glucose, 450 μL 2 M $CaCl_2$ solution. Weigh and add dry components to a clean glass bottle. Dissolve in H_2O. Add 10× HEPES and 2 M $CaCl_2$ solutions. Adjust the volume to approx. 480 mL with H_2O. Adjust the pH to 7.2–7.3 using NaOH. Make up the final volume to 500 mL with H_2O. Store at 4 °C and discard after 1 month.
18. Oxytocin stock solution: 1 mM, reconstituted in H_2O. Aliquot and store at −20 °C.
19. Oxytocin 2× solution: 150 nM, diluted in PSS-Ca^{2+}. Prepare fresh on the day of the experiment. Keep on ice until ready to use.
20. Confocal microscope, preferentially with a resonant scanner. We typically use a resonant scanning Olympus FV3000 confocal microscope or a resonant scanning Leica STELLARIS 8 confocal microscope.
21. ImageJ software (v1.52e, National Institutes of Health).

2.2 Mammary Cell Isolation and In Vitro Ca^{2+} Imaging

1. Female mice (aged 8–34 weeks) expressing GECI of choice in cell type of interest (*see* **Notes 3** and **4**).
2. Red blood cell lysis buffer (Sigma-Aldrich).
3. Dissociation media: Leibovitz's L-15 medium supplemented with 1× penicillin/streptomycin (100 units/mL penicillin, 100 ug/mL streptomycin) and 10% FBS.
4. Predigestion mix: Leibovitz's L-15 medium supplemented with 1× penicillin/streptomycin.
5. Digestion mix: Predigestion mix supplemented with ≥2250 units/mL of trypsin (from bovine pancreas) and 3 mg/mL Collagenase A (from *Clostridium histolyticum*). Prepare fresh the day of dissociation, filter-sterilize and store on ice until use. We recommend max. Two pairs of fourth and fifth mammary glands (from two mice) per 10 mL of digestion mix.
6. Fibroblast clearing media: DMEM high glucose (without L-glutamine or HEPES) supplemented with 1× GlutaMax, 1× penicillin/streptomycin and 10% FCS.
7. Mammary cell culture media: DMEM/F12 with HEPES and L-glutamine supplemented with 1× penicillin/streptomycin and 10% FBS.
8. Dissection tools: sharp surgical scissors and fine forceps.
9. Ice.
10. 70% ethanol.
11. gentleMACS™ Dissociator.

12. gentleMACS™ C Tubes.
13. 15 mL centrifuge tube.
14. 75 cm^2 tissue culture flask.
15. Optical grade (0.17 mm, #1.5) glass bottom sterile cell culture dishes.
16. Physiological salt solution with 1.8 mM Ca^{2+} (PSS-Ca^{2+}; see Subheading 2.1, **item 17**).
17. *Optional*, CellTracker Red CMTPX (#C34552 ThermoFisher) (*see* **Note 2**). For the stock solution, reconstitute to 6 mM in DMSO. Store at −20 °C. Avoid excessive freeze-thawing. For the 1.5 μM working solution, dilute 6 mM stock in mammary culture media (Subheading 2.2, **item 7**), and mix well by vortexing. Prepare working solution on the day of the experiment.
18. Epifluorescence or confocal microscope. We typically use either an Olympus FV3000 confocal microscope or a Leica STELLARIS 8 confocal microscope. However, confocal capabilities are not a requirement for imaging in vitro 2D samples.
19. *Optional*, pharmacological agents to stimulate Ca^{2+} events in cells, for example, 2× oxytocin solution (see Subheading 2.1, **item 19**).
20. Humidified cell culture incubator (37 °C, 5% CO_2).
21. Biosafety cabinet.

3 Methods

3.1 Mammary Gland Ex Vivo Ca^{2+} Imaging

When possible, prepare the microscope prior to euthanizing the mouse to maximize imaging time with viable tissue:

3.1.1 Mammary Tissue Preparation for Ex Vivo Imaging

1. Euthanize animals according to local regulations. We employ different methods of euthanasia for embryos, neonates, and pregnant or lactating mice (*see* **Note 5**). Start a laboratory timer to monitor time-post-cull (an important consideration for tissue viability).
2. Dissect the mammary glands as illustrated in Fig. 4a and described in detail in Stewart and Davis [4] (*see* **Note 6**).
3. Place the mammary gland in a large petri dish (Fig. 4b), and cut the gland into 3–5 mm^3 pieces without delay (Fig. 4c; *see* **Notes 7** and **8**).
4. Place approx. 15–20 tissue pieces in a 50 mL centrifuge tube containing 10–15 mL of pre-warmed sterile live tissue media (Subheading 2.1, **item 10**), as shown in Fig. 4d.

   Optional: incubate some tissue pieces in CellTracker Red working solution (Subheading 2.1, **item 13**) for 20 min (*see*

Note 9). Loosen the lid of the centrifuge tube to allow for some gas exchange.

5. Store tissue pieces in the humidified incubator at 37 °C and 5% CO_2. Tissue should ideally be used within 4–5 h post-dissection.
6. Place a mammary tissue piece in the center of an imaging dish (Fig. 4e), and anchor it in place using a piece of net and a washer as shown in Fig. 4f, g.
7. Add 200 μL of PSS-Ca^{2+} (Subheading 2.1, **item 17**) to the tissue piece, and immediately proceed to the confocal microscope (*see* **Notes 10** and **11**).

3.1.2 Ex Vivo Imaging of Mammary Tissue

Data presented in this chapter have been obtained using either a resonant scanning Olympus FV3000 confocal microscope or a resonant scanning Leica STELLARIS 8 confocal microscope. Imaging settings will depend on the confocal system available in your laboratory, and thus we provide only very general guidelines in this section:

1. Prepare the imaging system. Imaging settings routinely used by our laboratory include:
   (a) Use of long working distance water or silicone immersion objectives (25–30×).
   (b) Imaging GCaMP6f and CellTracker (if used) on the same track to improve acquisition speed.
   (c) Use of a resonant scanner and bidirectional scanning to improve acquisition speed.
2. Focus on the sample and identify a region of interest (ROI) for imaging (*see* **Note 12**).
3. Define the beginning and end of the volume/stack. Typically, we acquire 30–40 μm volumes (approx. 7–10× 3–5 μm optical slices) in approximately 3.5 s.
4. Image for 2–5 min prior to agonist addition to establish a baseline.
5. Add 200 μL of 2× oxytocin (Subheading 2.1, **item 19**) to the dish containing 200 μL of PSS-Ca^{2+} and continue imaging for 20–25 min.
6. After data acquisition is complete, images can be visualized using the Bio-Formats plug-in (ImageJ) [41, 42].

Expected time to complete procedure: 8–10 h, of which 30 min (**step 3**), 1 h 15 min (**step 10**) and 4–6 h (**step 15**) are incubation times:

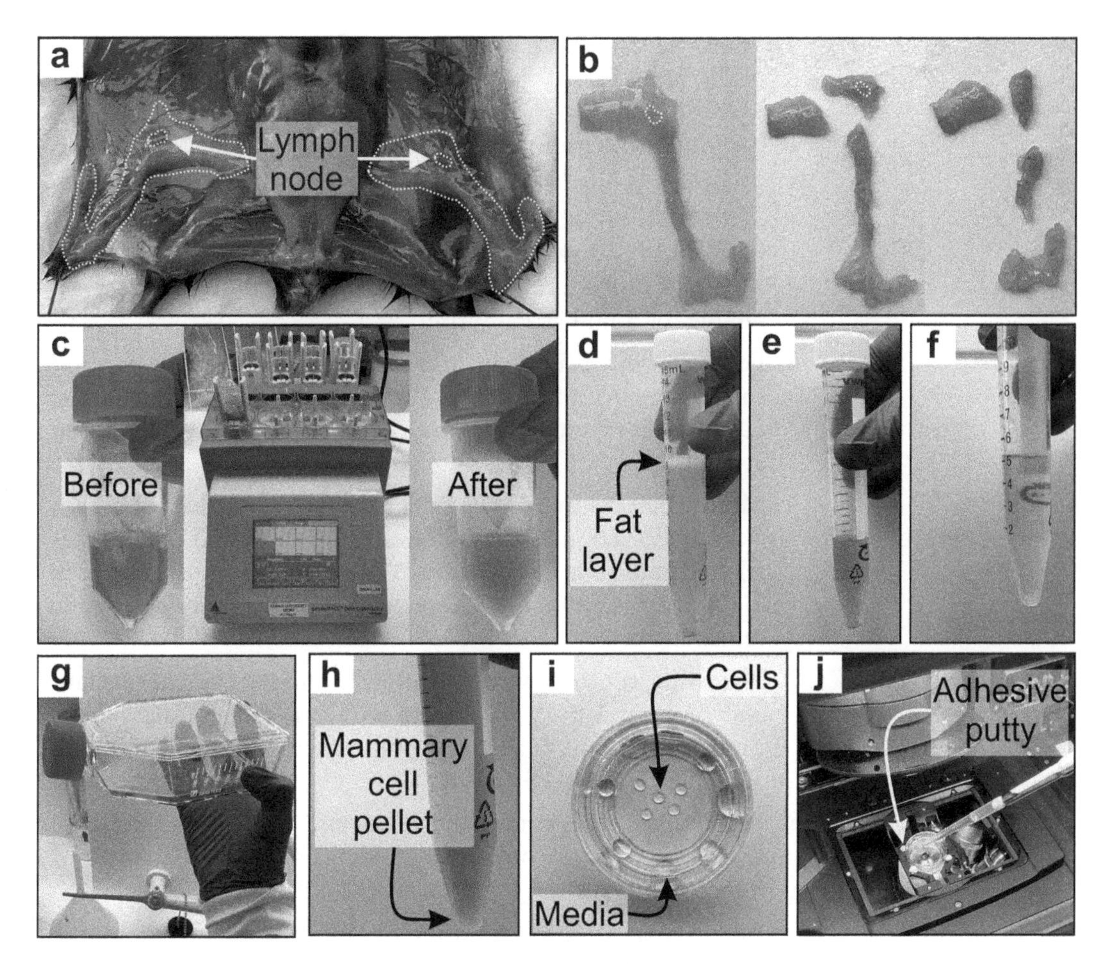

Fig. 5 Mammary tissue preparation for in vitro imaging. (**a**) Nulliparous mouse abdominal (fourth), and inguinal (fifth) mammary glands (white-dashed outline) with lymph nodes identified by yellow arrow and outline. (**b**) Dissected mammary glands should have lymph nodes identified (yellow-dashed line, panel 1) and removed (panel 2) prior to dicing roughly (panel 3). (**c**) Diced mammary glands are dissociated via gentleMACS™ Dissociator. (**d**) Dissociated mammary tissue post-centrifugation, prior to removal of fat layer and supernatant, and (**e**) after removal. (**f**) Following incubation in red blood cell lysis buffer, the cell suspension becomes colorless. (**g**) Fibroblasts are removed via short incubation in a cell culture flask. (**h**) Cell pellet at the end of isolation protocol should be clearly visible at the bottom of a 15 mL centrifuge tube. (**i**) Following isolation, mammary cells are counted and plated in imaging dishes via the droplet approach. (**j**) An example of liquid addition onto an imaging plate mounted on a confocal microscope. Adhesive putty placed gently around the imaging plate can minimize movement artifacts due to liquid addition

3.2 Primary Mammary Cell Isolation and In Vitro Ca^{2+} Imaging

3.2.1 Mammary Epithelial Cell Isolation

1. Euthanize animals according to local regulations. Dissect abdominal and inguinal mammary glands (Fig. 5a; *see* **Note 13**). Briefly dip tissue in 70% ethanol to help maintain sterility (particularly if dissecting on the bench).
2. Carefully remove the lymph nodes and roughly dice mammary tissue (Fig. 5b).

3. Transfer tissue to a gentleMACS™ C Tube containing 10 mL of digestion mix (Subheading 2.2, **item 5**). Place the gentleMACS™ C Tube in the gentleMACS™ Dissociator and run the pre-set program 37C_m_LDK_1 (Fig. 5c; *see* **Note 14**).
4. From this point forward, all work is completed in a biosafety cabinet using aseptic technique. At the completion of the gentleMACS™ program, transfer the contents of the gentleMACS™ C Tube to a 15 mL centrifuge tube. Centrifuge the sample for 5 min at 1500 rpm.
5. Following centrifugation, transfer the fat layer (Fig. 5d) and supernatant to a fresh 15 mL tube (tube 2). Resuspend the contents via pipetting and centrifuge tube 2 for a further 5 min at 1500 rpm.
6. Meanwhile, resuspend the pellet from the original tube with 1 mL dissociation media (Subheading 2.2, **item 3**), and transfer to a new 15 mL tube (tube 3). Wash the original tube with 2 mL of dissociation media, and transfer to tube 3 to collect any remaining cells.
7. Following centrifugation of tube 2, remove the supernatant and resuspend the small pellet in 1 mL dissociation media. Transfer to tube 3 and centrifuge for 5 min at 1500 rpm (Fig. 5e; *see* **Note 15**).
8. Remove the supernatant and resuspend the pellet in 5 mL of red blood cell lysis buffer. Incubate at room temperature for 5 min (Fig. 5f).
9. Add 10 mL of dissociation media and centrifuge for 5 min at 1500 rpm (*see* **Note 16**).
10. Remove the supernatant and resuspend the cells thoroughly but gently in 1 mL of fibroblast clearing media (Subheading 2.2, **item 6**) with a P1000 pipette. Transfer to a 75 cm^2 tissue culture flask. Wash the 15 mL tube with 12 mL of fibroblast clearing media to collect any remaining cells and transfer to the 75 cm^2 tissue culture flask. Gently shake the flask in a horizontal plane, and incubate in a humidified incubator for 1 h and 15 min at 37 °C and 5% CO_2 (Fig. 5g).
11. After the incubation, most of the fibroblasts will attach to the tissue culture plastic while epithelial cells will not. Shake the flask in a horizontal plane with moderate vigor, and transfer media to a new 15 mL tube. Rinse the tissue culture flask with another 13 mL of dissociation media, and transfer to another 15 mL tube. Centrifuge both cell suspensions for 5 min at 1500 rpm (Fig. 5h).
12. Remove the supernatants and combine pellets by resuspending mammary cell culture media (Subheading 2.2, **item 7**).

13. Count the cell suspension and adjust solution to a concentration of 5 million cells/mL.
14. Prepare imaging plates by pipetting approx. 5 μL droplets of cell suspension in the center of the dish. To maintain humidity within the dish, fill the rim of the dish with droplets of cell-free complete media (see Fig. 5i). Plating the cells in this way maximizes the amount of imaging dishes that can be created and assists in minimizing adhesion time for the cells.
15. Incubate the plated cells at 37 °C, 5% CO_2 in a humidified incubator for 4–6 h until cells have adhered. At the end of incubation, gently wash the dish with 500 μL mammary cell culture media to remove any material that has not adhered. Replenish the dish with mammary cell culture media.
16. Isolated cells should be imaged with 24 h of dissection.

3.2.2 *In Vitro Imaging of Isolated Mammary Epithelial Cells*

Data presented here have been obtained using either an Olympus FV3000 confocal microscope or a Leica STELLARIS 8 confocal microscope; however, confocal capabilities are not a requirement for imaging in vitro 2D samples:

1. *Optional*: Load the cells with CellTracker Red by incubating the plates in CellTracker working solution (Subheading 2.2, **item 17**) for at least 30 min at 37 °C and 5% CO_2. Following incubation, replace CellTracker loading media with mammary cell culture media (Subheading 2.2, **item 7**) until ready to image.
2. Prepare the imaging system. Imaging settings routinely used by our laboratory include:
   (a) Imaging via water or silicone objectives.
   (b) Imaging GCaMP6f and CellTracker (if used) on the same track to maximize acquisition speed.
   (c) Use of a resonant scanner and bidirectional scanning to improve acquisition speed.
   (d) A small volume (*z* stack) of approximately 4.5 μm, (3× 1.5 μm optical slices) in approx. 1 s.
   (e) Adaptive Focus Control (or an equivalent *z*-drift compensator).
3. Replace the media on one imaging dish with 1 mL PSS-Ca^{2+} (*see* **Note 10**).
4. Transfer the sample to the imaging system. To reduce the likelihood of the imaging plate moving during liquid addition, a small amount of adhesive putty can be gently placed around the plate on the stage (Fig. 5j).

5. Identify a region of interest and commence imaging. After approximately 40 s, gently add 1 mL of 2× agonist solution gently onto the imaging plate (Fig. 5j).

4 Notes

1. Tamoxifen has a dose-dependent effect on mammary gland development [4]. Tamoxifen dose and method of administration should be optimized in tamoxifen-dependent models to suit the experiment. Pregnant mice that have previously been administered tamoxifen should be monitored for dystocia around the time of littering.
2. Most GECIs have low fluorescence at resting cytosolic Ca^{2+} concentrations. CellTracker Red is useful for the identification of ROIs in tissue pieces. If imaged, the red channel can also be used to visualize the nature and extent of alveolar warping upon oxytocin stimulation. Other dyes in the CellTracker family (or other live cell stains) could be considered; however, the best results in our laboratory have been obtained using CellTracker Red CMTPX. This dye preferentially stains luminal cells (see Fig S1A, Ref [16]).
3. This protocol is suitable for mice of any age; however, mice <8 weeks old (where ductal morphogenesis is incomplete) may have a lower yield of mammary epithelial cells. Samples isolated from mice >20 weeks old may have greater variability in yield and purity.
4. This protocol has been optimized for isolation of cells from nulliparous mammary tissue. In our experience, cell isolation during lactation is problematic due to epithelial cell death. A method for isolating basal cells during gestation is outlined in Reference [16] and includes use of nano-patterned dishes for optimal basal cell attachment.
5. We euthanize pregnant and lactating mice by CO_2 inhalation for 10 min or 5 min, respectively. Embryos from pregnant mice are immediately removed from the mother's uterus and decapitated using sharp scissors. Due to their immature lung development, we typically euthanize 10–12-day-old neonates by intraperitoneal injection of a lidocaine–pentobarbital mixture, followed by swift decapitation with sharp scissors.
6. For lactation studies, experiments are ideally performed on lactation days 10–12 (peak lactation) on mice nursing six or more pups. When litter size <6, some mammary glands may not be frequently or adequately suckled and may appear white, blotchy, or streaky (Fig. 4a, milk stasis insert). This can signify

partial involution (or in some cases mastitis) and the glands should generally be avoided.

7. We often favor the fifth and upper mammary glands for pharmacological studies. These glands are thinner and thus allow faster diffusion of agonists/antagonists through the tissue to the ROI. Nevertheless, it is also reasonable and appropriate to use the fourth mammary gland for ex vivo imaging.
8. To avoid ripping-, tearing-, or crushing-like damage at the edge of tissue pieces, we prefer to dice mammary tissue between two #20 scalpel blades pulled swiftly in opposite directions. In our hands, this method causes less damage than cutting with dissection scissors or by pressing the gland under the force of a single blade. Do not allow the gland to dry out during cutting.
9. Tissue can remain incubated in CellTracker Red working solution for several h.
10. Samples should not remain in PSS-Ca^{2+} for more than 10 min prior to imaging.
11. Remove as much culture medium as possible from the tissue piece before adding the PSS-Ca^{2+}.
12. CellTracker Red can help to identify mammary alveoli down the eyepiece. Ideally alveoli within the field of view should be close to the edge of the tissue piece (approx. 2–3 alveolar diameters from the edge), to ensure that the agonist consistently diffuses to the ROI within several minutes of application. Damaged regions (typically characterized by high, uniform resting GCaMP6f fluorescence or a broken/damaged appearance) should be avoided. Alveoli on the outermost edge of the tissue are more subject to motion artifacts when the agonist is applied, particularly when using oxytocin, which causes basal cells to contract and alveolar structures to shift in x, y, and z planes. This can lead to the ROI moving out of the field of view or out of focus during acquisition. Even for the most experienced investigator, motion artifacts are a major challenge during imaging.
13. Thoracic mammary glands are less suitable for this protocol due to the propensity of these glands to be contaminated with surrounding skeletal muscle.
14. This is a pre-set, proprietary program the details of which have not been disclosed to us by the distributor. If adopting a different system, the program will need to be optimized for mammary tissue. The program involves heating to 37 °C for approximately 30 min with agitation.

15. **Steps 5–7** enable the recovery of maximal material, but could be skipped to save time if more than enough material are recovered at the end of step 4.
16. If pellet remains red following red blood cell lysis, **steps 8** and **9** could be repeated.

Acknowledgments

This work was supported by the Novo Nordisk Foundation (NNF20OC009705), the Carlsberg Foundation (CF21-0170), and the National Health and Medical Research Council of Australia (2003832). We thank Drs Teneale Stewart and Bethan Lloyd-Lewis, as well as Mr Alexander Stevenson, for their help developing and optimizing these protocols. We thank Ms Trine Lund Ruus and Dr Amita Gautam Ghadge for their animal and laboratory support. We would also like to thank the TRI Microscopy Core Facility (University of Queensland), the Katharina Gaus Light Microscopy Facility (University of New South Wales), and the Bioimaging Core Facility (Aarhus University). Figures were created in part with BioRender.com.

References

1. Macias H, Hinck L (2012) Mammary gland development. WIREs Dev Biol 1:533–557. https://doi.org/10.1002/wdev.35
2. Davis FM (2021) Cell and developmental biology of the mammary gland. Semin Cell Dev Biol 114:81–82. https://doi.org/10.1016/j.semcdb.2021.03.015
3. Victora CG, Bahl R, Barros AJD et al (2016) Breastfeeding in the 21st century: epidemiology, mechanisms, and lifelong effect. Lancet 387:475–490. https://doi.org/10.1016/S0140-6736(15)01024-7
4. Stewart TA, Davis FM (2020) Got milk? Identifying and characterizing lactation defects in genetically-engineered mouse models. J Mammary Gland Biol Neoplasia 25:255–272. https://doi.org/10.1007/s10911-020-09467-y
5. Hennighausen L, Robinson GW (2005) Information networks in the mammary gland. Nat Rev Mol Cell Biol 6:715–725. https://doi.org/10.1038/nrm1714
6. Lloyd-Lewis B, Harris OB, Watson CJ, Davis FM (2017) Mammary stem cells: premise, properties, and perspectives. Trends Cell Biol 27:556–567. https://doi.org/10.1016/j.tcb.2017.04.001
7. Gieniec KA, Davis FM (2022) Mammary basal cells: stars of the show. Biochim Biophys Acta (BBA) – Mol Cell Res. 1869:119159. https://doi.org/10.1016/j.bbamcr.2021.119159
8. Watson CJ, Khaled WT (2008) Mammary development in the embryo and adult: a journey of morphogenesis and commitment. Development 135:995–1003. https://doi.org/10.1242/dev.005439
9. Davis FM (2017) The ins and outs of calcium signalling in lactation and involution: implications for breast cancer treatment. Pharmacol Res 116:100–104. https://doi.org/10.1016/j.phrs.2016.12.007
10. Kelleher SL (2023) The ins and outs of mammary gland calcium and zinc transport: a brief review. JDS Commun 4:240–244. https://doi.org/10.3168/jdsc.2022-0291
11. Jenness R (1986) Lactational performance of various mammalian species. J Dairy Sci 69: 869–885. https://doi.org/10.3168/jds.S0022-0302(86)80478-7
12. Jenness R, Holt C (1983) Interrelationships of constituents and partition of salts in milk samples from eight species. Comp Biochem Physiol 77A:275–282. https://doi.org/10.1016/0300-9629(84)90060-4

13. Berridge MJ, Bootman MD, Roderick HL (2003) Calcium signalling: dynamics, homeostasis and remodelling. Nat Rev Mol Cell Biol 4:517–529. https://doi.org/10.1038/nrm1155
14. Clapham DE (2007) Calcium Signaling. Cell 131:1047–1058. https://doi.org/10.1016/j.cell.2007.11.028
15. Davis FM, Janoshazi A, Janardhan KS et al (2015) Essential role of Orai1 store-operated calcium channels in lactation. Proc Natl Acad Sci USA 112:5827–5832. https://doi.org/10.1073/pnas.1502264112
16. Stevenson AJ, Vanwalleghem G, Stewart TA et al (2020) Multiscale imaging of basal cell dynamics in the functionally mature mammary gland. Proc Natl Acad Sci USA 117:26822–26832. https://doi.org/10.1073/pnas.2016905117
17. VanHouten J, Sullivan C, Bazinet C et al (2010) PMCA2 regulates apoptosis during mammary gland involution and predicts outcome in breast cancer. Proc Natl Acad Sci USA 107:11405–11410. https://doi.org/10.1073/pnas.0911186107
18. Stewart TA, Davis FM (2020) A primary cell and organoid platform for evaluating pharmacological responses in mammary epithelial cells. ACS Pharmacol Transl Sci 3:63–75. https://doi.org/10.1021/acsptsci.9b00090
19. Jardé T, Lloyd-Lewis B, Thomas M et al (2016) Wnt and Neuregulin1/ErbB signalling extends 3D culture of hormone responsive mammary organoids. Nat Commun 7:13207. https://doi.org/10.1038/ncomms13207
20. Mourao L, Ciwinska M, Van Rheenen J, Scheele CLGJ (2022) Longitudinal intravital microscopy using a mammary imaging window with replaceable lid. J Vis Exp 63326. https://doi.org/10.3791/63326
21. Jacquemin G, Benavente-Diaz M, Djaber S et al (2021) Longitudinal high-resolution imaging through a flexible intravital imaging window. Sci Adv 7:eabg7663. https://doi.org/10.1126/sciadv.abg7663
22. Lloyd-Lewis B (2020) Multidimensional imaging of mammary gland development: a window into breast form and function. Front Cell Dev Biol 8:203. https://doi.org/10.3389/fcell.2020.00203
23. Chen T-W, Wardill TJ, Sun Y et al (2013) Ultrasensitive fluorescent proteins for imaging neuronal activity. Nature 499:295–300. https://doi.org/10.1038/nature12354
24. Dong TX, Othy S, Jairaman A et al (2017) T-cell calcium dynamics visualized in a ratiometric tdTomato-GCaMP6f transgenic reporter mouse. eLife 6:e32417. https://doi.org/10.7554/eLife.32417
25. Li J, Wang L, Chen Y et al (2020) Visible light excited ratiometric-GECIs for long-term in-cellulo monitoring of calcium signals. Cell Calcium 87:102165. https://doi.org/10.1016/j.ceca.2020.102165
26. Wagner K-U, Wall RJ, St-Onge L et al (1997) Cre-mediated gene deletion in the mammary gland. Nucleic Acids Res 25:4323–4330. https://doi.org/10.1093/nar/25.21.4323
27. Englund JI, Ritchie A, Blaas L et al (2021) Laminin alpha 5 regulates mammary gland remodeling through luminal cell differentiation and Wnt4-mediated epithelial crosstalk. Development 148:dev199281. https://doi.org/10.1242/dev.199281
28. Selbert S, Bentley DJ, Melton DW et al (1998) Efficient BLG-Cre mediated gene deletion in the mammary gland. Transgenic Res 7:387–398. https://doi.org/10.1023/A:1008848304391
29. Ilchuk LA, Stavskaya NI, Varlamova EA et al (2022) Limitations of tamoxifen application for in vivo genome editing using Cre/ERT2 system. Int J Mol Sci 23:14077. https://doi.org/10.3390/ijms232214077
30. Stewart TA, Hughes K, Hume DA, Davis FM (2019) Developmental stage-specific distribution of macrophages in mouse mammary gland. Front Cell Dev Biol 7:250. https://doi.org/10.3389/fcell.2019.00250
31. Dawson CA, Pal B, Vaillant F et al (2020) Tissue-resident ductal macrophages survey the mammary epithelium and facilitate tissue remodelling. Nat Cell Biol 22:546–558. https://doi.org/10.1038/s41556-020-0505-0
32. Hitchcock J, Hughes K, Pensa S et al (2022) The immune environment of the mammary gland fluctuates during post-lactational regression and correlates with tumour growth rate. Development 149:dev200162. https://doi.org/10.1242/dev.200162
33. Tolg C, Cowman M, Turley E (2018) Mouse mammary gland whole mount preparation and analysis. Bio-Protocol 8:e2915. https://doi.org/10.21769/BioProtoc.2915
34. Plante I, Stewart MKG, Laird DW (2011) Evaluation of mammary gland development and function in mouse models. J Vis Exp 2828. https://doi.org/10.3791/2828
35. Yip HYK, Papa A (2021) Generation and functional characterization of murine mammary organoids. STAR Protoc 2:100765. https://doi.org/10.1016/j.xpro.2021.100765

36. Petersen OW, Rønnov-Jessen L, Howlett AR, Bissell MJ (1992) Interaction with basement membrane serves to rapidly distinguish growth and differentiation pattern of normal and malignant human breast epithelial cells. Proc Natl Acad Sci USA 89:9064–9068. https://doi.org/10.1073/pnas.89.19.9064
37. Stokes DG, Liu G, Dharmavaram R et al (2001) Regulation of type-II collagen gene expression during human chondrocyte de-differentiation and recovery of chondrocyte-specific phenotype in culture involves Sry-type high-mobility-group box (SOX) transcription factors. Biochem J 360: 461–470. https://doi.org/10.1042/bj3600461
38. Smalley MJ (2010) Isolation, culture and analysis of mouse mammary epithelial cells. In: Ward A, Tosh D (eds) Mouse Cell Culture. Humana Press, Totowa, pp 139–170
39. Smalley MJ, Kendrick H, Sheridan JM et al (2012) Isolation of mouse mammary epithelial subpopulations: a comparison of leading methods. J Mammary Gland Biol Neoplasia 17:91–97. https://doi.org/10.1007/s10911-012-9257-1
40. Prater M, Shehata M, Watson CJ, Stingl J (2013) Enzymatic dissociation, flow cytometric analysis, and culture of normal mouse mammary tissue. In: Helgason CD, Miller CL (eds) Basic cell culture protocols. Humana Press, Totowa, NJ, pp 395–409
41. Linkert M, Rueden CT, Allan C et al (2010) Metadata matters: access to image data in the real world. J Cell Biol 189:777–782. https://doi.org/10.1083/jcb.201004104
42. Schindelin J, Arganda-Carreras I, Frise E et al (2012) Fiji: an open-source platform for biological-image analysis. Nat Methods 9: 676–682. https://doi.org/10.1038/nmeth.2019

Chapter 16

Live-Cell Calcium Imaging in 3D Intestinal Organoids

Eva Rath and Tamara Zietek

Abstract

Live-cell Ca^{2+} imaging is an important tool to detect activation of receptors by a putative ligand/drug and complements studies on transport processes, as intracellular Ca^{2+} changes provide direct evidence for substrate fluxes. Organoid-based systems offer numerous advantages over other in vitro systems such as cell lines, primary cells, or tissue explants, and in particular, intestinal organoid culture has revolutionized research on functional gastrointestinal processes. Calcium imaging using the fluorescent Ca^{2+} indicator Fura-2-AM can be applied to 3D intestinal organoids, which show an excellent dye-loading efficiency. Here we describe live-cell Ca^{2+} imaging in intestinal organoids, an important technique to improve research on malabsorption syndromes, secretory diarrhea, and metabolic disorders

Key words 3D-organoid culture, Calcium imaging, Drug transport, Incretin secretion, Intestinal organoids, Live-cell imaging, Nutrient absorption

1 Introduction

Calcium, one of the most widespread second messengers, can be imaged indirectly by fluorescent Ca^{2+} indicators such as Fura-2-AM. In response to stimuli, Ca^{2+} may either enter the cell from the external surrounding or be released from internal stores such as the endoplasmic reticulum. Consequently, intracellular Ca^{2+} changes reflect activation of receptors and transport processes and regulate intracellular translocation events such as exocytosis. Hence, live-cell imaging of calcium fluxes is commonly used in basic research and frequently applied in pharmacological drug screenings. In the intestinal epithelium, Ca^{2+} signaling is of particular interest in the context of nutrient and drug absorption processes as well as mucin/antimicrobial peptide secretion and gut hormone release including the diabetes treatment-relevant hormone GLP-1 [1–3].

Corresponding authors: Correspondence regarding organoid culture should be addressed to Eva Rath (eva.rath@tum.de) and that regarding imaging to Tamara Zietek (zietek@aerzte-gegen-tierversuche.de).

Caroline M. Gorvin (ed.), *Calcium Signaling: Methods and Protocols*, Methods in Molecular Biology, vol. 2861,
https://doi.org/10.1007/978-1-0716-4164-4_16,

Intestinal organoids resemble intestinal architecture and have been proven a suitable and versatile model for functional studies on the intestinal epithelium [4]. They contain all subtypes of intestinal epithelial cells including stem cells, antimicrobial peptide-producing Paneth cells, mucin-producing goblet cells, tuft cells, and hormone-producing enteroendocrine cells and constitute an excellent source of differentiated, fully functional enterocytes [5]. Furthermore, intestinal organoids can be generated from different parts of the gastrointestinal tract and retain the features of their tissue of origin, allowing for segment-specific studies [6]. By modifying medium composition, e.g., by withdrawal of Wnt factors [7, 8] or addition of inhibitors [9, 10], intestinal organoids can be enriched for certain rare cell populations. Facilitating experiments, standard molecular biological techniques for manipulation like transfection with CRISPR/Cas9 can be applied on intestinal organoids, for example, to easily identify the cell subtype of interest by fluorescent tagging [3]. Live-cell imaging of calcium fluxes has been done in primary isolated crypts (which are the starting material for intestinal organoid culture) [11], 3D organoid culture [1, 2], and 2D cell lawns derived from organoid cells [3]. Regarding availability and easy as well as economic experimental procedures, 3D organoid culture is the most straightforward approach. Similar to cell-based assays, 3D intestinal organoids can be readily loaded with the fluorescent Ca^{2+} indicator Fura-2-AM directly in the matrix they grow in, resulting in an even intracellular dye distribution across crypt as well as villus domains. This type of organoid culture is also compatible with perfusion systems, allowing simple addition of test substances and subsequent imaging and data analysis according to standard protocols.

2 Materials

2.1 Equipment

Appropriate (inverted) live-cell imaging microscope (×40 objective recommended), fluorescence camera, and software package for data analysis, perfusion system (optional) cell culture equipment (*see* **Note 1**).

2.2 Intestinal Organoid Culture

For details on the establishment and culture of intestinal organoids, please refer to standard protocols as given, for example, in [1, 2, 12, 13]. For calcium imaging, organoids can be either grown in 24-/48-well cell culture plates or on glass slides, depending if the intended measurement requires perfusion/media flow or not (e.g., in the case of reaction-to-stimuli tests). In any case, intestinal organoid culture requires particular media, dependent on the gastrointestinal location as well as species of origin. The composition of media is described in detail elsewhere [12, 13]; however, these media can also be obtained as ready-to-use medium. Furthermore, a matrix (such as Matrigel) in which organoids can grow three-dimensionally is required.

2.3 Solutions for Calcium Imaging

Prepare all solutions using sterile substances and deionized water and/or use sterile filtration (0.2 μm pore size) to obtain solutions suitable for cell culture.

2.3.1 Stock Solutions

1. Fura-2AM stock solution: *Caution, light sensitive!* Add 3.56 μL DMSO to a 50 μg Fura-2AM aliquot (MW 1001.86) to prepare a 14 mM stock solution. Carefully mix and store at −20 °C.
2. Pluronic stock solution: Add 1 g pluronic to 10 mL DMSO to obtain a 10% $^{v}/_{w}$ pluronic in DMSO solution, if necessary heat up to approx. 40 °C, and sonicate to dissolve. Store solution at room temperature. If precipitation is observed prior to use, pre-warm pluronic stock to 37 °C, and vortex or sonicate until it goes back into solution.
3. Eserine stock solution: *Caution, prepare under fume hood!* Add 24.212 μL chloroform to 1 mg eserine (MW 275.35) to prepare a 150 mM stock solution. Carefully dissolve and store at −20 °C.
4. Probenecid stock solution: Probenecid (MW 285.36) solubilized in 1M NaOH to obtain a 175 mM stock solution (*see* **Note 2**).

2.3.2 Working Solutions

1. Standard HEPES-saline buffer: Contains the following in (in mM and g/L) HEPES (MW 238.3, 10 mM, 2.383 g/L), NaCl (MW 58.44, 138 mM, 8.065 g/L), KCl (MW 74.55, 4.5 mM, 0.335 g/L), $NaHCO_3$ (MW 84.01, 4.2 mM, 0.3528 g/L), NaH_2PO_4 (MW 120, 1.2 mM, 0.144 g/L), $CaCl_2$ (1 M, 2.6 mM, 2.6 mL), $MgCl_2$ (1 M, 1.2 mM, 1.2 mL). Adjust to the desired pH with 1 M NaOH. Store at 4 °C (*see* **Note 3**).
2. Calcium-free HEPES-saline buffer with EGTA: Same as above [4] but without $CaCl_2$. $CaCl_2$.is replaced by an equal amount of $MgCl_2$. Add EGTA (MW 380.35, 1 mM, 380.35 g/L). Adjust to the desired pH with 1 M NaOH. Store at 4 °C.
3. Fura-2AM working solution: *Caution, light sensitive! Prepare freshly.* To prepare 1 mL Fura-2AM working solution, add 0.5 μL Fura-2AM stock solution, 1 μL pluronic stock solution, 2.5 μL eserine stock solution, and 11.4 μM probenecid stock solution, to 984.6 μL standard HEPES-saline buffer to obtain the following final concentrations: 7 μM Fura-2AM, 0.01% pluronic, 375 μM eserine, 2 mM probenecid. Adjust to the desired pH with 1 M NaOH or HCl (if probenecid is used).

2.3.3 Calibration/Positive Control/Inhibitors

1. Calibration (*see* **Note 4**): Signal elicited by minimal Ca^{2+} concentrations—use calcium-free HEPES-saline buffer with EGTA supplemented with 5 μM ionomycin. Signal elicited by maximal Ca^{2+} concentrations: use standard HEPES-saline buffer supplemented with 5 μM ionomycin.
2. Positive control: prepare a 500 μM ATP stock solution in standard HEPES-saline buffer. To obtain an end concentration of 50 μM ATP in a 48-well plate well containing 270 μL buffer, add 30 μL ATP stock solution.
3. Inhibitors: The following substances can be directly added to the Fura-2AM working solution in the concentrations indicated to specify calcium signaling: 10 μM nifedipine, 50 μM thapsigargin, 20 μM BAPTA-AM, 5 μM ionomycin (*see* **Note 5**).

3 Methods

3.1 Organoid Culture

Intestinal organoids derived from (induced) stem cells, freshly isolated crypts, or after passing can be used (*see* **Note 6**).

1. Spin down intestinal crypts in a centrifuge (300 g, 5 min, 4 °C). Remove supernatant (the pellet should be quite firm, and the tube can be turned upside down on a paper towel to get rid of remaining PBS).
2. Resuspend pellet in 5 mL of cold PBS. Pipette 10 μL of the suspension on a glass slide, and count the number of crypts under a microscope. Repeat three times. Multiply the average of the 3 individual counts with 500 to calculate the total number of crypts contained in 5 mL of suspension.
3. Take the appropriate volume to seed 600 crypts per well, and transfer into a labeled 50 mL conical tube. Spin down the crypts in a centrifuge (300 g, 5 min, 4 °C). Remove supernatant.
4. Using chilled pipette tips, add 30 μL Matrigel per well to be seeded. Carefully pipette up and down ten times to thoroughly resuspend the pellet. Avoid introducing any bubbles. Keep on ice.
5. Using chilled pipette tips, carefully apply 30 μL of the suspension directly on the center of a well from a pre-warmed 48-well plate. The suspension should form a hemispherical droplet (*see* **Note 7**). Incubate the plate at 37 °C and allow the Matrigel to solidify for approximately 5–10 min (*see* **Note 8**).
6. Overlay the crypt/organoid-containing Matrigel droplet with 300 μL of intestinal organoid medium (*see* **Note 9**).

7. Change medium every 2–3 days and check for growth and contaminations daily (*see* **Note 10**).

3.2 Calcium Imaging

1. Take cell culture plates out of the incubator, and locate organoid structures/cells of interest suitable for imaging (*see* **Note 11**).
2. Remove medium. Carefully wash the organoid-containing Matrigel-dome one time with 500 μL pre-warmed standard HEPES-saline buffer (pH 7.4).
3. Add 300 μL Fura-2AM working solution. Incubate for 45 min at 37 °C in the dark.
4. Wash five times with 500 μL pre-warmed standard HEPES-saline buffer. After removal of the fluid from the last washing step, add 300 μL pre-warmed standard HEPES-saline buffer.
5. Incubate for 30 min at 37 °C in the dark to allow de-esterification of the dye (*see* **Note 12**).
6. Transfer the organoids into a chamber on the stage of an inverted fluorescent microscope.
7. Optional: Set up perfusion system.
8. Set up microscope according to the manufacturer's instructions. 20× or 40× objective is recommended. The Ca^{2+}-bound form of Fura2 gets excited at 340 nm, the Ca^{2+}-unbound form of Fura2 gets excited at 380 nm, and emitted light is measured around 510 nm. Set up camera and software according to the manufacturer's instructions. Set up exposure time (default time: around 100 ms).
9. Check loading efficiency and equal dye distribution of cells/organoids (*see* **Note 13**).
10. Define regions of interest (ROIs) for normalization (background/noise subtraction) and calibration purposes, and define ROIs for target cells/organoid structures.
11. Record images every 5–10 s.
12. Run calibration using first calcium-free HEPES-saline buffer with EGTA supplemented with ionomycin and subsequently standard HEPES-saline buffer supplemented with ionomycin.
13. Add test substances. If directly applied into the well, let the tip of the pipette touch the upper rim of the well to allow the fluid to slowly flow into the buffer (*see* **Note 14**).
14. Analyze recorded images using an appropriate software suite (*see* **Note 1**) and calculating the fluorescence ratio $F(\lambda_{ex}340\text{ nm})/F(\lambda_{ex}380\text{ nm})$ of ROIs (*see* **Note 15**).
15. Live-cell Ca^{2+} imaging data are commonly presented in the form of pseudocolor images (Fig. 1) indicating fluorescence

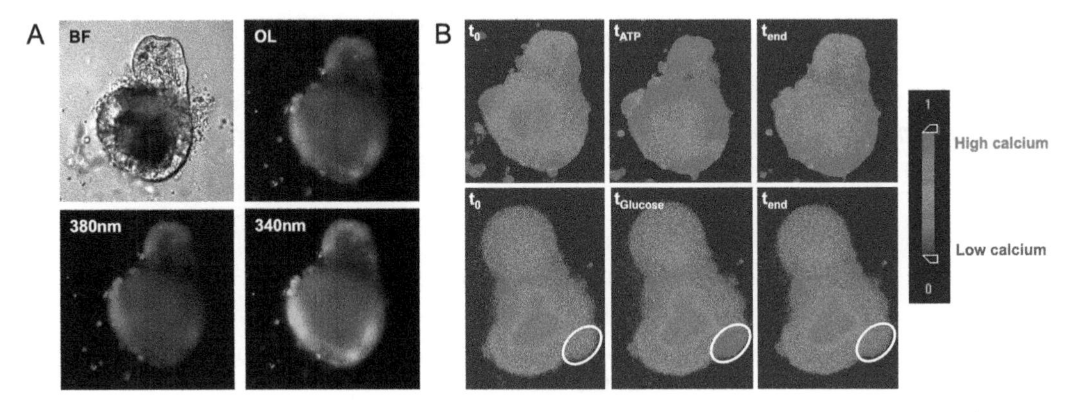

Fig. 1 (**a**) Bright field image of a 3D intestinal organoid, fluorescence images depicting 340 nm (Ca^{2+} bound Fura2) and 380 nm (Ca^{2+} unbound Fura2), and the corresponding overlay (OL). (**b**) Live-cell imaging of intracellular calcium fluxes in intestinal organoids depicted as pseudocolor images. Calcium responses to ATP (upper panel) and glucose (lower panel). A ROI (white circle) reacting to glucose stimulation is indicated. Ratio of fluorescence intensities $F(\lambda_{ex}340\ nm)/\ F(\lambda_{ex}380\ nm)$ reflects changes in intracellular calcium levels according to the color legend. (Adapted from Ref. [1])

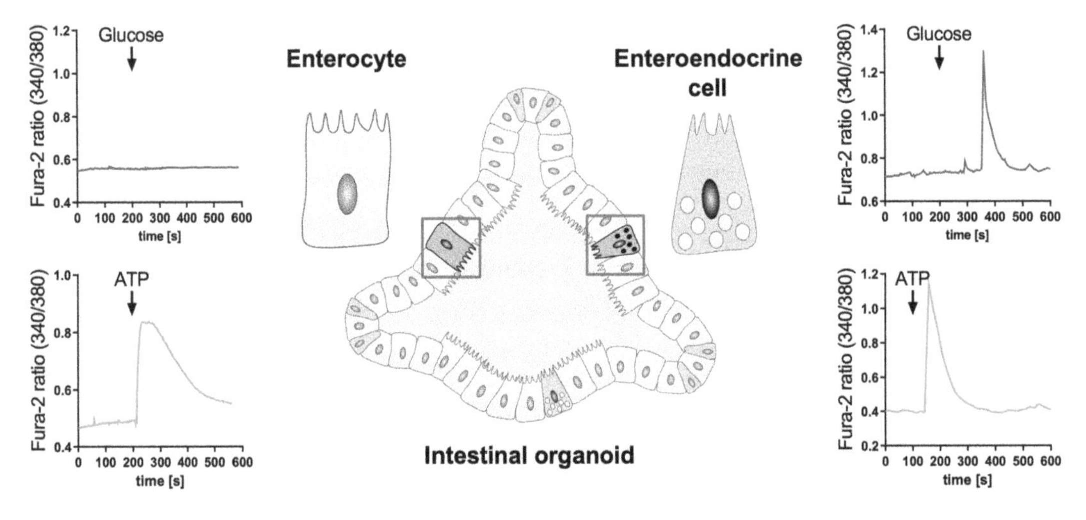

Fig. 2 Live-cell imaging of intracellular calcium fluxes in intestinal organoids depicted as traces and schematic drawing of a 3D intestinal organoid. Responses to glucose (upper panel) and ATP (lower panel) are depicted for enterocytes (left) and enteroendocrine cells (right). Enteroendocrine cells employ Ca^{2+} as a messenger in response to glucose sensing to initiate exocytosis of GLP-1-containing granula

ratio or as traces, where the average ratio values within regions of interest (ROIs) are plotted with respect to time (Fig. 2).

4 Notes

1. We use a Leica DMI6000 B inverted fluorescence microscope (Leica Microsystems) and the Leica Application Suite

Advanced Fluorescence software for data analysis, and a custom-made perfusion system.

2. The nonionic, low-toxicity detergent pluronic facilitates AM-ester solubilization. The esterase blocker eserine improves cytoplasmic dye loading. The organic anion transporter inhibitor probenecid prevents leakage, thus prolonging cellular retention of the dye.
3. Standard pH of the saline solution is 7.4. However for some cases like investigation of signaling associated with H^+-dependent transport processes, pH 6.5 might give better results. Depending on the process to be investigated, 5 mM glucose can be added to the standard saline solution to avoid starving effects.
4. For calibration, the signals of the maximum and minimum amount of bound Ca^{2+} indicator need to be determined. This can be done within the same experiment.
5. To ensure specificity and scrutinize calcium signaling, the calcium channel blocker nifedipine can be used and/or combinations of the SERCA-inhibitor thapsigargin which promotes Ca^{2+} release from the endoplasmic reticulum; the Ca^{2+}ionophore ionomycin, which mainly facilitates Ca^{2+} entry from the extracellular space; and the cell-permeable calcium chelator BAPTA-AM.
6. Please refer to standard protocols for intestinal crypt isolation and intestinal organoid passaging [13, 12, 14]. Of note, intestinal organoids directly derived from primary crypts show reduced expression of several differentiation markers within days in culture [2]. When working with organoids, the individual genetic and sex background needs to be considered.
7. For applying the crypt-Matrigel suspension on the plate, pipette tips with a wider opening can be used to improve handling.
8. If a perfusion system should be used, place sterile coverslips into a well from a cell culture plate or Petri dish and apply the crypt-Matrigel suspension on the coverslip. Let solidify and add medium. Culture organoids for the desired time. The coverslips need to be lifted carefully (e.g., using a plastic inoculation loop) before placing them into the perfusion system.
9. Standard intestinal organoid medium including the commercially available ready-to-use medium fosters stemness and reduces terminal differentiation into epithelial subpopulations. To improve differentiation or steer differentiation into certain subpopulations, Wnt factors can be withdrawn and specific inhibitors added.

10. Dependent on the mode of passaging (mechanically or using dissociation reagents) and thus size of seeded "crypts," the approximate time in culture before Ca^{2+} imaging is 3–4 days for small intestinal organoids and 5–6 days for slower-growing colonic organoids. These times need to be adapted individually, particularly if additional experimental procedures are applied, for example, transfection or steered differentiation [9, 3, 10].
11. This is of particular importance if you aim for a certain cell population. Notes and images should be taken of the localization of important parameters such as fluorescently labeled cells or morphologically identified cell populations, and a logical system, for example, using clock positions, may be useful.
12. Time for de-esterification of the dye can be adapted and needs to be balanced with cellular mechanisms reducing cellular dye content. In general, 15–45 min is suitable. If the live-cell imaging microscope does not possess a heated chamber/incubator, the last 15 min of incubation time should be at room temperature to avoid cold shock effects on cells. Take care to keep the plate protected from light (e.g., cover with tinfoil).
13. To avoid artifacts due to unequal cell thickness, uneven dye distribution, leakage of dye, or photo bleaching, ratioing methods should be applied.
14. Adding test substrates to 3D intestinal organoids is the most straightforward approach for functional studies. Yet, intestinal organoids grow in Matrigel with the luminal compartment facing inside and the basolateral compartment facing outside. Thus, when adding effectors, these first reach the basolateral side of organoids, targeting basolateral transporters or receptors. However, small molecules (<4 kDa) rapidly pass the epithelium of organoids and enter the organoid lumen [1]. As it is not possible in this system to target the apical and basolateral side separately, genetic manipulation and the use of inhibitors might help to decipher molecular processes.
15. Experiments should be performed several times to confirm results.

References

1. Zietek T, Rath E, Haller D, Daniel H (2015) Intestinal organoids for assessing nutrient transport, sensing and incretin secretion. Sci Rep 5(1):1–10. https://doi.org/10.1038/srep16831
2. Zietek T, Giesbertz P, Ewers M, Reichart F, Weinmüller M, Urbauer E, Haller D, Demir IE, Ceyhan GO, Kessler H, Rath E (2020) Organoids to study intestinal nutrient transport, drug uptake and metabolism – Update to the human model and expansion of applications. Front Bioeng Biotechnol 8: 577656. https://doi.org/10.3389/fbioe.2020.577656
3. Goldspink DA, Lu VB, Miedzybrodzka EL, Smith CA, Foreman RE, Billing LJ, Kay RG, Reimann F, Gribble FM (2020) Labeling and characterization of human GLP-1-secreting

L-cells in primary ileal organoid culture. Cell Rep 31(13):107833. https://doi.org/10.1016/j.celrep.2020.107833

4. Zietek T, Boomgaarden WAD, Rath E (2021) Drug screening, oral bioavailability and regulatory aspects: a need for human organoids. Pharmaceutics 13(8):1280. https://doi.org/10.3390/pharmaceutics13081280
5. Sato T, Vries RG, Snippert HJ, van de Wetering M, Barker N, Stange DE, van Es JH, Abo A, Kujala P, Peters PJ, Clevers H (2009) Single Lgr5 stem cells build crypt-villus structures in vitro without a mesenchymal niche. Nature 459(7244):262–265. https://doi.org/10.1038/nature07935
6. Middendorp S, Schneeberger K, Wiegerinck CL, Mokry M, Akkerman RD, van Wijngaarden S, Clevers H, Nieuwenhuis EE (2014) Adult stem cells in the small intestine are intrinsically programmed with their location-specific function. Stem Cells 32(5):1083–1091. https://doi.org/10.1002/stem.1655
7. Lindeboom RG, van Voorthuijsen L, Oost KC, Rodriguez-Colman MJ, Luna-Velez MV, Furlan C, Baraille F, Jansen PW, Ribeiro A, Burgering BM, Snippert HJ, Vermeulen M (2018) Integrative multi-omics analysis of intestinal organoid differentiation. Mol Syst Biol 14(6):e8227. https://doi.org/10.15252/msb.20188227
8. Foulke-Abel J, In J, Yin J, Zachos NC, Kovbasnjuk O, Estes MK, de Jonge H, Donowitz M (2016) Human enteroids as a model of upper small intestinal ion transport physiology and pathophysiology. Gastroenterology 150(3):638–649 e638. https://doi.org/10.1053/j.gastro.2015.11.047
9. Petersen N, Reimann F, van Es JH, van den Berg BM, Kroone C, Pais R, Jansen E, Clevers H, Gribble FM, de Koning EJ (2015) Targeting development of incretin-producing cells increases insulin secretion. J Clin Invest 125(1):379–385. https://doi.org/10.1172/JCI75838
10. de Lau W, Kujala P, Schneeberger K, Middendorp S, Li VS, Barker N, Martens A, Hofhuis F, DeKoter RP, Peters PJ, Nieuwenhuis E, Clevers H (2012) Peyer's patch M cells derived from Lgr5(+) stem cells require SpiB and are induced by RankL in cultured "miniguts". Mol Cell Biol 32(18):3639–3647. https://doi.org/10.1128/MCB.00434-12
11. Pelaez-Llaneza N, Jones V, Kam C, Lee A, Parris A, Tattan S, Loader M, Champion J, Russam G, Miller B, Juge N, Wharton R, Speakman C, Kapur S, Hernon J, Stearns A, Shaik I, Pal A, Tsigka A, Batsoulis D, Williams M (2022) Guardian crypt-base-goblet-cells protect the human colonic stem cell niche by triggering cholinergic calcium signal-dependent MUC2 secretion and luminal flushing. bioRxiv:2022.2001.2004.474646. https://doi.org/10.1101/2022.01.04.474646
12. Mahe MM, Sundaram N, Watson CL, Shroyer NF, Helmrath MA (2015) Establishment of human epithelial enteroids and colonoids from whole tissue and biopsy. J Vis Exp (97). https://doi.org/10.3791/52483
13. Zietek T, Rath E (2018) Chapter 3 – Intestinal organoids: Mini-guts grown in the laboratory. In: Davies JA, Lawrence ML (eds) Organs and organoids. Academic, pp 43–71
14. Mahe MM, Aihara E, Schumacher MA, Zavros Y, Montrose MH, Helmrath MA, Sato T, Shroyer NF (2013) Establishment of gastrointestinal epithelial organoids. Curr Protoc Mouse Biol 3(4):217–240. https://doi.org/10.1002/9780470942390.mo130179

Chapter 17

Calcium Imaging and Analysis in Beta Cells in Acute Mouse Pancreas Tissue Slices

Eva Paradiž Leitgeb, Viljem Pohorec, Lidija Križančić Bombek, Maša Skelin Klemen, Maja Duh, Marko Gosak, Jurij Dolenšek, and Andraž Stožer

Abstract

Ca^{2+} ions play a central role in the stimulus-secretion coupling cascade of pancreatic beta cells. The use of confocal microscopy in conjunction with the acute pancreas tissue slice technique offers valuable insights into changes in the intracellular calcium concentration following stimulation by secretagogues. This allows the study of beta cells on a single cell level, as well as their behavior on a multicellular scale within an intact environment. With the use of advanced analytical tools, this approach offers insight into how single cells contribute to the functional unit of islets of Langerhans and processes underlying insulin secretion. Here we describe a comprehensive protocol for the preparation and utilization of acute pancreas tissue slices in mice, the use of high-resolution confocal microscopy for observation of glucose-stimulated calcium dynamics in beta cells, and the computational analysis for objective evaluation of calcium signals.

Key words Acute pancreas tissue slice, Beta cell, Calcium imaging, Calcium oscillation, Islet of Langerhans, Network

1 Introduction

The Ca^{2+} ion is a crucial secondary messenger in many excitable and non-excitable cells, playing a central physiological role in processes such as excitation–contraction coupling and stimulus-secretion coupling [1, 2]. In pancreatic beta cells, a variety of metabolic primary messengers, neurotransmitters, and hormones, employing partly overlapping signaling pathways, converge by eliciting oscillations of intracellular Ca^{2+} concentration ($[Ca^{2+}]_{IC}$), triggering secretion of insulin, and in turn orchestrating postprandial storage and interprandial utilization of energy-rich nutrients [3, 4]. The

Co-first authors: Eva Paradiž Leitgeb and Viljem Pohorec.
Co-corresponding authors: Jurij Dolenšek and Andraž Stožer.

Caroline M. Gorvin (ed.), *Calcium Signaling: Methods and Protocols*, Methods in Molecular Biology, vol. 2861,
https://doi.org/10.1007/978-1-0716-4164-4_17,

spatiotemporal patterns of $[Ca^{2+}]_{IC}$ dynamics become disrupted during development of type 1 and type 2 diabetes, and therefore imaging studies are crucial to better understand both normal and pathological beta cell physiologies [5, 6].

During the last four decades, the introduction of fluorescent indicators and genetic probes has enabled studies of $[Ca^{2+}]_{IC}$ dynamics with camera-based or confocal Ca^{2+} imaging setups in dispersed beta cells [7, 8] and in isolated islets of Langerhans in vitro [9, 10], in tissue slices in situ [11, 12], and in islets transplanted into the anterior chamber of the eye [13, 14] as well as in islets in vivo [15, 16]. Inspired by the classical work of Warburg on liver slices a century ago [17] and more recent work on slices from other tissue types, especially neuronal [18], neuroendocrine [19], and lung tissue [20], the pancreas tissue slice approach was introduced 20 years ago and combines many advantages from both extremes of the methodological spectrum [21]. More specifically, compared with isolated cells and islets, it does not involve any considerable mechanical or enzymatic stress, and there is no selective damage to the outermost islet layers that contain predominantly non-beta cells in mouse islets [22]. Additionally, beta cells in the core areas of islets can be imaged in slices, and there is no need for cultivation following isolation, which can have important effects on cell identity and physiology [23–25]. Furthermore, in slices, beta cells are embedded in the surrounding tissue and thus retain mechanical support and paracrine interactions [26]. The tissue slice method also enables studying beta cells during the perinatal period and in models with deranged islet architecture when the yield using isolation procedures is low [27, 28]. Compared with studies in vivo, the major advantages of tissue slices are a more rapid preparation; optimal handling of biological material and control over experimental conditions, such as the concentration of secretagogues; and compatibility with other opto-, electrophysiological, and morphological methods, with transfection approaches, as well as with tissue from other species [11, 25, 29].

During the last decade, Ca^{2+} imaging in mouse tissue slices has yielded important insights into the complex intra- and intercellular patterns of $[Ca^{2+}]_{IC}$ signals in beta cells during stimulation with various secretagogues [30–32] in different strains of mice [33], under both normal and pathological conditions [28]. Since it enables signal acquisition with high spatiotemporal resolution over prolonged periods of time, it has also provided the necessary dataset for studying beta cell functional connectivity with tools from network science [34–36]. To facilitate a more widespread use of this promising pipeline, in this chapter we describe in detail the methodological steps required to prepare acute pancreas tissue slices, precision cut from a mouse pancreas using a vibratome, and to use them for live cell imaging of Ca^{2+} dynamics in beta cells embedded in their native environment in islets of Langerhans employing laser-scanning confocal microscopy. We also outline the analytical steps for time-series export, processing, determination of classical physiological parameters of beta cell function, such as frequency, duration, relative active time, and identification of first-responder and wave-initiating cells, as well as for construction of functional networks and determination of network parameters, such as node degree, clustering coefficient, and identification of hub cells.

2 Materials

2.1 Preparation of Extracellular Solution (ECS) with 6 mM Glucose

1. Fill a 1000 mL glass laboratory bottle with approximately 800 mL of MilliQ (ultrapure-Type I) water and add the 125 mM NaCl, 26 mM $NaHCO_3$, 6 mM glucose, 6 mM lactic acid, 3 mM myo-inositol, 2.5 mM KCl, 2 mM Na-pyruvate, 1.25 mM NaH_2PO_4, and 0.5 mM ascorbic acid in the amounts specified. Remember to handle all chemicals with care and to follow appropriate safety protocols (*see* **Note 1**).
2. Thoroughly mix the ingredients by stirring the solution with a magnetic stirrer until they are completely dissolved.
3. Once all the ingredients are dissolved, add 2 mM $CaCl_2$ and 1 mM $MgCl_2$. Fill up to 1000 mL with MilliQ water and check the osmolarity of the solution using an osmometer. Follow the manufacturer's instructions for calibrating and using the osmometer. The osmolarity of the ECS should fall within the range of 300–320 mOsm (*see* **Note 2**).
4. To maintain a physiological pH value of 7.4 throughout the slicing and experimental processes, it's essential to continuously bubble the extracellular solution (ECS) with carbogen gas, consisting of 95% oxygen (O_2) and 5% carbon dioxide (CO_2), at barometric pressure (*see* **Note 3**).

2.2 Preparation of HEPES Buffer Solution (HBS) with 6 mM Glucose

1. Measure each chemical compound accurately.
2. Fill a 1000 mL glass laboratory bottle with approximately 800 mL of MilliQ water, and add the 150 mM NaCl, 10 mM HEPES, 6 mM glucose, and 5 mM KCl in the amounts specified. Remember to handle all chemicals with care and to follow appropriate safety protocols.
3. Thoroughly mix the ingredients by stirring the solution with a magnetic stirrer until they are completely dissolved.
4. Once all the ingredients are dissolved, add 2 mM $CaCl_2$ and 1 mM $MgCl_2$. Fill up to 1000 mL using MilliQ water, and check the osmolarity of the solution using an osmometer. Follow the manufacturer's instructions for calibrating and using the osmometer. The osmolarity of the ECS should fall within the range of 300–320 mOsm.
5. Adjust the pH value to 7.4 using 1 M NaOH (*see* **Note 4**).

2.3 Reagents Required for Pancreas Tissue Slices

1. SeaPlaque® Agarose (Lonza Bioscience, Basel, Switzerland).
2. Water bath.
3. Disposable ice cube bags.
4. 2 C-shaped stabilizing lead rings.
5. 2–3 Petri dishes, 100 mm.
6. 2 Petri dishes, 30 mm.
7. Stereomicroscope.
8. Specimen dissecting board.
9. 4–5 paper pushpins.
10. Curved hemostat forceps.
11. Graefe forceps, straight.
12. Delicate suture tying forceps.
13. Fine scissors—ToughCut®, 11.5 cm (FST, Heidelberg, Germany).
14. Fine scissors—ToughCut®, 9 cm (FST, Heidelberg, Germany).
15. Syringe, Luer lock, 5 mL.
16. Syringe needle, 30 G.
17. Razor blade.
18. Super glue.
19. Crushed ice.
20. Microwave oven.
21. Vibratome.

2.4 Preparation of 1.9% Agarose

1. Add 0.475 g of low-melting-point agarose into an Erlenmeyer flask containing 25 mL of ECS with 6 mM glucose.
2. Heat the mixture in a microwave oven until the agarose completely dissolves.
3. Place the Erlenmeyer flask in a water bath pre-warmed to 40 °C to cool it down, and maintain its liquid state until injection. Use lead rings for stabilization (*see* **Note 5**).

2.5 Preparation of ECS Ice Cubes

1. Fill a disposable ice cube bag with ECS containing 6 mM glucose, and freeze it ahead of time to ensure that the ice cubes are readily available on the day of the experiment. For the preparation of acute pancreas tissue slices, please *see* Subheading 3.1.

2.6 Preparation of Dye Loading Solution for Ca^{2+} Imaging

1. Calbryte™ 520 AM (AAT Bioquest, Pleasanton, California, USA).
2. Poloxamer (e.g., Pluronic F-127, 20% solution in DMSO, Invitrogen by ThermoFisher Scientific).

3. Screw cap tube, 15 mL.
4. Petri dish, 35 mm.
5. Vortex.
6. Ultrasonic bath chamber.
7. Orbital shaker.

2.6.1 Preparation of 6 μM Calbryte 520 AM Solution with 0.11% DMSO, and 0.037% Poloxamer

1. Dissolve one 50 μg vial of the cell-permeable Ca^{2+} indicator Calbryte 520 AM in 6.667 mL of HBS containing 6 mM glucose, 7.5 μL DMSO, and 2.5 μL of the poloxamer in a 15 mL screw cap tube (*see* **Note 6**).
2. Use a 1 mL pipette to aspirate and expel the solution in the tube repeatedly for 20 s, vortex the solution for 30 s, and optionally submerge the tube in an ultrasonic bath chamber for 30 s to improve solubilization.
3. Aliquot 3.333 mL of the dye solution into a 5 mL Petri dish (*see* **Note 7**).

2.6.2 Equipment for Calcium Imaging

1. Confocal microscope with 20X objective (numerical aperture [NA] between 0.77 and 1.00) and 488 Argon laser.
2. Temperature controller.
3. Stage controller.
4. Recording chamber.
5. Harp (U-shaped slice anchor with nylon support mesh).
6. Perifusion system with peristaltic pumps and flexible silicone tubing.
7. Source of carbogen and tubing for carbogen bubbling.
8. ECS solution containing varying concentrations of glucose or other secretagogue for cell stimulation.

3 Methods

3.1 Preparation of Pancreatic Tissue Slices

3.1.1 Preparation of Agarose and ECS Prior to Injection

1. Load a 5 mL syringe with the liquid agarose extracted from the Erlenmeyer flask kept in the warm bath. Eliminate any air bubbles present within the syringe. On top of the syringe, fix a cap-protected 30 G needle (*see* **Note 8**).
2. Transfer the flask containing the liquid agarose and syringe back to the warm water bath. Position the needle downward and ensure the entire agarose volume is submerged. Use the C-shaped lead ring for stabilization.
3. Put a bottle of ECS into a bucket of ice and start continuously bubbling it with carbogen.
4. Place two 100 mm Petri dishes on ice. You will fill them with bubbled ECS just prior to pancreas extraction.

3.1.2 Pancreas Injection with Agarose

1. Euthanize the mouse in strict adherence to all applicable national, regional, and international guidelines governing the handling of experimental animals. Endeavor to minimize any distress or suffering experienced by the animal.
2. Place the mouse on a specimen dissecting board lying on its back, and secure it in place by pinning its extremities on the board using drawing paper pushpins.
3. Perform a laparotomy to access the abdominal cavity (Fig. 1a). Carefully reposition the intestines to the left side of the mouse to expose the common bile duct. Use forceps to gently elevate the duodenal section of the intestine and locate the major duodenal papilla—also known as the papilla of Vater. Using a curved hemostat forceps, clamp the common bile duct at the duodenal papilla (Fig. 1a, b) to prevent any leakage of agarose from the duct into the duodenum (*see* **Note 9**).
5. Using small sharp forceps, carefully clear the common bile duct of any surrounding membranous and adipose tissue while ensuring not to sever it from pancreatic tissue.

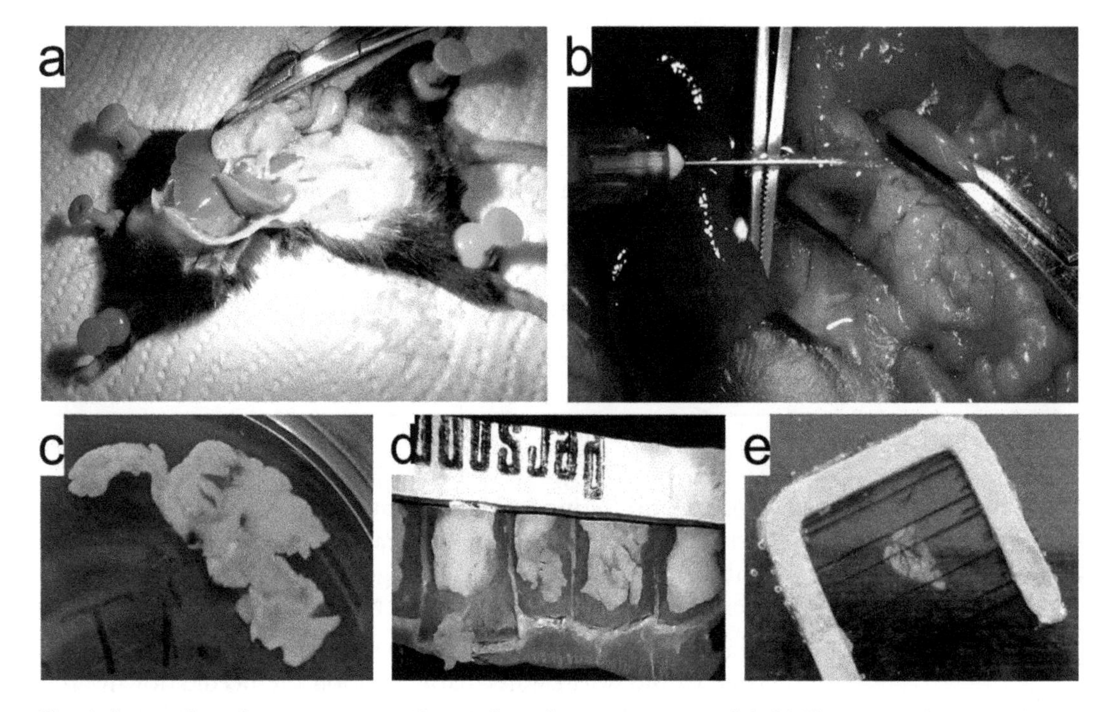

Fig. 1 Preparation of acute pancreas tissue slices from a mouse model. (**a**) The mouse is secured to a dissecting board, and the abdominal cavity is accessed via laparotomy post euthanasia. (**b**) The common bile duct is supported with delicate suture tying forceps and heated agarose is injected into it proximally. (**c**) The extracted pancreas is placed into a Petri dish with ice-cold ECS. (**d**) The tissue slices embedded into agarose blocks are transferred on a sample plate of the vibratome for sectioning. (**e**) Tissue slices are carefully retrieved and transferred to a Petri dish after sectioning. An anchor is used to secure it for imaging

6. With the delicate suture tying forceps in your nondominant hand, reach under the duct and position the duct perpendicularly on top of the forceps (Fig. 1b). Slightly lift the forceps with the duct lying on top of it so the duct is slightly taut.
6. With the dominant hand, take the agarose-filled syringe from the water bath, and administer the liquid agarose into the proximal portion of the common bile duct (Fig. 1b). Apply significant pressure on the syringe piston as the agarose is highly viscous. Continue filling the pancreas until it appears whitish or for a duration of at least 20–30 s (*see* **Note 10**).
7. Withdraw the syringe and apply ice-cold ECS onto the pancreas to solidify the agarose and cool the tissue.
8. Pour some ice-cold bubbled ECS into the two prepared 100 mM Petri dishes.
9. Carefully extract the pancreas from the abdominal cavity (*see* **Note 11**).
10. Transfer the pancreas into the prepared 100 mm Petri dish. Rinse the pancreas and remove any adipose tissue that might be floating on the surface of the ECS.
11. Transfer the pancreas into the second 100 mm Petri dish with the ice-cold ECS (Fig. 1c).
12. Identify the optimally injected portion of the pancreas, distinguished by its whitish coloration and firmer texture, and prepare tissue blocks measuring 0.1–0.2 cm^3 using fine scissors (*see* **Note 12**).
13. Fill a 35 mm Petri dish with 3–4 mL of the liquid agarose, transfer tissue blocks into it, and promptly return on ice to chill and harden.
14. After about 5 min as the agarose has hardened, use a razor blade to cut individual agarose cubes each containing a single piece of tissue. Ensure that each tissue piece is surrounded by a continuous layer of agarose.
15. Using forceps gently position the agarose blocks containing tissue pieces onto the sample plate of a vibratome and fix them using super glue (Fig. 1d).

3.1.3 Tissue Slicing

1. Fill the chamber of the vibratome with ice-cold, continuously bubbled ECS with 6 mM glucose. Arrange crushed ice around the cutting chamber and place two ice cubes into the cutting chamber (*see* **Note 13**).
2. Attach the razor blade onto the vibratome and secure the sample plate containing the tissue blocks.
3. Adjust the blade travel speed to 0.9 mm/s, frequency to 70 Hz, and section thickness to 140 μm.

4. Pause the cutting process after each step to retrieve the slices using a fine paintbrush, and transfer them into a 100 mm Petri dish containing HBS with 6 mM glucose (Fig. 1e) (*see* **Note 14**).

3.1.4 Dye Loading

1. Transfer up to 15 acute tissue slices, and incubate them in dye loading solution at room temperature exposed to ambient air on an orbital shaker at 40 rotations per minute for 50 min.
2. After 50 min move the stained tissue slices into a 100 mm Petri dish containing fresh HBS with 6 mM glucose until recording (*see* **Note 15**).

3.2 Calcium Imaging

3.2.1 Preparation of the Microscope for Calcium Imaging

1. Mount a suitable recording chamber to the microscope stage. The recording chamber should be compatible with a form of a perifusion system and temperature control that allows an inflow and outflow of approximately 1–2 mL/min of solution at 37 °C.
2. The chamber itself should be filled with 1–2 mL of perifusate, and the level should be kept stable to avoid drifts while recording. Exchange of solutions can be achieved either via a computer-controlled valve control system or by manually switching the solutions.
3. For visualization of whole islets, groups of acini, or larger ducts, choose an objective with a suitable magnification (20X and 25X; numerical aperture [NA] between 0.77 and 1.00). Higher magnifications are required for visualization of intracellular events.
4. In the microscope acquisition software, set the excitation and emission parameters at 488 nm and 500–700 nM, respectively.

3.2.2 Performing the Imaging

1. Place a single tissue slice into the prepared recording chamber (Subheading 2.5), and fix it with a harp to avoid drifts in recordings.
2. Use the eyepiece and stage controller to locate the structure of interest (islet, acinus, or duct) using bright-field illumination. When located, use the live recording feature in the imaging software to position the structure of interest in the center and adjust the zoom as required. The focal plane should be confined to a depth of at least approximately 15 μm below the cut surface to avoid potentially damaging cells (*see* **Note 16**).
3. Adjust the imaging parameters to optimize the signal-to-noise ratio (adjust the pinhole and laser power). Set the preferred acquisition settings (resolution of the image, averaging/binning/accumulation of lines/frames/pixels) balancing between desired spatial and temporal resolution and possibility of photobleaching). Keep the laser power and dwell time as

minimal as possible while still achieving desired detection. In addition to minimizing laser dwell time, phototoxicity can be reduced by a rest interval between two sequential illuminations of the sample (*see* **Note 17**).

4. Capturing a high-resolution image (1024 × 1024, line accumulation/averaging >12) is good practice that allows an overview of the captured structure and identification of individual cells for later analysis and data presentation.
5. The sampling frequency depends on the durations of phenomena that are observed. For capturing initial Ca^{2+} transients in response to a stimulus or slower Ca^{2+} oscillations (lasting minutes), a sampling frequency of 1–2 Hz or even lower is arguably sufficient. However, capturing faster oscillations (lasting seconds) or recordings intended for analyzing the similarity between signals of individual cells or for detection of Ca^{2+} waves require higher acquisition rates (10 Hz or higher). A resonant scanner capable of faster scanning speeds is used for these types of recordings.
6. Save recording data at the end of each experiment, and create backup copies, with one copy placed on media off-site.

3.3 Calcium Analysis

Understanding the intricacies of cellular signaling mechanisms, their alterations during disease development, and how they are affected by pharmacological interventions requires a firm characterization of the measured cellular signals. The advancement in multicellular imaging, enabling simultaneous recordings of signals from a large number of cells with high spatial and temporal resolution, has highlighted the necessity for developing suitable computational frameworks whose aim is to objectively evaluate cellular dynamics, potentially leading to method standardization. In this section, we introduce a computational pipeline for analyzing signals from beta cells obtained through multicellular high-resolution confocal microscopy in tissue slices. The first part focuses on signal preprocessing and extraction of basic cellular signaling parameters, while the second part delves into quantifying collective cellular activity and intercellular waves.

3.3.1 Characterizing Calcium Dynamics

1. Load and display time-series data. This can be done using either software of the confocal microscope or third-party software (*see* **Note 18**).
2. Check visually for motion artefacts (e.g., drifts in either xy- or z-plane) with a fast playthrough of image sequence. Discard the data if motion artifacts occur in the z-plane, e.g., cells drift out of the focal plane of recording. If motion artifacts are limited to the xy-plane, resort to xy-motion correction using either official microscope software or third-party software (*see* **Note 19**).

3. Select regions of interest (ROIs) that correspond to individual beta cells (for pros and cons of choice of software, *see* **Note 18**). Individual cells can be discriminated visually, using a combination of (i) cell boundaries on high-resolution images captured before or after experimentation and (ii) differences in calcium dynamics in individual beta cells. Search for latencies in response onsets and temporal delays due to the presence of calcium waves, best displayed by fast playthrough of image sequence while selecting ROIs.
4. Save and export time stamps, average intensity, and coordinates for selected ROIs for further analysis.
5. Compensate for photobleaching of the signal. Decreased amplitude of calcium oscillations due to photobleaching (Fig. 2a) may produce artifacts when oscillations are detected based on amplitude. Fit through segments of your data corresponding to baseline (unstimulated) $[Ca^{2+}]_i$ intensity at the beginning and end of recording using Eq. (1):

$$\mathrm{Fit}(t) = A + B * f(t) + C * e^{-D*f(t)}, \tag{1}$$

where $f(t)$ and $\mathrm{Fit}(t)$ denote absolute and fitted fluorescence at time points t. Search for parameters A, B, and C in

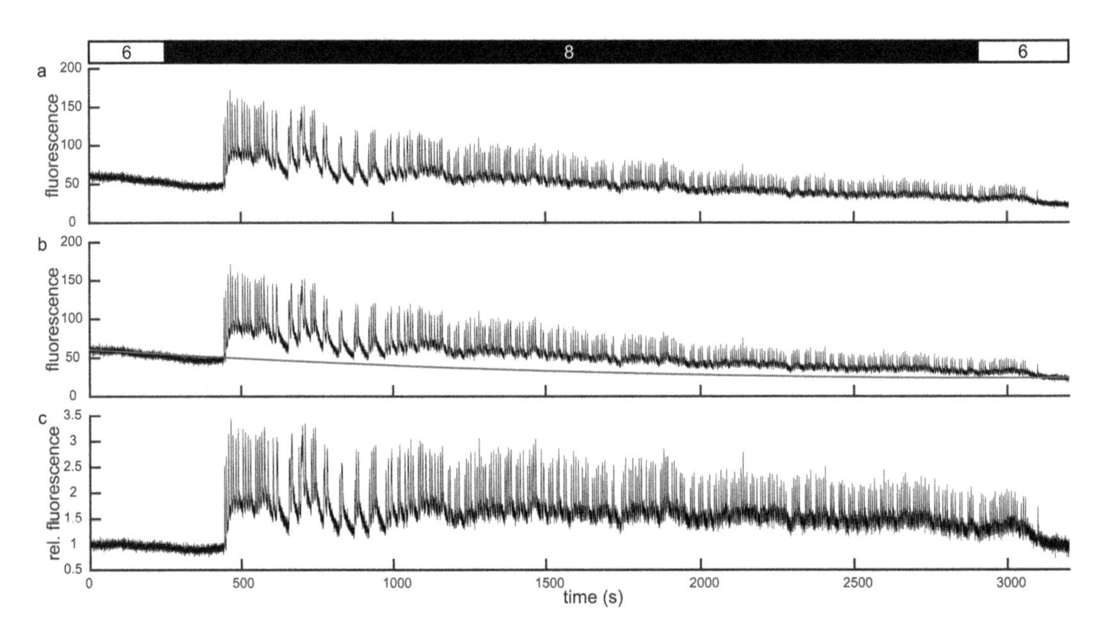

Fig. 2 Visualization of beta cell $[Ca^{2+}]_i$ and correction for photobleaching. (**a**) Raw data from a ROI corresponding to a beta cell during stimulation with 8 mM glucose (stimulation protocol indicated on the figure, in mM). Note the effect of photobleaching and decreased amplitude of $[Ca^{2+}]_i$ oscillations. (**b**, **c**) Correction for photobleaching. Segments at the beginning and at the end of the recording are used to construct a curve based on a combination of a linear and exponential decay curve (**b**, magenta). Dividing the raw data (**b**, black) with photobleaching estimation curve (**b**, magenta) yields relative fluorescence (**c**) that corrects for relative decrease in oscillation amplitude

Eq. (1) using the least square method and extrapolate the result to your time stamps (Fig. 2b, magenta curve). Calculate compensated fluorescence (Fig. 2c) using Eq. (2):

$$\frac{f(t)}{\mathrm{Fit}(t)}. \tag{2}$$

Repeat the procedure (*see* **Note 20**) for individual ROIs, then save and plot the data.

6. Time-series data can be used to analyze the following: (i) heterogeneity in response onsets and deactivations or (ii) to binarize individual oscillations:

   (i) To measure heterogeneity in the first phase of response to glucose stimulation during activation, display time-series data of beta cell population, and zoom in between extremes of response latencies (Fig. 3). Similarly, if response deactivations are to be analyzed, zoom in to the segment displaying the last oscillations. Use visual determination of response onsets for individual beta cells (Fig. 3a, red dot). Display data as individual response latencies (Fig. 3b), plot color-coded scatterplots to demonstrate the relationship with physical position within islets (Fig. 3c) or pool the data to demonstrate heterogeneity (Fig. 3d).

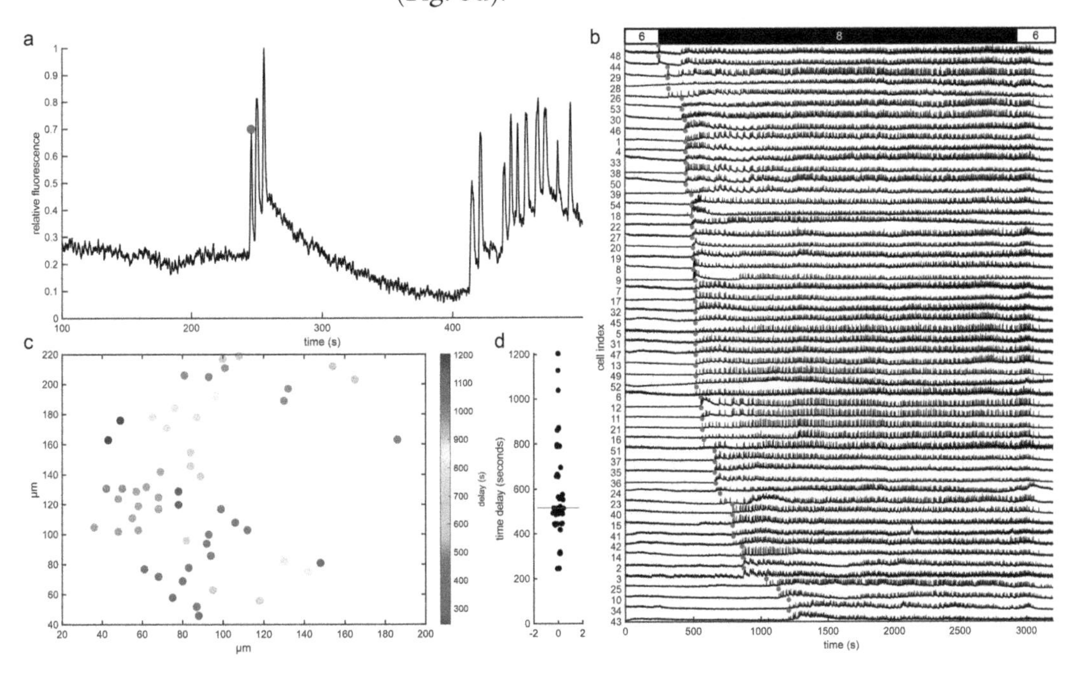

Fig. 3 Visualization and characterization of response onsets in beta cells. (**a**) Zooming in on the first phase of response to 8 mM glucose in a beta cell allows visual determination of response onset (red dot). (**b-d**) Response heterogeneity within islets can be visualized for (**b**) individual beta cells, (**c**) plotted as a function of physical position, or (**d**) pooled for the entire islet

(ii) For binarization of $[Ca^{2+}]_i$ oscillations, apply data smoothing with moving averages filter, and adjust window size not to distort the shape of oscillations (*see* **Note 21**). Detect individual oscillations by searching for local maxima and select timepoints at half-maximal amplitude as start- and endpoints for individual oscillation (Fig. 4a; *see* **Note 22**). Store start- and endpoints of individual oscillations, and plot these as raster plots for visualization (Fig. 4b). Use binarized data to calculate oscillation frequencies (as number of oscillations within selected time window) and durations (Fig. 4d, e). Calculate active time (AT) per ROI as summed activity relative to time window (Fig. 4c, f) and interoscillation interval (Fig. 4g). Plot heterogeneity in these parameters by pooling data from ROIs (Fig. 4d–g).

3.3.2 Characterizing Multicellular Activity

Beta cell signals within intact islets show oscillatory patterns that are not only different from those in individual cells but are also remarkably complex. The collective beta cell activity is marked by their functional heterogeneity, along with the influence of cell–cell contacts, electrical coupling, and paracrine signals. Despite their inherent heterogeneity and multimodal nature of oscillatory dynamics, intercellular coupling acts as a homogenizing force, enabling coordinated responses through the propagation of intercellular waves. Given its critical role in tightly regulated insulin secretion and its known alterations in diabetes, quantitatively assessing multicellular rhythmicity in islets and investigating its establishment under the influence of various functional subpopulations becomes paramount. In this subsection, we present several methodological approaches for extracting intercellular waves from multicellular imaging experiments, along with visualizations and analyses enabled by this extraction. Furthermore, we introduce the concept of network analysis for evaluating collective cellular activity, which has become a popular tool in recent years in the islet community. The analysis of intercellular activity is described in the following five steps:

1. First, it is important to visually inspect collective activity, which can be effectively accomplished using a raster plot (Fig. 5a). The raster plot displays the temporal evolution of binarized activity of all cells in the tissue slice and depicts the dynamic patterns of individual beta cell activities in much greater detail than merely analyzing the average signal of all cells.
2. Based on the raster plot and average signal, we can distinguish between two phases of activity:

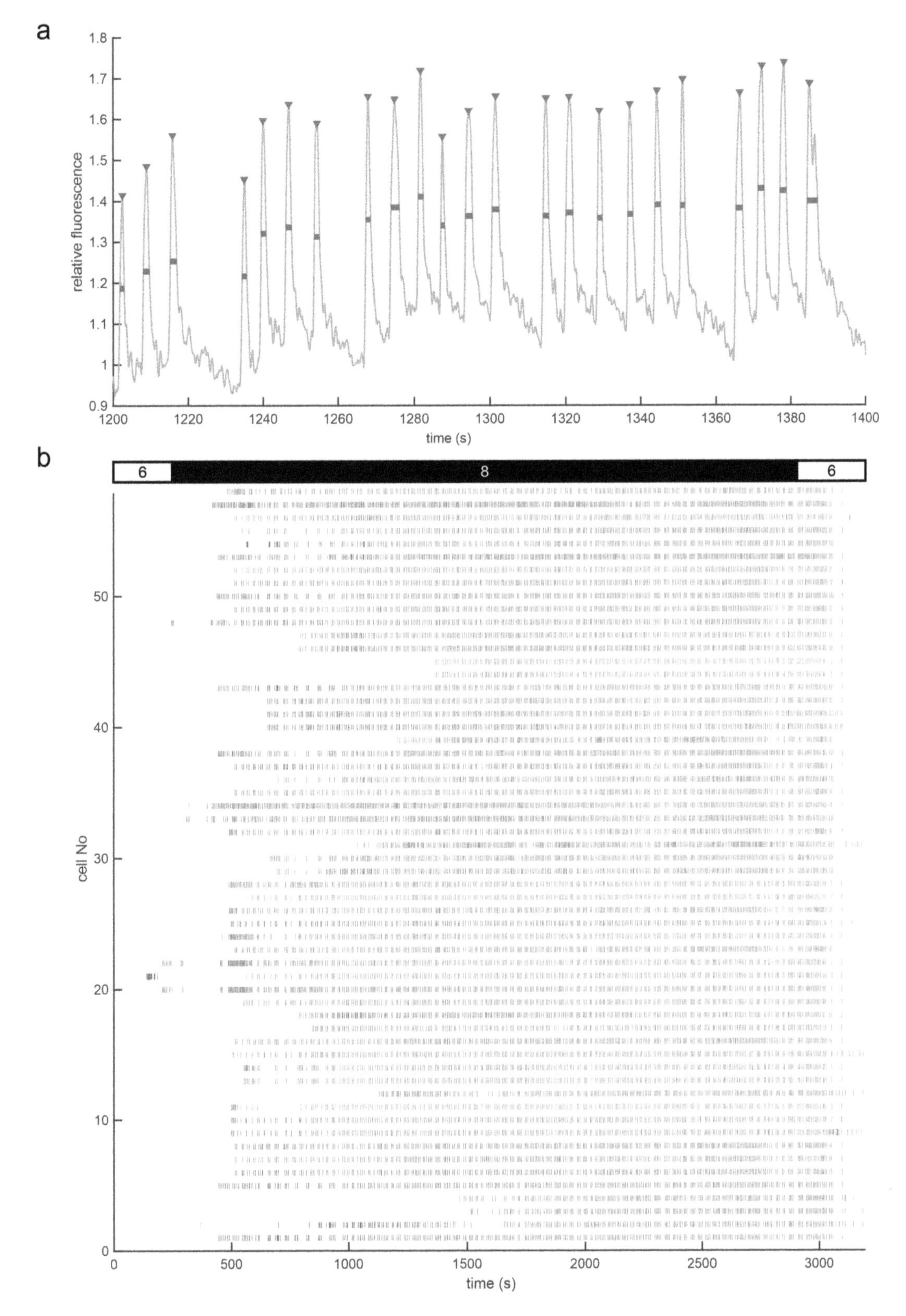

Fig. 4 Visualization and characterization of beta cell sustained oscillatory activity. (**a**) Binarization of individual fast oscillations (triangles denote local maxima and lines show timepoints at half amplitude). (**b–d**) Beta cell activity can be presented as raster plots of binarized data. Data from an islet during stimulation with 8 mM glucose (as indicated in the figure, in mM) is shown. (**d**) Visual presentation of active time. Blue lines present

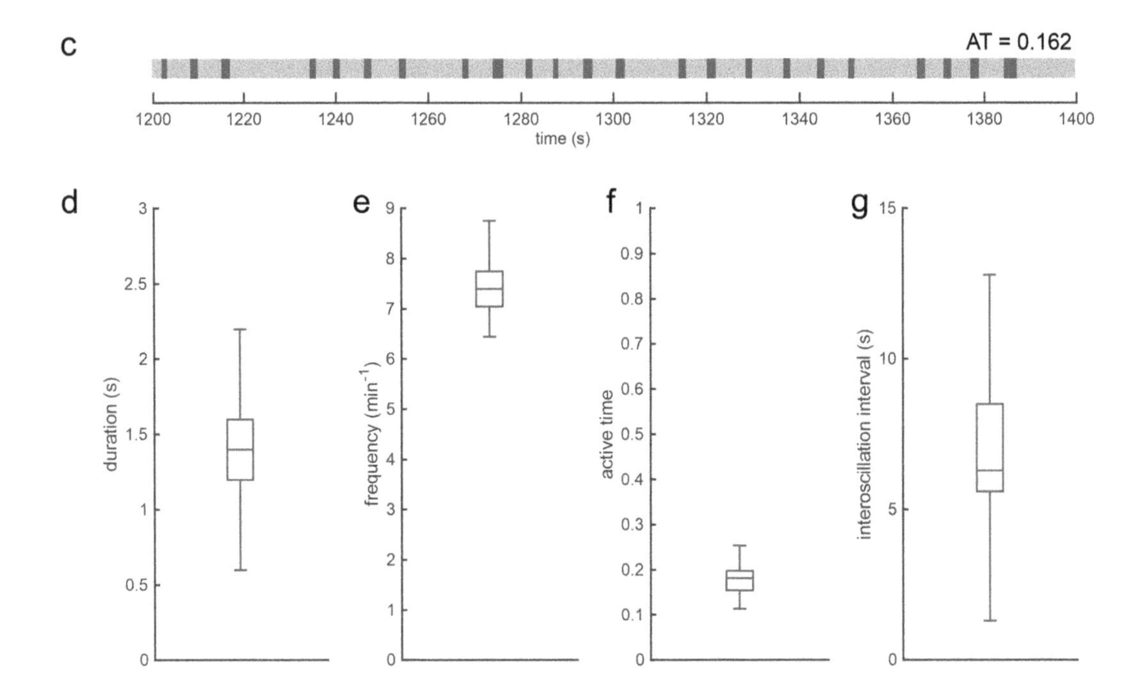

Fig. 4 (continued) binarized data from panel **a**; gray area presents selected time window. Active time (AT) is calculated as summed activity relative to selected time window. (**d–g**) Oscillation properties presented as pooled data for (**d**) duration, (**e**) frequency, (**f**) active time, and (**g**) interoscillation interval

(i) the activation phase, during which cells are recruited and gradually activated. This phase is characterized by small and heterogeneous waves that cover smaller portions of the islets. (ii) The plateau phase, characterized by stable cell activity reflected in more global waves that often spread across the majority of cells. For the analysis of cellular activity, the plateau phase is typically used, while the transient activation phase with variable activity is skipped.

3. Using the space-time clustering algorithm [37], we can efficiently distinguish individual waves and identify the sequence of cell activations within them (Fig. 5b). The colors of individual points illustrate the sequence of activations (red for first and blue for last) within individual waves, providing insight into the robustness of wave progression and also identifying cells that more frequently trigger them (i.e., wave initiator cells), as detailed in **step 5**.

4. Intercellular waves not only do not always propagate in the same direction but also involve different proportions of cells. This can be quantified by displaying the distribution of sizes of individual waves (Fig. 5c). We can see that during the plateau phase, the wave sizes range from approximately 50–100% of cells, with an average of 75% of cells. The proportion of cells

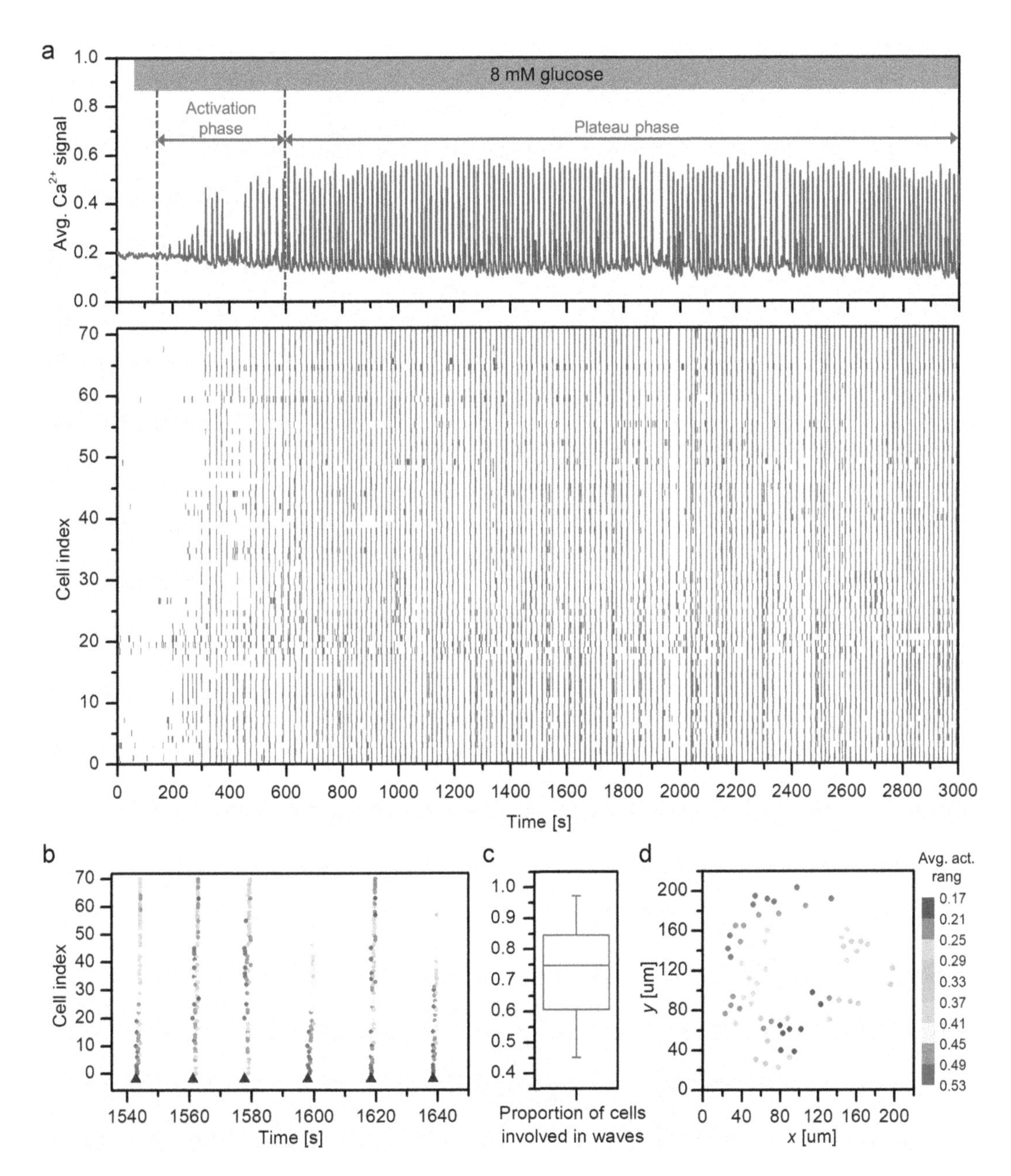

Fig. 5 Visualizing multicellular activity and intercellular Ca^{2+} wave analysis. (**a**) Average signal of fast oscillatory activity with indicated 8 mM glucose stimulation interval (upper panel) and the corresponding raster plot (lower panel) showing the binarized activity of all cells in the islet. (**b**) Raster plot of activation times of cells within six subsequent Ca^{2+} waves. The colors of dots represent the activation rank of individual cells within each wave, with red and blue colors denoting the first and last activated cell, respectively. Triangles indicate the start times of individual waves. (**c**) Boxplot representing the proportion of cells involved in all waves in this recording. The box determines the 25th and 75th percentile, whiskers extend to the minimum and maximum values, and the horizontal line within the box indicates median value. (**d**) Islet cells color-coded in accordance with their average relative rank of activation, as specified by the color bar

involved in individual waves is a glucose-dependent parameter [37] and can potentially also be modulated by pharmacological interventions and disease states [38, 39].

5. It is known that waves in islets often originate from regions with elevated excitability levels [40, 35]. Subpopulations of these cells are termed as wave initiator cells; they exhibit specific metabolic profiles and play a crucial role in the regulation of multicellular activity. There are several methods for their identification, and one of the more robust and illustrative ones is calculating the average rank of individual cells, which can be presented as color-coded relative activation ranks (Fig. 5d) where we can observe that most waves start in the region located in the upper left corner and propagate toward the lower central region.

A more advanced method to quantify multicellular activity in islets involves constructing functional beta cell networks. In these networks, individual beta cells are represented as nodes, and connections between them are established based on the temporal similarity of measured cellular dynamics, typically Ca^{2+} signals [34, 41, 36]. This approach evaluates pairwise synchronicity between cellular activity patterns and is defined as follows:

1. Processed traces (typically representing the extracted fast oscillatory component) from all cells (Fig. 6a) are utilized to calculate the Pearson correlation coefficient matrix, indicating the degree of correlation between all pairs of cells (Fig. 6b).
2. The correlation matrix is then converted into a binary connectivity matrix through a thresholding procedure. This procedure considers only those cell pairs whose correlations exceed a specified threshold R_{th} as connected (Fig. 6c).
3. The resulting connectivity matrix defines the functional network (Fig. 6d). The extracted network can then be assessed with conventional methods for network analysis.
4. We define the degree of a node (i.e., a cell) by counting its connections. By averaging the degrees of all nodes, we calculate the network's average degree, a measure of node importance. Plotting a degree distribution allows us to characterize the network. Beta cell networks belong to the category of broad-scale networks, indicating the presence of hub cells. These hubs, with a significantly higher-than-average number of connections, play a crucial role in coordinating intercellular waves [41, 35]. In the box plot (Fig. 6e), the upper whisker represents these highly connected cells, revealing that the most connected cells are linked to up to 25% of cells within the islet.
5. In the table shown in Fig. 6e, network parameters are summarized. A detailed description of their significance and calculation

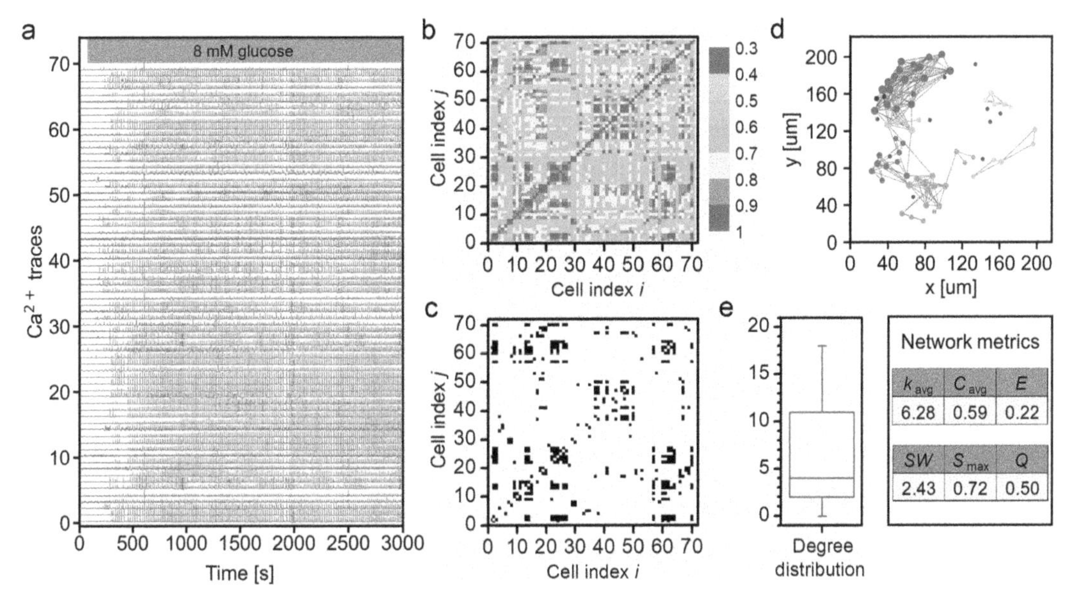

Fig. 6 Functional connectivity networks in mouse islets: from the recorded Ca^{2+} dynamics to pairwise correlation matrices and networks. (**a**) Recorded Ca^{2+} signals of all cells in the islet during stimulation with 8 mM glucose concentrations. (**b**) Pearson correlation coefficient matrix computed based on filtered and smoothed fast oscillatory signals of all cell pairs in the islet. (**c**) Extracted binary connectivity matrix based on the thresholded correlation matrix; only cell pairs i and j whose correlations exceeded $R_{th} = 0.85$ were considered connected. (**d**) The corresponding functional beta cell network designed based on fast-component only signals. The network was designed with fixed average network node degrees ($k_{avg} \approx 6$). Colored dots indicate physical locations of cells within islets, while gray lines represent functional connections between them. Colors of cells indicate different communities. (**e**) Node degree distributions for the network presented in panel **d**. The box determines the 25th and 75th percentile, whiskers may extend to the minimum and maximum values, and the full square and horizontal line within the box indicate the average and the median value, respectively. The table on the right shows the calculated network parameters: the average network node degree (k_{avg}), average clustering coefficient (C_{avg}), global efficiency (E), the small-world (SW) coefficient, the relative largest component (S_{max}), and modularity (Q)

methods is provided in Ref. [36], whereas we only briefly present them here. To evaluate the functional segregation of the beta cell network, we compute the average clustering coefficient and modularity. The latter reflects the presence of communities, depicted in Fig. 6d with different colors of cells. Relatively high values of the clustering coefficient and modularity indicate a relatively high level of segregation. For assessing functional integration, we calculate the global efficiency and the relative largest component. These measures also exhibit high values. High clustering together with high efficiency indicates small-world properties, as reflected in the small-world parameter SW, which is significantly greater than 1. The presence of small-world network characteristics is shared by beta cell networks with many other real networks [42].

3.4 Complementary Approaches

Beyond the comprehensive evaluation of Ca^{2+} dynamics in beta cells outlined in the previous sections, acute pancreas tissue slices provide a versatile platform for assessing a wide array of physiological parameters and cellular processes. This includes:

1. Functional discrimination between different types of endocrine cells [43–45], and non-endocrine tissue based on their distinct Ca^{2+} signals [46, 47] (*see* **Note 23**).
2. Measurement of hormone release in dynamic hormone secretion studies [48, 29].
3. Viability assessment, e.g., utilizing a LIVE/DEAD stain [49, 26].
4. Identification of different cell types through immunohistochemical staining techniques [25].
5. Electrophysiological characterization of islet cells with the patch-clamp method, enabling measurement of transmembrane electrical currents, potential changes, and exocytosis [50–54, 12].

4 Notes

1. To stimulate beta cells, solutions with increased concentrations of glucose can be prepared. All solutions should be prepared in advance and can be kept in the refrigerator for up to 2 weeks.
2. As an alternative, you can prepare a 10X stock solution using the following ingredients: 1250 mM NaCl, 260 mM $NaHCO_3$, 30 mM myo-inositol, 25 mM KCl, 20 mM Na-pyruvate, 12.5 mM NaH_2PO_4, and 5 mM ascorbic acid. When the extracellular buffer is required, mix 100 mL of the stock solution with 800 mL of MilliQ water, and add 2 mL of 1 M $CaCl_2$, 1 mL of 1 M $MgCl_2$, 0.455 mL of lactic acid, and 1.08 g of glucose. Fill up to 1000 mL with MilliQ water and check the osmolarity of the solution using an osmometer.
3. It is recommended to use tubing made of inert materials such as silicone and polytetrafluoroethylene (PTFE) to prevent any reactions with the carbogen gas.
4. If carbogen is unavailable, this buffer can be used instead of ECS.
5. If preparing the agarose in advance, store it in the refrigerator. Before use, warm the agarose in the microwave oven until it liquefies again. Transfer the Erlenmeyer flask into a water bath pre-warmed to 40 °C to cool it down and keep it liquid. Reuse the agarose up to five times using this method.

6. Pluronic acid may solidify over time, requiring heating to return it to a liquid state before use.
7. Cover the Petri dish containing the dye solution with aluminum foil prior to and during the dye loading to shield it from light exposure.
8. Avoid pushing any agarose into the needle to prevent it from hardening and blocking the needle. Since injection of agarose is one of the crucial steps in slice preparation, use larger animal specimens with larger ducts for practice (e.g., NMRI mice instead of Bl6J mice).
9. Instead of clamping the common bile duct at the papilla of Vater, one might consider clamping across the duodenum proximally and distally from the papilla. One or two straight or curved hemostats may be used depending on the skill of the person performing this task.
10. This step is the most crucial and delicate in slice preparation. The volume of agarose injected from the syringe is not a good marker of satisfactory injection as backflow at the injection site and leakage into the duodenum usually exceeds the volume injected into the pancreatic ductal tree. It is crucial to note that successful injections can be accomplished with minimal observable changes in syringe volume.
11. It is optimal to start the pancreas extraction procedure from the splenic portion of the pancreas. Begin by carefully flipping the gut and the injected pancreas to the right side of the mouse. Subsequently, extract the spleen, and delicately separate the pancreatic tissue from surrounding organs and structures within the abdominal cavity. Endeavor to avoid exerting pressure on the pancreas with the forceps.
12. Pay special attention to remove any connective and fatty tissue as well as blood vessels as they might cause tissue tears during slicing procedure.
13. Maintaining the ECS at low temperature throughout the slicing procedure minimizes enzymatic activity and facilitates smooth cutting. Additional crushed ice and/or ice cubes should be added if necessary.
14. Similar technique, with some species-specific adjustments, can also be applied for production of human, rat, and porcine pancreas tissue slices [25, 48, 55].
15. To prolong the survival time of the slices, replace the HBS every 2–3 h.
16. To distinguish islets from the surrounding tissue, especially in mouse models with altered tissue architecture or in tissue slices obtained from non-mouse pancreas, the detection of backscattered laser light can be used.

17. In case of unacceptable rates of photobleaching during acquisition, stop the recording, and decrease the laser power to a point that signal-to-noise ratio is still sufficient for later analysis. In case of mechanical drift, check for temperature changes and changes in solution volume in the recording chamber.
18. Bio-formats (https://www.openmicroscopy.org/bio-formats/) offers an open Java library with an application interface compatible with ImageJ, Matlab, and others. Performing analysis with third-party software may surpass often limited functionalities of the confocal microscope-related firmware; however, it requires additional training. Moreover, loading, displaying, and processing data is time-consuming for large datasets. These alternatives may outperform firmware, or, if computing power with strong graphic cards is available, developing custom software that utilizes GPU to read and analyze large data may speed up the process.
19. Several applications offer image registration to correct drifts, e.g., Manual Drift Correction Plugin in ImageJ (https://imagej.net/plugins/manual-drift-correction) based on preset image landmarks or automatic intensity-based image registration in Matlab (https://www.mathworks.com/help/images/intensity-based-automatic-image-registration.html).
20. Creating batch processing for fitting procedure may save time. In case your data consist of plateau oscillatory activity without data for substimulatory $[Ca^{2+}]_i$, using whole recording as input for Eq. (1) may successfully correct for bleaching.
21. Oversized window size of sliding average filter can significantly distort the shape of fast oscillations resulting in overestimation of their duration. Appropriate smoothing can, on the other hand, improve signal-to-noise ratio and oscillation detection.
22. Alternatively, start- and endpoints can be detected as thresholded timepoints of fluorescence derivative or as their inflection points. Although these methods can present a more precise description of fast oscillation duration, they can prove to be less accurate in noisier data. In such cases, a half-amplitude-based approach is robust in detecting relative changes (e.g., interspecies, due to pharmacological intervention); their accuracy in absolute terms may be improved by selecting timepoints at larger amplitude (e.g., 90%).
23. For the functional discrimination of individual cellular types, it is crucial to select an appropriate stimulus that elicits a cell-specific response. For example, low glucose, adrenalin, or kainite can be used for alpha cells [45, 43, 44], ghrelin for delta cells [56, 43], acetylcholine or cholecystokinin for acinar tissue [46], and bile acids for ductal cells [57, 47].

5 Conclusions

The use of acute pancreas tissue slices for calcium imaging provides a valuable tool for the study of physiological and pathophysiological phenomena in various pancreatic cells offering insight into intra- and intercellular mechanisms of pancreatic function and dysfunction in situ. The methods described within this chapter offer only a sample of what can be studied with the use of the acute pancreas tissue slice in conjunction with other methodologies. Future research utilizing this platform can expand on current knowledge in the context of diabetes as well as other metabolic and pancreatic disorders, potentially leading to more effective treatment and management strategies for these conditions.

References

1. Petersen OH, Michalak M, Verkhratsky A (2005) Calcium signalling: past, present and future. Cell Calcium 38(3–4):161–169. https://doi.org/10.1016/j.ceca.2005.06.023
2. Berridge MJ (2012) Calcium signalling remodelling and disease. Biochem Soc Trans 40(2): 297–309. https://doi.org/10.1042/bst20110766
3. Gilon P, Chae HY, Rutter GA, Ravier MA (2014) Calcium signaling in pancreatic beta-cells in health and in type 2 diabetes. Cell Calcium 56(5):340–361. https://doi.org/10.1016/j.ceca.2014.09.001
4. Skelin Klemen M, Dolensek J, Slak Rupnik M, Stozer A (2017) The triggering pathway to insulin secretion: functional similarities and differences between the human and the mouse beta cells and their translational relevance. Islets 9(6):109–139. https://doi.org/10.1080/19382014.2017.1342022
5. Evans-Molina C (2024) The ailing β-cell in diabetes: insights from a trip to the ER: the 2023 outstanding scientific achievement award lecture. Diabetes 73(4):545–553. https://doi.org/10.2337/dbi23-0030
6. Skelin Klemen M, Kopecky J, Dolenšek J, Stožer A (2023) Human beta cell functional adaptation and dysfunction in insulin resistance and its reversibility. Nephron 148(2):78–84. https://doi.org/10.1159/000534667
7. Rorsman P, Abrahamsson H, Gylfe E, Hellman B (1984) Dual effects of glucose on the cytosolic Ca2+ activity of mouse pancreatic beta-cells. FEBS Lett 170(1):196–200. https://doi.org/10.1016/0014-5793(84)81398-8
8. Grapengiesser E, Gylfe E, Hellman B (1988) Glucose-induced oscillations of cytoplasmic Ca2+ in the pancreatic beta-cell. Biochem Biophys Res Commun 151(3):1299–1304. https://doi.org/10.1016/s0006-291x(88)80503-5
9. Valdeolmillos M, Santos RM, Contreras D, Soria B, Rosario LM (1989) Glucose-induced oscillations of intracellular Ca2+ concentration resembling bursting electrical activity in single mouse islets of Langerhans. FEBS Lett 259(1): 19–23. https://doi.org/10.1016/0014-5793(89)81484-X
10. Gilon P, Henquin JC (1992) Influence of membrane potential changes on cytoplasmic Ca2+ concentration in an electrically excitable cell, the insulin-secreting pancreatic B-cell. J Biol Chem 267(29):20713–20720
11. Stožer A, Dolenšek J, Rupnik MS (2013) Glucose-stimulated calcium dynamics in islets of Langerhans in acute mouse pancreas tissue slices. PLoS One 8(1):e54638. https://doi.org/10.1371/journal.pone.0054638
12. Dolensek J, Stozer A, Skelin Klemen M, Miller EW, Slak Rupnik M (2013) The relationship between membrane potential and calcium dynamics in glucose-stimulated beta cell syncytium in acute mouse pancreas tissue slices. PLoS One 8(12):e82374. https://doi.org/10.1371/journal.pone.0082374
13. Chen C, Chmelova H, Cohrs CM, Chouinard JA, Jahn SR, Stertmann J, Uphues I, Speier S (2016) Alterations in β-cell calcium dynamics and efficacy outweigh islet mass adaptation in compensation of insulin resistance and prediabetes onset. Diabetes 65(9):2676–2685. https://doi.org/10.2337/db15-1718
14. Jacob S, Köhler M, Tröster P, Visa M, García-Prieto CF, Alanentalo T, Moede T, Leibiger B,

Leibiger IB, Berggren PO (2020) In vivo Ca (2+) dynamics in single pancreatic β cells. FASEB J 34(1):945–959. https://doi.org/10.1096/fj.201901302RR

15. Fernandez J, Valdeolmillos M (2000) Synchronous glucose-dependent [Ca(2+)](i) oscillations in mouse pancreatic islets of Langerhans recorded in vivo. FEBS Lett 477(1–2):33–36. https://doi.org/10.1016/s0014-5793(00)01631-8
16. Reissaus CA, Piñeros AR, Twigg AN, Orr KS, Conteh AM, Martinez MM, Kamocka MM, Day RN, Tersey SA, Mirmira RG, Dunn KW, Linnemann AK (2019) A versatile, portable intravital microscopy platform for studying beta-cell biology in vivo. Sci Rep 9(1):8449. https://doi.org/10.1038/s41598-019-44777-0
17. Warburg O, Minami S (1923) Versuche an Überlebendem Carcinom-gewebe. Klin Wochenschr 2(17):776–777. https://doi.org/10.1007/BF01712130
18. Stuart GJ, Dodt HU, Sakmann B (1993) Patch-clamp recordings from the soma and dendrites of neurons in brain slices using infrared video microscopy. Pflugers Arch 423(5):511–518. https://doi.org/10.1007/BF00374949
19. Moser T, Neher E (1997) Rapid exocytosis in single chromaffin cells recorded from mouse adrenal slices. J Neurosci 17(7):2314–2323. https://doi.org/10.1523/jneurosci.17-07-02314.1997
20. Parrish AR, Gandolfi AJ, Brendel K (1995) Precision-cut tissue slices: applications in pharmacology and toxicology. Life Sci 57(21):1887–1901. https://doi.org/10.1016/0024-3205(95)02176-j
21. Speier S, Rupnik M (2003) A novel approach to in situ characterization of pancreatic beta-cells. Arch Eur J Physiol 446(5):553–558. https://doi.org/10.1007/s00424-003-1097-9
22. Dolenšek J, Rupnik MS, Stožer A (2015) Structural similarities and differences between the human and the mouse pancreas. Islets 7(1):e1024405. https://doi.org/10.1080/19382014.2015.1024405
23. Gilon P, Jonas JC, Henquin JC (1994) Culture duration and conditions affect the oscillations of cytoplasmic calcium concentration induced by glucose in mouse pancreatic islets. Diabetologia 37(10):1007–1014
24. Rupnik M (2009) The physiology of rodent beta-cells in pancreas slices. Acta Physiol (Oxf) 195(1):123–138. https://doi.org/10.1111/j.1748-1716.2008.01927.x
25. Marciniak A, Cohrs CM, Tsata V, Chouinard JA, Selck C, Stertmann J, Reichelt S, Rose T, Ehehalt F, Weitz J, Solimena M, Slak Rupnik M, Speier S (2014) Using pancreas tissue slices for in situ studies of islet of Langerhans and acinar cell biology. Nat Protoc 9(12):2809–2822. https://doi.org/10.1038/nprot.2014.195
26. Stožer A, Dolenšek J, Križančić Bombek L, Pohorec V, Slak Rupnik M, Klemen MS (2021) Confocal laser scanning microscopy of calcium dynamics in acute mouse pancreatic tissue slices. J Vis Exp 170:e62293. https://doi.org/10.3791/62293
27. Meneghel-Rozzo T, Rozzo A, Poppi L, Rupnik M (2004) In vivo and in vitro development of mouse pancreatic beta-cells in organotypic slices. Cell Tissue Res 316(3):295–303. https://doi.org/10.1007/s00441-004-0886-6
28. Postić S, Pfabe J, Sarikas S, Ehall B, Pieber T, Korošak D, Slak Rupnik M, Huang Y-C (2023) Tracking Ca2+ dynamics in NOD mouse islets during spontaneous diabetes development. Diabetes 72(9):1251–1261. https://doi.org/10.2337/db22-0952
29. Panzer JK, Garcia PA, Pugliese A (2024) Generating human pancreatic tissue slices to study endocrine and exocrine pancreas physiology. J Vis Exp 205:e66468. https://doi.org/10.3791/66468
30. Stožer A, Klemen MS, Gosak M, Bombek LK, Pohorec V, Rupnik MS, Dolenšek J (2021) Glucose-dependent activation, activity, and deactivation of beta cell networks in acute mouse pancreas tissue slices. American Journal of Physiology-Endocrinology and Metabolism 321(2):E305–E323. https://doi.org/10.1152/ajpendo.00043.2021
31. Skelin Klemen M, Dolenšek J, Križančić Bombek L, Pohorec V, Gosak M, Slak Rupnik M, Stožer A (2023) The effect of forskolin and the role of Epac2A during activation, activity, and deactivation of beta cell networks. Front Endocrinol 14:1225486. https://doi.org/10.3389/fendo.2023.1225486
32. Paradiž Leitgeb E, Kerčmar J, Križančić Bombek L, Pohorec V, Skelin Klemen M, Slak Rupnik M, Gosak M, Dolenšek J, Stožer A (2024) Exendin-4 affects calcium signalling predominantly during activation and activity of beta cell networks in acute mouse pancreas tissue slices. Front Endocrinol (Lausanne) 14:1315520. https://doi.org/10.3389/fendo.2023.1315520
33. Pohorec V, Križančić Bombek L, Skelin Klemen M, Dolenšek J, Stožer A (2022)

Glucose-stimulated calcium dynamics in beta cells from male C57BL/6J, C57BL/6N, and NMRI mice: a comparison of activation, activity, and deactivation properties in tissue slices. Front Endocrinol 13:867663. https://doi.org/10.3389/fendo.2022.867663

34. Stožer A, Gosak M, Dolenšek J, Perc M, Marhl M, Rupnik MS, Korošak D (2013) Functional connectivity in islets of Langerhans from mouse pancreas tissue slices. PLoS Comput Biol 9(2):e1002923. https://doi.org/10.1371/journal.pcbi.1002923
35. Šterk M, Dolenšek J, Skelin Klemen M, Križančić Bombek L, Paradiž Leitgeb E, Kerčmar J, Perc M, Slak Rupnik M, Stožer A, Gosak M (2023) Functional characteristics of hub and wave-initiator cells in β cell networks. Biophys J 122(5):784–801. https://doi.org/10.1016/j.bpj.2023.01.039
36. Stožer A, Šterk M, Paradiž Leitgeb E, Markovič R, Skelin Klemen M, Ellis CE, Križančić Bombek L, Dolenšek J, MacDonald PE, Gosak M (2022) From isles of Königsberg to islets of Langerhans: examining the function of the endocrine pancreas through network science. Front Endocrinol 13:922640. https://doi.org/10.3389/fendo.2022.922640
37. Gosak M, Stožer A, Markovič R, Dolenšek J, Perc M, Rupnik MS, Marhl M (2017) Critical and supercritical spatiotemporal calcium dynamics in beta cells. Front Physiol 8. https://doi.org/10.3389/fphys.2017.01106
38. Gosak M, Yan-Do R, Lin H, MacDonald PE, Stožer A (2022) Ca2+ oscillations, waves, and networks in islets from human donors with and without type 2 diabetes. Diabetes 71(12):2584–2596. https://doi.org/10.2337/db22-0004
39. Šterk M, Križančić Bombek L, Skelin Klemen M, Slak Rupnik M, Marhl M, Stožer A, Gosak M (2021) NMDA receptor inhibition increases, synchronizes, and stabilizes the collective pancreatic beta cell activity: insights through multilayer network analysis. PLoS Comput Biol 17(5):e1009002. https://doi.org/10.1371/journal.pcbi.1009002
40. Westacott MJ, Ludin NWF, Benninger RKP (2017) Spatially organized β-cell subpopulations control electrical dynamics across islets of Langerhans. Biophys J 113(5):1093–1108. https://doi.org/10.1016/j.bpj.2017.07.021
41. Johnston Natalie R, Mitchell Ryan K, Haythorne E, Pessoa Maria P, Semplici F, Ferrer J, Piemonti L, Marchetti P, Bugliani M, Bosco D, Berishvili E, Duncanson P, Watkinson M, Broichhagen J, Trauner D, Rutter Guy A, Hodson David J (2016) Beta cell hubs dictate pancreatic islet responses to glucose. Cell Metab 24(3):389–401. https://doi.org/10.1016/j.cmet.2016.06.020
42. Watts DJ, Strogatz SH (1998) Collective dynamics of 'small-world' networks. Nature 393:440. https://doi.org/10.1038/30918
43. Hamilton A, Vergari E, Miranda C, Tarasov AI (2019) Imaging calcium dynamics in subpopulations of mouse pancreatic islet cells. J Vis Exp 153. https://doi.org/10.3791/59491
44. Pohorec V, Zadravec N, Turk M, Dolenšek J, Stožer A (2023) Alpha cell stimulus-secretion coupling and intercellular interactions in health and type 2 diabetes. Acta Medico-Biotechnica 16(1):21. https://doi.org/10.18690/actabiomed.246
45. Panzer JK, Cohrs CM, Speier S (2020) Using pancreas tissue slices for the study of islet physiology. In: King AJF (ed) Animal models of diabetes: methods and protocols. Springer, New York, pp 301–312. https://doi.org/10.1007/978-1-0716-0385-7_20
46. Marolt U, Paradiž Leitgeb E, Pohorec V, Lipovšek S, Venglovecz V, Gál E, Ébert A, Menyhárt I, Potrč S, Gosak M, Dolenšek J, Stožer A (2022) Calcium imaging in intact mouse acinar cells in acute pancreas tissue slices. PLoS One 17(6):e0268644. https://doi.org/10.1371/journal.pone.0268644
47. Gál E, Dolenšek J, Stožer A, Pohorec V, Ébert A, Venglovecz V (2019) A novel in situ approach to studying pancreatic ducts in mice. Front Physiol 10:938. https://doi.org/10.3389/fphys.2019.00938
48. Panzer JK, Caicedo A (2023) Protocol to generate and utilize pancreatic tissue slices to study endocrine and exocrine physiology in situ from mouse and human tissue. STAR Protoc 4(3):102399. https://doi.org/10.1016/j.xpro.2023.102399
49. Qadir MMF, Álvarez-Cubela S, Weitz J, Panzer JK, Klein D, Moreno-Hernández Y, Cechin S, Tamayo A, Almaça J, Hiller H, Beery M, Kusmartseva I, Atkinson M, Speier S, Ricordi C, Pugliese A, Caicedo A, Fraker CA, Pastori RL, Domínguez-Bendala J (2020) Long-term culture of human pancreatic slices as a model to study real-time islet regeneration. Nat Commun 11(1):3265. https://doi.org/10.1038/s41467-020-17040-8
50. Huang YC, Rupnik M, Gaisano HY (2011) Unperturbed islet α-cell function examined in mouse pancreas tissue slices. J Physiol 589(Pt 2):395–408. https://doi.org/10.1113/jphysiol.2010.200345
51. Speier S, Yang SB, Sroka K, Rose T, Rupnik M (2005) KATP-channels in beta-cells in tissue

slices are directly modulated by millimolar ATP. Mol Cell Endocrinol 230(1–2):51–58. https://doi.org/10.1016/j.mce.2004.11.002

52. Huang YC, Rupnik MS, Karimian N, Herrera PL, Gilon P, Feng ZP, Gaisano HY (2013) In situ electrophysiological examination of pancreatic α cells in the streptozotocin-induced diabetes model, revealing the cellular basis of glucagon hypersecretion. Diabetes 62(2): 519–530. https://doi.org/10.2337/db11-0786
53. Skelin M, Rupnik M (2011) cAMP increases the sensitivity of exocytosis to Ca^2+ primarily through protein kinase A in mouse pancreatic beta cells. Cell Calcium 49(2):89–99. https://doi.org/10.1016/j.ceca.2010.12.005
54. Skelin Klemen M, Dolenšek J, Stožer A, Slak Rupnik M (2014) Measuring exocytosis in endocrine tissue slices. In: Thorn P (ed) Exocytosis methods. Humana Press, Totowa, pp 127–146. https://doi.org/10.1007/978-1-62703-676-4_7
55. Liang T, Dolai S, Xie L, Winter E, Orabi AI, Karimian N, Cosen-Binker LI, Huang YC, Thorn P, Cattral MS, Gaisano HY (2017) Ex vivo human pancreatic slice preparations offer a valuable model for studying pancreatic exocrine biology. J Biol Chem 292(14): 5957–5969. https://doi.org/10.1074/jbc.M117.777433
56. DiGruccio MR, Mawla AM, Donaldson CJ, Noguchi GM, Vaughan J, Cowing-Zitron C, van der Meulen T, Huising MO (2016) Comprehensive alpha, beta and delta cell transcriptomes reveal that ghrelin selectively activates delta cells and promotes somatostatin release from pancreatic islets. Mol Metab 5(7): 449–458. https://doi.org/10.1016/j.molmet.2016.04.007
57. Molnár R, Madácsy T, Varga Á, Németh M, Katona X, Görög M, Molnár B, Fanczal J, Rakonczay Z Jr, Hegyi P, Pallagi P, Maléth J (2020) Mouse pancreatic ductal organoid culture as a relevant model to study exocrine pancreatic ion secretion. Lab Investig 100(1): 84–97. https://doi.org/10.1038/s41374-019-0300-3

Chapter 18

Real-Time Imaging of Calcium Dynamics in Human Sperm After Precise Single-Cell Stimulation

Melissa A. White, Roy Cohen, and Alexander J. Travis

Abstract

Calcium signaling is a critical regulator of sperm activation and function during the processes of capacitation and fertilization. Here, we describe a combined method for calcium imaging of single, live human sperm in response to stimuli administered with a precisely targeted delivery technique. This protocol is an adaptation of techniques developed for studies of murine sperm [1, 2], and enables real-time monitoring of human sperm calcium dynamics with high spatiotemporal resolution and concurrent detection of acrosome exocytosis (AE), a functional endpoint of sperm capacitation and requirement for physiological fertilization.

The described imaging technique provides a valuable tool for exploration of calcium regulation in human sperm, which is essential to answer important questions and knowledge gaps regarding the link between calcium dynamics, AE, and fertilization. The versatility of this technique can be amplified through use of various indicator dyes or integration with pharmacological strategies such as pre-treating sperm with inhibitors or activators targeting specific receptors, channels, or intracellular signaling pathways of interest. Beyond fundamental inquiries into sperm physiology, this method can also be applied to assess the impact of potential contraceptive compounds on calcium signaling, AE, and membrane integrity.

Key words Acrosome exocytosis, Calcium channels, Calcium signaling, Calcium transients, Calcium waves, Lipid regulation, Real-time imaging, Single-cell stimulation

1 Introduction

Sperm are transcriptionally and translationally quiescent. Consequently, most regulation of sperm physiological processes is achieved posttranslationally via a number of signaling mechanisms, involving various channels (e.g., CatSper [3, 4]), or intracellular pathways (e.g., soluble adenylyl cyclase, cyclic adenosine monophosphate, and protein tyrosine phosphorylation [5–7]). Correct spatial and temporal activation of these signaling pathways is critical for sperm to gain and maintain fertilizing ability. Immediately after ejaculation, sperm are unable to fertilize eggs. Instead, as the sperm progress up the female reproductive tract, they encounter and respond to various stimuli that initiate the process of sperm

Caroline M. Gorvin (ed.), *Calcium Signaling: Methods and Protocols*, Methods in Molecular Biology, vol. 2861, https://doi.org/10.1007/978-1-0716-4164-4_18,

functional maturation, or capacitation [8–10]. Although these stimuli vary among species, for most mammalian sperm, including human sperm, the key stimuli are removal of sterols from the plasma membrane and influx of bicarbonate and calcium [7]. Capacitation is required for sperm to undergo acrosomal exocytosis (AE) and subsequent fertilization.

One of the most important signaling molecules regulating capacitation and AE is calcium (Ca^{2+}). Ca^{2+} is a common secondary messenger molecule which relays signals received by the cell's surface to target intracellular effector proteins. The regulation of AE and fertilization by calcium is complex, involving interactions with multiple signaling cascades. Since Ca^{2+} can neither be created nor destroyed by the cell, Ca^{2+} influx, and the concentration of intracellular calcium $[Ca^{2+}]_i$ is tightly controlled via Ca^{2+}-ATPases, Ca^{2+} channels, Na^+/Ca^{2+} exchangers, and compartments that sequester Ca^{2+} within the cell [3, 4, 11–15]. Previous studies have shown that local fluctuations in $[Ca^{2+}]_i$ can occur in highly specific regions of the cell, illustrating the capacity of Ca^{2+} to govern diverse processes across distinct spatial and temporal contexts [1]. Understanding the intricate dynamics of $[Ca^{2+}]_i$ is critical to understanding how AE and fertilization occurs. Furthermore, the identification of compounds that disrupt sperm $[Ca^{2+}]_i$ stores or signaling may aid in the development of novel contraceptives.

Here, we describe an approach for live imaging of single spermatozoa, for the purpose of investigating cytosolic Ca^{2+} dynamics in response to a precise, single-cell stimulation delivery system. This approach is adapted for human sperm from the single-cell live imaging technique used with transgenic murine sperm as described in Cohen (2023) [2]. This method combines high frame rate and high magnification confocal imaging with precise stimulation of sperm preloaded with the fluorescent, cell-permeant Ca^{2+} indicator Fluo4-AM. We demonstrate how this technique can be used to monitor Ca^{2+} dynamics spatially and temporally in the heads of non-capacitated sperm stimulated with ionophore. Because of the intrinsically high intra-acrosomal Ca^{2+} concentration, this method also enables real-time monitoring of AE, and can be used to investigate the relationship between cytosolic Ca^{2+} influx and AE. The indication for AE comes from the rapid loss of fluorescence signal in the Fluo4-loaded cell [1]. Therefore, with this technique, one can monitor how different stimuli impact both Ca^{2+} influx and AE.

While live-cell imaging combined with precise stimulation offers versatility, the main limitation is its relatively low throughput, and requirement for costly imaging and dispensing systems. Nevertheless, despite its drawbacks, this method has the potential to significantly advance our understanding of the spatial and temporal regulation of Ca^{2+} influx in human sperm and how it regulates AE.

2 Materials

1. Wide-orifice transfer pipettes (*see* **Notes 1** and **2**).
2. Wide-orifice pipette tips (*see* **Notes 1** and **2**).
3. 5 mL polypropylene tubes.
4. 15 mL conical tubes.
5. 37 °C incubator or water bath.
6. Tabletop centrifuge with 15 mL conical swinging bucket rotor (*see* **Note 3**).
7. Hemocytometer.
8. MatTek dishes: 35 mm petri dish, 20 mm Microwell.
9. Poly-D-Lysine (*see* **Note 4**).
10. Capillaries: Borosilicate glass, O.D. 2.0 mm, I.D. 1.56 mm, 10 cm length (Sutter Instruments) (*see* **Note 5**).
11. Picospritzer (*see* **Note 6**).
12. Olympus FV3000 laser scanning confocal microscope with live-cell incubator (*see* **Note** 7).
13. Enhance S-Plus cell isolation media (Vitrolife).
14. Phenol-red-free Human Tubal Fluid (HTF) buffer: KCl 4.68 mM, $MgSO_4*7H_2O$ 200 μM, KH_2PO_4 370.3 μM, $CaCl_2*2H_2O$ 2.04 mM, Glucose 2.77 mM, Na-Pyruvate 331.8 μM, Na-lactate 24 mM, HEPES Na-salt 21 mM. Adjust osmolarity to 270 mOsm with NaCl, pH to 7.4 (recipe from Irvine Scientific).
15. Ionomycin.
16. (2-Hydroxypropyl)-beta-cyclodextrin (*see* **Note 8**).
17. Fluo4-AM (*see* **Note 9**).

3 Methods

Fresh semen specimens were obtained from healthy men of prime reproductive age (21–40 years) by masturbation after at least 48 h of sexual abstinence. All human subject research adhered to protocols approved by the Cornell University Institutional Review Board (ID 1809008283).

3.1 Sperm Washing

This human sperm washing protocol is adapted from Moody et al. (2017) [16]:

1. Allow fresh human semen to liquefy at 37 °C for 30–60 min (*see* **Note 10**).

2. In a 15 mL conical tube, carefully layer 1 mL semen on top of 1 mL pre-warmed (37 °C) Enhance S-Plus cell isolation medium (Vitrolife) using a wide-orifice transfer pipette. Centrifuge at 300× *g* for 10 min in a swinging bucket rotor at 37 °C (*see* **Note 3**).
3. After the first spin, the sperm will be in a pellet at the bottom of the tube. The seminal plasma will remain above the Enhance. Using a wide-orifice transfer pipette, completely remove and discard the seminal plasma layer. Transfer the remaining Enhance and sperm pellet to a new 15 mL conical tube using a new wide-orifice pipette (*see* **Note 1**).
4. Add 4 mL pre-warmed (37 °C) phenol-red-free modified Human Tubal Fluid (PRF-mHTF), gently mix by titration with a wide-orifice pipette tip, and spin at 600× *g* for 10 min in a swinging bucket rotor at 37 °C. At the end of the spin, sperm will form a pellet at the bottom of the tube (*see* **Note 1**).
5. Using a wide-orifice pipette, remove and discard as much supernatant as possible without disturbing the sperm pellet (*see* **Note 1**).
6. Gently resuspend sperm in 100–200 μL PRF-mHTF using a wide-orifice pipette (*see* **Note 1**).
7. Calculate concentration of sperm using a hemocytometer.

8a. For capacitated sperm, 15 mM bicarbonate and 5 mM 2-hydroxypropyl-beta-cyclodextrin (2-OHCD) should be added to PRF-mHTF. Incubate 3×10^6 sperm in 300 μL capacitating media in a 5 mL polypropylene tube, for at least 3 h at 37 °C (*see* **Note 8**).

8b. For non-capacitated sperm, incubate 3×10^6 sperm in 300 μL PRF-mHTF for at least 3 h at 37 °C in a 5 mL polypropylene tube (*see* **Note 8**).

3.2 Poly-D-Lysine (PDL) Coating of Coverslip Dishes for Live-Cell Imaging

1. In the center of a coverslip dish, place 0.5 μL (0.5 mg/mL) PDL.
2. Use a 10 μL pipette tip to gently smear and disperse the PDL over one face of the coverslip (*see* **Note 4**).
3. Dry the coverslip for 10 min at 37 °C. Dishes should be kept at 37 °C until use.

3.3 Capillary Pulling and Loading for Puffing

1. Using a micropipette puller (Sutter instruments P-97, Heat, 330; Pull, 250; Velocity, 250; Time, 70), pull borosilicate glass capillaries (O.D. 2 mm, I.D. 1.56) (*see* **Note 5**).
2. Using fine-tip forceps, break off 1–2 mm from the tip of a glass capillary, creating a ~ 5 μM opening. Fill the capillary with 5× the final desired concentration of reagent used to stimulate

sperm. This will compensate for diffusion between the tip of the capillary and the target cell.

3. Gently flick the capillary to remove air bubbles, and mount on micromanipulator.

3.4 Preparing the Picospritzer Dispenser for Sperm Puffing (See Note 6)

1. Remove residual air from the tip of the capillary by activating a 1-s pulse at 40 psi. A droplet should be visible at the tip of the capillary.
2. Set the Picospritzer to generate a 10-s pulse at 5 psi.

3.5 Loading Sperm with Fluo4-AM

1. Add 1 μL of Fluo4-AM (1 mg/mL in DMSO), a cell-permeant fluorescent Ca^{2+} indicator to one 5 mL incubation tube containing 3×10^6 sperm in 300 μL. Incubate for 5 min at 37 °C (*see* **Note 9**).
2. Centrifuge sperm at 500× *g* for 2 min at 37 °C in a swinging bucket rotor.
3. Remove supernatant and resuspend in an equal amount of PRF-mHTF.

3.6 Imaging (See Note 7)

1. Add 2.7 mL PRF-mHTF with 10 mM $CaCl_2$ to a pre-warmed PDL-treated coverslip dish.
2. Add 300 μL Fluo4-AM loaded and washed sperm to coverslip dish.
3. Securely mount coverslip dish on microscope. The image acquisition setup is shown in Fig. 1.
4. Identify a field of cells to image (*see* **Note 11**).
5. Using a micromanipulator, lower the capillary ~100 μM away from the target cells.
6. Excite Fluo4 with 488 nM laser with a 500–600 nM bandpass filter (*see* **Note 12**).
7. Start image acquisition on microscope, and take 10–15 s of baseline readings.
8. After taking baseline readings, activate the dispensing system to deliver a 10-s pulse at 5 psi (*see* **Note 13**).
9. Continue imaging for 10–15 min.
10. Raise the capillary, and repeat the imaging procedure with a different field of cells (*see* **Note 14**).

3.7 Image Analysis

All image analysis was conducted in ImageJ (1.52a) using the Bio-Formats package to import and read OIR confocal image files.

1. Open the image file and if needed split the acquisition channels (typical experiment will include wide-field and fluorescence (Fluo-4) channels; Image > Color > Split Channels).

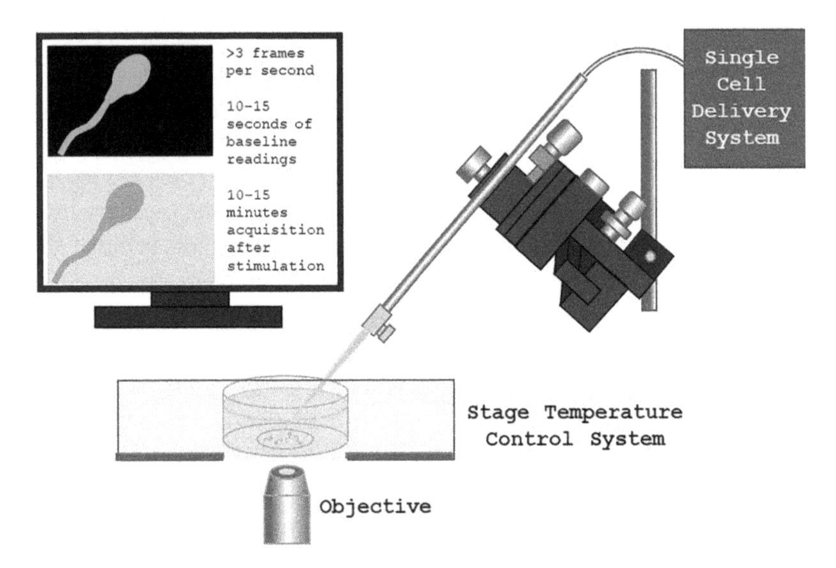

Fig. 1 System configuration. Live human sperm placed in a PDL-coated coverslip are placed in an incubation chamber at 37 °C. A borosilicate capillary attached to a delivery system is positioned near the sperm cell in the center of the imaging area. Fluorescence (Fluo-4) and wide-field channels are continuously recorded at a high frame rate before, during, and after stimulation for up to 15 min

2. Quickly scroll through the bright-field images to confirm that sperm heads remained fixed throughout the imaging process and that the imaging field remained unobstructed.
3. Open the region of interest (ROI) manager (Analyze > Tools > ROI Manager).
4. Using the circle selection tool, draw an ROI within the portion of the sperm head in which you are interested, such as the acrosomal region. Spatial determination of response location can be determined by moving the ROI around the head area and identifying the peak fluorescence response.
5. To plot fluorescence intensity as a function of frame/time, use the Z-profiler plugin (Plugins > Stacks > Z-Profiler) or Time Series Analyzer (Plugins > Stacks > Time Series Analyzer).
6. Save the spreadsheet data for further analysis in Microsoft Excel. Also save the ROI.
7. Once open in Excel, a variety of parameters can be extracted from the data. The specific parameters calculated will largely depend on the experimental design and overall goal of the research. We have described below some elementary data analyses:
   (a) In general, the shape and pattern of the response should be assessed. Descriptions of the specific patterns of single-cell Ca^{2+} influx in human sperm have been described elsewhere [17, 18]. Briefly, transient responses are

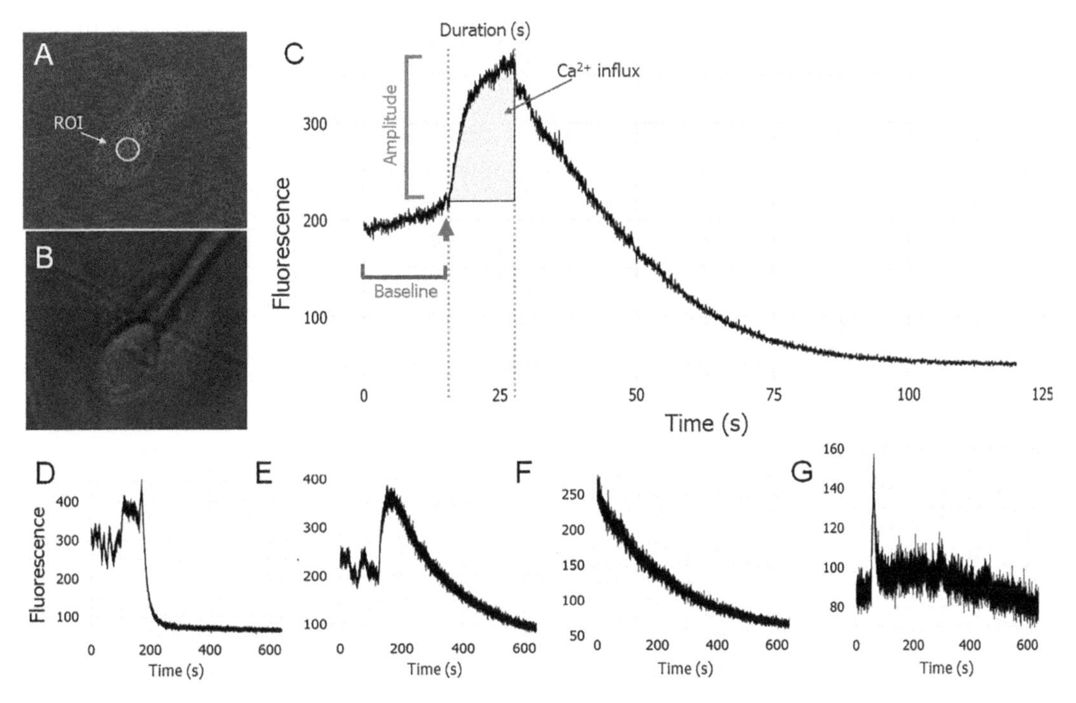

Fig. 2 Representative data and basic analysis. (**a**) A single image from the fluorescent channel (Fluo-4) showing the signal coming from the sperm's head, where brighter colors represent higher calcium concentrations. Yellow arrow points to the region of interest (ROI) where the fluorescence intensity is measured (trace shown in **c**). (**b**) Bright-field image of the sperm shown in **b**. (**c**) Representative trace of the Fluo-4 signal measured in human sperm shown in **b** (measured at the indicated ROI) stimulated with ionophore (15 s into the recording (red arrow)). Some of the basic analysis parameters are shown here, including baseline signal, fluorescence signal amplitude, rise duration, and Ca^{2+} Influx. (**d**) Representative data trace showing a calcium rise followed by a rapid signal loss that is characteristic of AE. (**e**) Representative data trace showing a calcium wave. (**f**) Representative data trace showing the signal coming from a sperm that did not get activated by the ionophore stimulation. (**g**) Representative trace of a Ca^{2+} transient

typically observed as fast-rising peaks followed by rapid decay to baseline levels (Fig. 2g), and are different from Ca^{2+} waves (Fig. 2e) that demonstrate much slower decay rates, if at all.

(b) The baseline Ca^{2+} level (F_0) can be calculated by averaging the fluorescence intensity values before addition of the stimulus.

(c) The amplitude of the response can be determined by subtracting the baseline fluorescence value from the maximum value.

(d) Latency from stimulation to beginning of Ca^{2+} response is an important metric that is typically measured.

(e) The rise duration refers to the amount of time it takes to reach the maximum response.

(f) The peak time can be calculated either from the beginning of the recording or from the time point where the Ca^{2+} rise is first observed.

(g) The Ca^{2+} influx can be calculated from integrating the area under the signal increase.

3.8 Representative Results

In Fig. 2, we show representative analyses of non-capacitated sperm stimulated with 10 μM ionomycin. Ionomycin is a Ca^{2+} ionophore that will stimulate a non-physiological Ca^{2+} rise in human sperm. As shown, in Fig. 2, ionomycin will typically trigger a rapid influx of Ca^{2+}, followed by signal decay that in some cases is a result of AE.

4 Notes

1. Sperm are extremely vulnerable to damage from shear forces. Therefore, throughout the protocol, all pipetting of live sperm should be done with wide-orifice pipettes. It is particularly important to utilize a fresh wide-orifice pipette after the first wash so that contaminating seminal plasma is not inadvertently added back to the washed cells.
2. Samples with hyperviscous seminal plasma should not be passed through syringe needles because of their sensitivity to membrane damage from shearing. Passage through long, narrow-gauge needles or Pasteur pipettes will succeed in breaking down the viscosity, but will impair sperm viability/motility over time [16]. It has also been shown that treatment of seminal plasma with enzymes such as chymotrypsin and bromelain can initially improve motility but over time will reduce viability and interfere with capacitation [16].
3. A swinging bucket rotor must be used when centrifuging live sperm to minimize damaging interactions between the sperm membrane and the sides of the centrifuge tubes.
4. We have found that the best way to disperse the PDL droplet is with a gentle but vigorous cross-hatching motion.
5. Capillaries can be pulled ahead of time and stored in a covered box.
6. While priming the Picospritzer or other air pressure-based, single-cell dispensing system, it is necessary to wear proper eye protection.
7. Our images were taken on an Olympus FV3000 confocal laser scanning microscope equipped with a Tokai Hit STXF stage-top incubator. Given the temperature-dependent nature of membrane lipid dynamics and the crucial role of plasma membrane stability in sperm function, it is strongly recommended

to conduct imaging at 37 °C. Microscope settings were optimized for high frame rate acquisition: 33 frames per second using a 100x objective and additional 6–7× with galvanometer scanning. These magnification and zoom settings were sufficient to image 1–3 live cells per image. A minimum frame rate of three frames per second is recommended. If available, Z-position drift correction should be activated to prevent shifting of the focal plane. A diagram of the imaging setup is shown in Fig. 1.

8. Other sterol acceptors, such as bovine serum albumin, can be used to stimulate capacitation. The concentration of the sterol acceptor as well as the capacitation incubation time can be varied according to the experimental design. Longer times will result in higher percentages of sperm that capacitate but will also lead to increased percentages of sperm that undergo spontaneous AE and will be lost for analysis.
9. We have noticed that human sperm tend to take up Fluo4-AM very quickly. Overloading sperm with dye may cause oversaturation of images and may introduce intracellular buffering artifacts.
10. It is vital to completely remove the seminal plasma layer if capacitated sperm are required for the study, because seminal plasma contains semenogelin that inhibits capacitation [19].
11. When identifying optimal sperm to image, it is important to note that human sperm can exhibit significant morphological heterogeneity, even within a single specimen. Sperm targeted for imaging should be free of gross morphological defects [20]. Additionally, imaged sperm should exhibit fixed heads (no lateral movement) while maintaining tail motility to verify viability. Excess lateral head movement will lead to inaccurate fluorescence quantification.
12. While acquiring images, laser power should be kept to a minimum as excessive laser exposure can cause phototoxicity. Also, bright-field images should be collected in tandem with fluorescence. The bright-field channel is important for the assessment of sperm morphology, motility, and ensuring an unobstructed imaging field.
13. It is necessary to include unstimulated and sham-stimulated controls to control for spontaneous Ca^{2+} influx, and potential mechanical effects induced by liquid flowing over the cells respectively.
14. Refrain from imaging cells that were within range of previous stimulations. Three 10–15-min images can be acquired per dish.

References

1. Cohen R, Buttke DE, Asano A et al (2014) Lipid modulation of calcium flux through CaV2.3 regulates acrosome exocytosis and fertilization. Dev Cell 28:310–321. https://doi.org/10.1016/j.devcel.2014.01.005
2. Cohen R, Sosnicki DM, White MA et al (2023) Real-time imaging of acrosomal calcium dynamics and exocytosis in live mouse sperm. J Vis Exp 2023. https://doi.org/10.3791/65962
3. Lishko PV, Botchkina IL, Kirichok Y (2011) Progesterone activates the principal Ca2+ channel of human sperm. Nature 471:387–392. https://doi.org/10.1038/nature09767
4. Strünker T, Goodwin N, Brenker C et al (2011) The CatSper channel mediates progesterone-induced Ca2+ influx in human sperm. Nature 471:382–387. https://doi.org/10.1038/nature09769
5. Balbach M, Ghanem L, Rossetti T et al (2021) Soluble adenylyl cyclase inhibition prevents human sperm functions essential for fertilization. Mol Hum Reprod 27:1–13. https://doi.org/10.1093/molehr/gaab054
6. Akbari A, Pipitone GB, Anvar Z et al (2019) ADCY10 frameshift variant leading to severe recessive asthenozoospermia and segregating with absorptive hypercalciuria. Hum Reprod 34:1155–1164. https://doi.org/10.1093/humrep/dez048
7. Osheroff JE, Visconti PE, Valenzuela JP et al (1999) Regulation of human sperm capacitation by a cholesterol efflux-stimulated signal transduction pathway leading to protein kinase A-mediated up-regulation of protein tyrosine phosphorylation. Mol Hum Reprod 5(11): 1017–1026
8. Travis AJ, Kopf GS (2002) The role of cholesterol efflux in regulating the fertilization potential of mammalian spermatozoa. J Clin Invest 110:731–736. https://doi.org/10.1172/jci200216392
9. Langlais J, Kan FWK, Granger L et al (1988) Identification of sterol acceptors that stimulate cholesterol efflux from human spermatozoa during in vitro capacitation. Gamete Res 20: 185–201. https://doi.org/10.1002/mrd.1120200209
10. Molina LCP, Luque GM, Balestrini PA et al (2018) Molecular basis of human sperm capacitation. Front Cell Dev Biol 6:72
11. Harper C, Wootton L, Michelangeli F et al (2005) Secretory pathway Ca2+-ATPase (SPCA1) Ca2+ pumps, not SERCAs, regulate complex [Ca2+]i signals in human spermatozoa. J Cell Sci 118:1673–1685. https://doi.org/10.1242/jcs.02297
12. Costello S, Michelangeli F, Nash K et al (2009) Ca2+-stores in sperm: their identities and functions. Reproduction 138:425–437
13. Schuh K, Cartwright EJ, Jankevics E et al (2004) Plasma membrane Ca2+ ATPase 4 is required for sperm motility and male fertility. J Biol Chem 279:28220–28226. https://doi.org/10.1074/jbc.M312599200
14. Krasznai Z, Krasznai ZT, Morisawa M et al (2006) Role of the Na+/Ca2+ exchanger in calcium homeostasis and human sperm motility regulation. Cell Motil Cytoskeleton 63:66–76. https://doi.org/10.1002/cm.20108
15. Okunade GW, Miller ML, Pyne GJ et al (2004) Targeted ablation of plasma membrane Ca2+-ATPase (PMCA) 1 and 4 indicates a major housekeeping function for PMCA1 and a critical role in hyperactivated sperm motility and male fertility for PMCA4. J Biol Chem 279: 33742–33750. https://doi.org/10.1074/jbc.M404628200
16. Moody MA, Cardona C, Simpson AJ et al (2017) Validation of a laboratory-developed test of human sperm capacitation. Mol Reprod Dev 84:408–422. https://doi.org/10.1002/mrd.22801
17. Sánchez-Cárdenas C, Servín-Vences MR, José O et al (2014) Acrosome reaction and Ca2+ imaging in single human spermatozoa: new regulatory roles of [Ca2+]i. Biol Reprod 91: 67. https://doi.org/10.1095/biolreprod.114.119768
18. Bedu-Addo K, Barratt CLR, Kirkman-Brown JC, Publicover SJ (2007) Patterns of [Ca2+]i mobilization and cell response in human spermatozoa exposed to progesterone. Dev Biol 302:324–332. https://doi.org/10.1016/j.ydbio.2006.09.040
19. De Lamirande E (2007) Semenogelin, the main protein of the human semen coagulum, regulates sperm function. Semin Thromb Hemost 33:60–68
20. WHO laboratory manual for the examination and processing of human semen, 8th edn

Chapter 19

Calcium Imaging in *Drosophila*

David V. Gazzo and Jeremiah J. Zartman

Abstract

Ex vivo calcium imaging in *Drosophila* opens an expansive amount of research avenues for the study of live signal propagation through complex tissue. Here, we describe how to isolate *Drosophila* organs of interest, like the developing wing imaginal disc and larval brain, culture them for extended periods, up to 10 h, and how to image the calcium dynamics occurring within them using genetically encoded biosensors like GCaMP. This protocol enables the study of complex calcium signaling dynamics, which is conserved throughout biology in such processes as cell differentiation and proliferation, immune reactions, wound healing, and cell-to-cell and organ-to-organ communication, among others. These methods also allow pharmacological compounds to be tested to observe effects on calcium dynamics with the applications of target identification and therapeutic development.

Key words Calcium imaging, Calcium dynamics, *Drosophila*, Ex vivo, Microfluidics

1 Introduction

Why is calcium (Ca^{2+}) such a frequently measured signal in biology? With its innate chemical properties allowing it to be readily bio-reactive and its natural abundance, Ca^{2+} is the foremost second messenger that biological life utilizes for many, if not most, of its activities. Because of its unique combination of valency, ionic radius, polarizability, hydration energy, and charge density, it interacts with many proteins in various complex and diverse ways. High calcium concentrations are also toxic to cells and will form insoluble phosphate salts with a cell's main energy source, ATP. This requires maintaining it at low cytosolic basal levels in the nM range. A sophisticated control system for Ca^{2+} sequestration thus enables Ca^{2+} to be a complex and highly regulated signaling mediator [1].

Ca^{2+} has been found to facilitate signaling across numerous cellular processes and does so through several pathways. This includes tissue development, cell fate, cell division, fertilization, metabolism, mobilization, and cognition, among others [2–5]. How has it become so diverse? The frequency, amplitude, and

Caroline M. Gorvin (ed.), *Calcium Signaling: Methods and Protocols*, Methods in Molecular Biology, vol. 2861,
https://doi.org/10.1007/978-1-0716-4164-4_19,

spatial/temporal patterning can all influence how Ca^{2+} messages are conveyed [5, 6]. Additionally, the Ca^{2+} "toolkit" is expansive with many components contained within it [3, 6]. This allows signals to be augmented and shaped in further diverse ways and has allowed Ca^{2+} to become a key second messenger found in all known biological systems.

To fully comprehend the complexity of calcium (Ca^{2+}) signaling within and between tissues, the cultivation of live organs is needed. With the calcium imaging chambers described here, complex Ca^{2+} signals can be imaged in real time using explanted micro-organs of *Drosophila melanogaster*. *Drosophila* is an excellent model organism to use when studying signaling pathways in developmental biology. This is because multiple phenomena like signal transduction and organ growth can be observed within their thin transparent organs [7–11]. Additionally, *Drosophila* can be used to recapitulate many genetic diseases found in humans, are inexpensive, and have a wide range of genetic tools created to facilitate complex control of protein and genetic expression levels [12, 13].

Here, we describe the fabrication and implementation of simple and advanced calcium imaging chambers for use in deciphering complex Ca^{2+} signals in real time using explanted micro-organs of *D. melanogaster*. To preserve organ viability, an improved culture media using Grace's media supplemented with 20-hydroxyecdysone (20E) is employed [7, 8, 10, 11]. These chambers allow for short and extended imaging sessions, which can also be adapted for use with other systems like zebrafish and 3D tissue cultures. The first and more simple imaging chamber allows for easy setup for capturing rapid responses within the first 2–3 h of imaging, while the advanced chamber incorporates microfluidics, which facilitates a more controlled environment allowing for long-term imaging (10–12 h) [14–17]. The use of advanced chambers will also allow for the addition and removal of compounds of interest for the detection of Ca^{2+} stimulation. Ca^{2+} plays a critical role in developing organs, and investigations into its dynamics can now be easily implemented in any laboratory with these devices and imaging techniques.

2 Materials

2.1 Culture Media

1. Grace's Insect Medium: Prepare Grace's Insect Medium with 5 mM BIS-TRIS. Adjust pH to 6.6 with 1M KOH. Filter sterilize with 0.22 μm bottle filters, twice. Store at 4 °C for no more than 1 month or freeze as small aliquots at −20 °C (*see* **Note 1**). On the day of experiment, add fetal bovine serum and penicillin–streptomycin for final amounts of 5% and 1%, respectively, as well as 20-hydroxyecdysone for a final concentration of 20 nM.

2.2 Materials for Simple Imaging Chambers

1. #1 Glass bottom Petri dishes.
2. Millicell Standing Cell Culture Inserts (*Sigma: PI8P01250*).
3. Blade for cutting.
4. Kimwipes.
5. Embryo oil.

2.3 Materials for Advanced Microfluidic Imaging Chamber

1. Craft Cutter and Software (Silhouette Cameo) (*see* **Note 2**).
2. Cutting Mat for Craft cutter (Silhouette) (*see* **Note 3**).
3. 3 or 5 mil Thermal Laminating Pouches.
4. Rubber roller.
5. 24 × 60 mm #1 cover glass.
6. Laminator.
7. 6 and 8 mm Furniture Bumpers with adhesive.
8. 1 and 2 mm hole punches.
9. Revolving punch pliers.
10. Scissors.
11. Forceps.
12. Double-sided tape.
13. Polyetheretherketone (PEEK) tubing and syringe adapters (ZEUS).
14. 1 mL Luer lock syringes.
15. Syringe pump (for small volume control).

2.4 Material for Dissections

1. Dissection microscope.
2. A siliconized glass plate (*see* **Note 4**).
3. Sigmacote for siliconizing plates (Sigma: SL2).
4. Two dissection forceps (Fine Science Tools Inc.: Dumont #5 Forceps).
5. 100 μL micropipette with disposable tips.
6. 2 μL micropipette with disposable tips.
7. Four Petri dishes or a six-well plate for washing larvae.

2.5 Solutions

1. Phosphate-buffered saline, PBS (1 L): 8 g NaCl, 0.2 g KCl, 1.44 g Na_2HPO_4, 0.24 g Kh_2PO_4

3 Methods

3.1 Fabrication of Simple Imaging Chamber

1. Prepare the Millicell Standing Cell Culture Insert by removing the bottom legs with a blade (*see* **Note 5**).

2. Siliconize the glass bottom Petri dish (*see* **Note 6**).
3. Your device is now ready for sample loading.

3.2 Fabrication of Advanced Imaging Chamber

1. If creating a new design, start with hand drawings.
2. Once drawn, create the design on the computer (any software can be used), and surround it with a border (*see* **Note 7**).
3. Clip a screenshot of the design.
4. Open the Silhouette software.
5. Paste the design into Silhouette.
6. Open the trace panel and trace the design (*see* Fig. 1a).
7. To help size the design, use the grided view (Click on *view → show grid*).
8. Each square is 1 inch so change the design size by clicking and dragging points.
9. Measure channel width with the line drawing tool (*see* Fig. 1b) (*see* **Note 8**).
10. The overall size should be somewhat smaller than a 24 × 60 mm cover glass as this will be the base of the device (*see* **Note 9**).
11. Create circles between 1.5 and 2.0 mm in diameter, and place them at each entry and exit point.
12. Copy the entire design and paste it nearby.

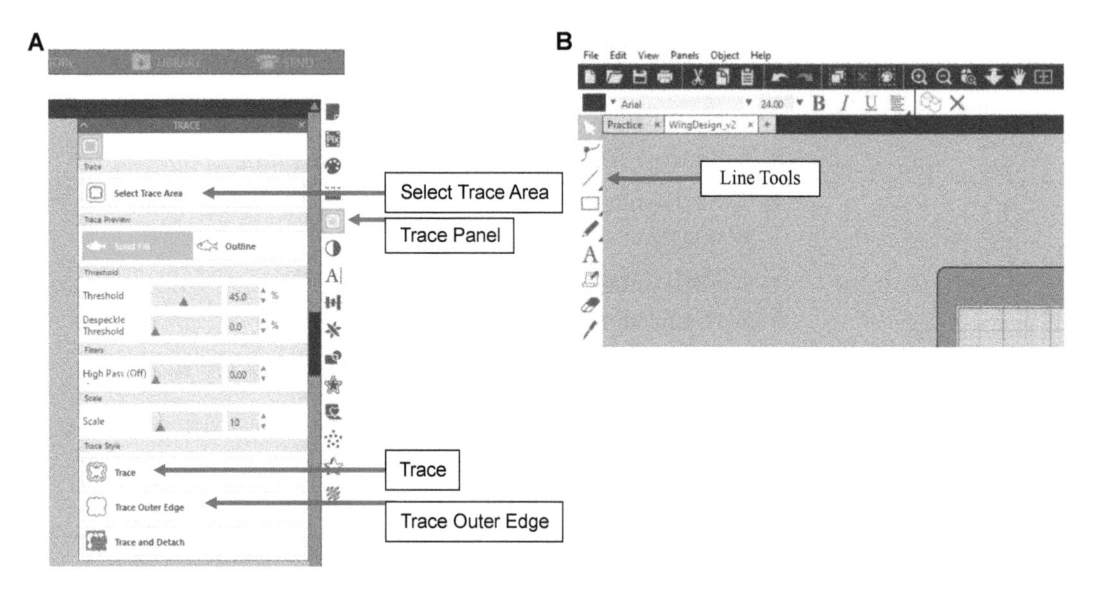

Fig. 1 Tracing designs in Silhouette software. (**a**) In the "Trace Panel" choose "Select Trace Area." Click and drag over the design (the yellow highlight will become the design). To create the trace, select "Trace" or "Trace Outer Edge." (**b**) To measure channel widths, use the "Line Tools" selection and draw a line that spans the length of the channel. The length (and width) of this line will be displayed

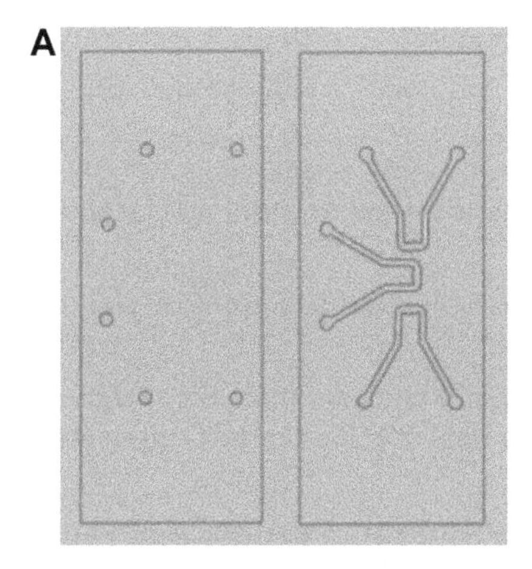

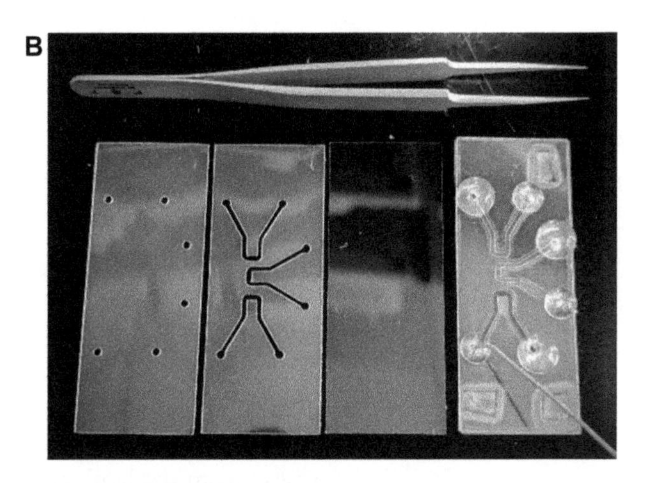

Fig. 2 Example of advanced microfluidic imaging chamber. (**a**) Silhouette outline of chamber design, (**b**) which consists of one glass layer and two laminate layers. The top layer (left) consists of inlets and outlets, while the middle layer contains the channels. At the bottom of the imaging chamber is a glass coverslip (24 × 60 mm #1). The fully assembled device (right) is sealed from the bottom with the glass slide and from the top, the in−/outlet channels. Bumpers then allow for easy access and the addition of solution and tissue

13. Remove all channels leaving only the entry and exit circles (this will be the top layer of the design).
14. Box your design so that it will be cut out (*see* example in Fig. 2a).
15. When placing your design in the grid, **Do Not** place it in the top left corner as the Cameo cutter tests its readiness here.

3.2.1 Cutting Design

1. With gloves (to prevent oil transfer from hands), replace the protective cover on the cutting mat with a piece of laminate with the shiny side of the laminate face down (*see* **Note 10**).
2. Using a roller, remove any air bubbles from underneath the laminate, fully securing it to the cutting mat.
3. Line the cutting mat up with the blue (line) guide on the CAMEO and hit "Load cut mat."
4. In your design file, click "Send".
5. In this menu set the force, speed, material, passes, and blade depth (*see* **Note 11**).
6. Connect your computer to the Cameo cutter (*see* **Note 12**).
7. Remove anything from the back of the machine as it will use this space to pull the cutting mat through.
8. Make sure the design that you have created in silhouette will be cut on an area where you have placed laminate.
9. The Cameo cutter should indicate it's "Ready" and "TEST" should be selectable.
10. Test the readiness by selecting "TEST."

11. Click "Send" in the "Send window".
12. Once cutting is complete hit "Unload" to unload the cutting mat.

3.2.2 Laminating Design

1. Turn on the laminator and allow it time to warm up (*see* **Note 13**).
2. Remove the cut design from the laminate using a pair of tweezers.
3. With small pieces of double-sided tape, assemble the design in the proper layered order with the coverslip on the bottom (see Fig. 2b).
4. Make sure to have all adhesive, matte, sides facing inwards, away from the laminator (*see* **Note 14**).
5. Once assembled, run the cut design attached to the coverslip through the laminator (*see* **Note 15**).
6. Perforate furniture bumpers (*see* **Note 16**):
   (a) Punch holes in larger bumpers with the revolving punch pliers.
   (b) Using biopsy punches, create holes in the smaller bumpers.
7. Remove cut material.
8. Place the perforated furniture bumpers over the exit and entry holes (*see* **Note 17**).
9. Your device is now complete and ready for sample loading (see Fig. 2b).
10. Assemble tubing and syringe with the polyetheretherketone (PEEK) tubing adapter.
11. Then begin by cleaning the tubing that will be used to supply media to the microfluidic device by flushing the tubing with 70% isopropyl alcohol (IPA) and air, alternating between them.
12. Do this three times, ending with air, by adding solutions to the syringe using a 1 mL pipette after removing the syringe plunger.
13. Rinse a final time with PBS or Grace's media.
14. Place aside and retrieve for final assembly.

3.3 Dissection of *Drosophila* Tissue

1. Obtain wandering larvae that contain a genetic background with a calcium biosensor such as a variant of GCaMP (*see* **Note 18**).
2. Prepare the dissection microscope by shifting the working area to the black surface as this will help make the larva more visible and dissection easier.

3. Place a siliconized black glass dissection plate over the microscope stage.
4. Prepare four small Petri dishes with two containing PBS, one with 70% isopropyl alcohol (IPA), and one with a small amount of deionized water (DI) (*see* **Note 19**).
5. Retrieve larvae using blunt forceps and rinse in first dish of PBS, making sure to remove any debris.
6. Transfer to 70% IPA and let sit for ~3 min, swirling occasionally.
7. After sterilizing in 70% IPA, rinse in second dish of PBS.
8. Then place larvae in the final dish with DI water until ready to dissect (*see* **Note 20**).
9. Many larvae can be "stored" like this at a time so repeat **steps 5–8** as many times as needed.
10. Clean the dissection plate with a Kimwipe and 70% IPA.
11. Place a drop of Grace's medium (~30 μL) onto the glass plate (*see* **Note 21**).
12. Using the blunt tweezers, place a larva in the drop.
13. Under the microscope, locate the posterior, and using the blunt tweezers grab the larva around the middle (closer to the posterior) (*see* **Note 22**).
14. With the sharp tweezers, grab the larva next to the blunt ones and tear the posterior part away. The posterior half can be discarded (e.g., into an adjacent "garbage" drop).
15. Now grab the torn end with the blunt tweezers to stabilize the larva so that the sharp tweezers can be inserted into the anterior end (*see* **Note 23**).
16. With the blunt tweezers, invert the larva by pushing it onto the sharp tweezers.
17. This will expose the larva tissue that contains the imaginal discs.
18. Remove the sharp tweezers after the larva has been inverted.
19. To locate the discs, begin to remove some of the tissue like the gut and fat.
20. After some tissue removal, locate desired tissues.
21. These can be found by first finding the trachea as this is a bright white line of tissue that runs along the entire larva, is easy to identify, and has the wing imaginal discs attached to it at the anterior end (along with the thoracic leg and haltere discs).
22. If targeting the brain, it is found at the anterior end of the larvae, surrounded by additional developing tissue like the eye discs.

23. When the desired tissue is identified, isolate it while paying attention not to stretch, scratch, or tear it.
24. Once isolated, wash tissues by transferring them through a series of drops using a 2 μL micropipette.
25. Tissues can be stored in a drop of medium until all dissections are complete (*see* **Note 24**).

3.4 Preparation of Imaging Chambers for Sample Loading

3.4.1 Simple Imaging Chamber

1. Add 20 μL of Grace's medium to the center of the glass bottom Petri dish.
2. Transfer dissected and isolated tissues to the drop of medium using a 2 μL micropipette.
3. Orient the tissue as required for the microscope you will be imaging with (*see* **Note 25**).
4. Place the prepped Millicell Standing Cell Culture Insert ("phantom" legs down) over the tissue, paying attention to not use lateral movements, which can disrupt the orientation of tissues (*see* **Note 26**).
5. Add 100–150 μL of Grace's medium to the top of the Millicell Standing Cell Culture Insert.
6. Place enough (100–150 μL) embryo oil over the Grace's medium to completely cover it.
7. Place a rolled-up Kimwipe around the inner walls of the Petri dish, and wet it with 400 μL of PBS or surround the outside of the imaging chamber with embryo oil (100–150 μL) (*see* **Note 27**).
8. The chamber is now ready to transfer to the microscope (*see* **Note 28**).

3.4.2 Advanced Microfluidic Imaging Chamber

1. Retrieve a microfluidic device (*see* **Note 29**).
2. Partially fill the larger exit channel (large bumper) with Grace's media.
3. Place the desired tissue in the larger opening using a 2 μL micropipette.
4. Use a wire to push the tissue down and slightly into the channel, orienting it properly (*see* **Note 25**).
5. With a 20 μL micropipette, slowly suck the wing disc down into the channel by pipetting volume up from the entrance channel (small bumper) until it has reached a desired location in the design (*see* **Note 30**).
6. Once the microfluidic device has been prepared, retrieve the cleaned tubing.

7. With a new syringe (and attached needle), suck up ~1 mL of Grace's media, and remove as many air bubbles as you can in the syringe by taping the tip and expelling some media.
8. Remove the needle and replace the syringe that is attached to the tubing with this new one loaded with media (*see* **Note 31**).
9. Wash about 0.1–0.2 mL of media through the tubing, ensuring all air is removed.
10. With the syringe and microfluidic device prepared, mount the syringe into the syringe pump, which should be placed next to the microscope.
11. Using the menu on the syringe pump, push out a small drop of media, and then insert the tubing into the entrance of the entrance channel (small bumper) on the microfluidic device.
12. Set the flow rate on the syringe pump (*see* **Note 32**).
13. Place the imaging chamber onto the microscope and begin imaging.

3.5 Live Calcium Imaging of Drosophila Tissue

It is recommended to utilize a spinning disc confocal microscope for improved acquisition speed with conserved resolution. The chosen interval and light exposure levels are essential for proper image acquisition and downstream analysis. This will largely depend on the signaling pathway that will be observed and biosensors being used (*see* **Note 18**). Generally, for long-term imaging (8+ h), intervals of 5–6 min or larger are recommended to avoid damage to cellular tissue. Shorter intervals of 10–30 s can be used but will limit the length of imaging to about 2 h before the tissue's health starts to become impaired. Additionally, laser power is recommended to be maintained at the lowest detectable setting as is allowed by the system. This will also help prevent tissue damage.

It is also important to understand that the working distance of spinning confocal microscopes are limited to boost resolution. Therefore, the imaging plane chosen is significant when acquiring data. Either a single z-slice should be imaged or a z-stack of limited range. Larger z-stacks can be acquired with larger step sizes; however, this is not recommended as the resolution of the z-axis will be severely limited. The imaging plan should be chosen based on the dynamics to be observed and the tissue oriented appropriately (*see* **Note 25**).

3.6 Calcium Analysis

Following the imaging of calcium dynamics is a workflow of image analysis around the acquired fluorescent signals. This usually involves denoising, motion correction if drift is present, segmentation and identification of regions of interest (ROI) (cells experiencing calcium spikes), and signal extraction. This is then ultimately followed by the analysis of the calcium transients and their interpretation. Many platforms have been created for this over a variety of cell types [18–23]. Here we utilize and demonstrate analysis with CaImAn—an open-source library that can be used in either MATLAB or Python for calcium image analysis [18]. We have adapted it here for use in the developing wing imaginal disc instead of its original use in neurons (Fig. 3). This tool takes time-lapse videos of fluorescent capture (Fig. 3a) and begins by correcting for motion, which is followed by ROI identification (Fig. 3b), and finally data extraction (Fig. 3c).

Modifications will likely be required and depends on the platform chosen. Key parameters that must be accounted for in the analysis include the interval rate of capture and cell size, along with the dynamics of the calcium indicator. When selecting an image analysis method or platform, one must also factor in the amount of prior coding experience needed. Many of the platforms, like CaImAn, allow for the specification of the system to sample needs but can require an advanced understanding of computer languages. Others, like EZcalcium, include a graphical user interface and can be navigated by those with little coding experience [19].

In conclusion, the post-acquisition analysis workflow involves several decisions to quantify and interpret the results. Several open-access image analysis methods are available. However, regardless of the chosen platform, customization and understanding of coding languages may be necessary, although user-friendly options are available. Thus, selecting the appropriate platform requires consideration of specific experimental needs and the level of coding expertise available.

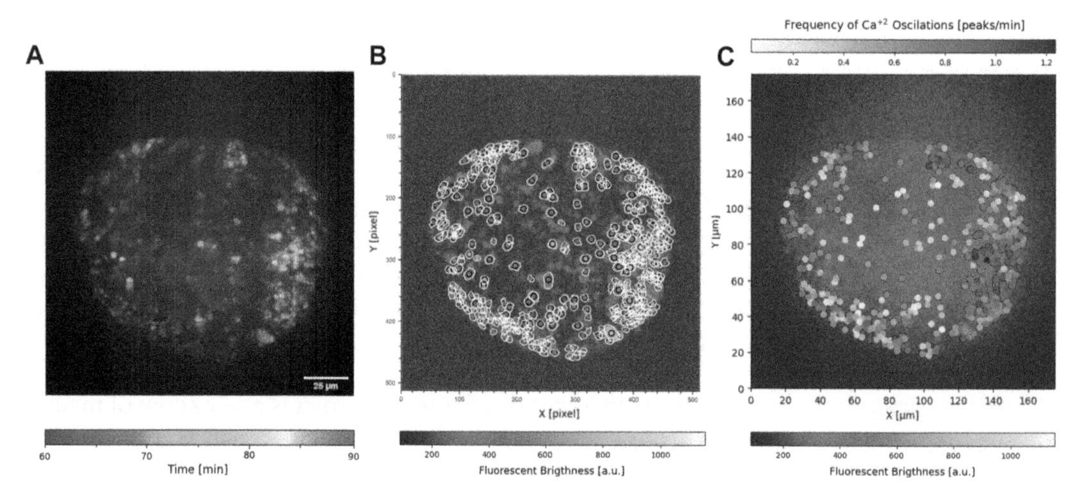

Fig. 3 Calcium analysis using CalmAn on a developing wing imaginal disc treated with 1 mM Yoda1. (**a**) Temporal color-code of time-lapse series over 30 min of imaging with an interval of 3 s. (**b**) Segmentation mask and ROI identification by CalmAn following motion correction. (**c**) Calcium dynamics extracted and presented as frequency of oscillation

4 Notes

1. Check for precipitates or small floating particles throughout storage time as this will indicate when the medium has expired and/or been contaminated. Usually, 100 mL of Grace's media is made up at a time to reduce waste. Dissolve 4.59 g Grace's Insect Medium (45.9 g/L) with 0.1 g BIS-TRIS (5 mM) in Milli-Q water and adjust pH to 6.6 with 1M KOH. KOH is used as K^+ is more physiologically relevant than Na^+ (NaOH) [25, 26]. 20-Hydroxyecdysone is prepared fresh weekly and stored as a 1000x stock solution in ethanol at −20 °C. Other organ culture media are available and will lead to various levels of stimulation of calcium. This includes WM1 (Wing Disc Medium 1), with FEX (fly extract), which results in the formation of intercellular calcium waves [2, 15, 16].
2. Silhouette Cameo will come with a cutting mat; however, it is recommended to have replacements for unavoidable wear and tear.
3. The Silhouette software can be found at Silhouette Software Downloads (https://www.silhouetteamerica.com/software). Laminating Pouches, laminate rollers, laminators, furniture bumpers, biopsy punches, and revolving punch pliers can be purchased from many craft stores or on Amazon.
4. Alternatively, this can be a siliconized plastic plate like the bottom of a Petri dish; however, it will not perform as well.
5. Completely remove any leg material, making sure that the surface is smooth. This will ensure a nice seal around the edges.
6. Though not required, it is recommended that Petri dishes are siliconized before each use as this ensures that drops of media bead well.
7. Using six-point thick lines is recommended for initial creation. Leave space in one of the corners where tape can be placed when assembling. It may be beneficial to place tape in two corners if the design is large. This will prevent the layers from shifting when being laminated.
8. Channels should **never be greater than 1 mm**. When larger than this, the top and bottom layers will bend and touch each other in the channel when assembled.
9. A glass coverslip is used as the bottom layer of designs for imaging purposes, but if this is ever not required, laminate can also act as a bottom layer.
10. Pieces of laminate come as pouches, but you only need a single piece, so separate them. Placing the shiny side of the laminate face down on the cutting mat has been found to work best.

11. Force, 25; Speed, 4; Material, Sticker paper-clear; Passes, 1; and Blade Depth, 5 are used for most designs; however, these settings are adjustable depending on the thickness of the laminate you are cutting.
12. Hard connection (plug the Cameo cutter into your computer), external memory (copy files onto a flash drive and plug into the Cameo cutter), or Bluetooth (depends on the Cameo model).
13. For any device with four or more layers, use the 5-mil setting; otherwise, use 3-mil (1-mil is short for a thousand of an inch). Material thickness determines channel thickness, so stack materials if thicker channels are required. Remove anything from the back of the machine as anything placed in the laminator will come out of the back (depending on the model).
14. Do not overlap tape throughout layers as this will make the device too thick and will delaminate. **Make sure to place the adhesive, matte side down so that it will not stick to the laminator**. Only the shiny side should ever be facing the laminator. If the adhesive matte side faces up toward the laminator, then it will stick, and that part of the laminator will no longer be usable.
15. It is recommended to run the design through the laminator two times to ensure complete lamination. This will prevent any leaks.
16. Hole sizes on bumpers depend on the desired use. Larger holes on entry bumpers can be used to create a reservoir of solution, whereas smaller holes will allow for tight connections with tubing.
17. For exit channels use the large bumpers (8 mm), and for the entry channels, use the small bumpers (6 mm).
18. Calmodulin-based genetically encoded fluorescent calcium indicators (GCaMP and its derivatives) have been developed across several generations to enhance their ability to rapidly detect Ca^{2+} with varying binding affinities, allowing for dynamic observation of calcium signaling. The different GCaMP variants are categorized based on their decay kinetics after calcium binding, each suited for specific experimental needs. **GCaMPs** is characterized by a **slow (s) decay** after calcium detection, making it ideal for experiments where sensitive signal detection is benefical (e.g., BDSC #'s 92593, 92594, 92595). In contrast, **GCaMPm** exhibits a **medium (m) decay** of signal (e.g., BDSC #'s 92590, 92591, 92592), while **GCaMPf** is designed with a **fast (f) decay**, making it the most dynamic variant for capturing rapid Ca^{2+} signaling events (e.g., BDSC #s 92587, 92588, 92589)[25]. Each variant is specifically suited for different experimental contexts depending on the required temporal dynamics of calcium detection.

Currently, the latest generation, **GCaMP8** (BDSC #'s listed above), is available in all three decay kinetics (s, m, and f) and provides a versatile set of tools for precise calcium imaging. For the most dynamic and accurate representation of rapid Ca^{2+} signals, the GCaMP8f variant is recommended, however, each variant has its use and should be used accordingly.

19. Make sure to use solutions at room temperature to minimize stress on the larvae.
20. The DI water is simply used to facilitate easy transport of larvae into the final dish. This way, larvae will not stick to the forceps.
21. Multiple drops can be placed along the dissection plate so that dissections can be conducted in rapid order.
22. The larva will be breathing through its posterior spiracles so they will naturally orient this side to be at or slightly above the surface. Lightly grab and slowly squeeze to prevent it from popping the stomach and releasing all its contents. This will prevent clouding and contamination of solution.
23. The anterior end is the head of the larva and will look like a mouth trying to eat things. Insert the tweezers into the mouth.
24. It is recommended to dissect 2–3 specimens to ensure a viable specimen is retrieved for live imaging. Practice is key here.
25. Tissue orientation is important and will depend on the type of microscope you are using, upright or inverted microscope. The general rule of thumb is to orient a tissue so that the brightest part of the tissue is facing the objective. This way, light will not "bleed through" from below, impairing resolution while imaging. For example, the developing wing imaginal disc is oriented with the peripodial cells facing the objective as they tend to have the brightest signal.
26. This can be difficult, and practice will be required. Tissue orientation will be inevitably disturbed, so add multiple samples to the chamber in the hopes that at least one has the proper orientation.
27. Embryo oil and/or wetted Kimwipe will prevent evaporation. Embryo oil is recommended.
28. An even simpler imaging chamber can be created by only using a glass bottom Petri dish with a larger drop of Grace's medium (100–200 μL). This is only recommended for short imaging sessions of 30 min or less as evaporation will begin to affect accurate data acquisition.
29. Devices can be cleaned and reused; however, this is not recommended. If cleaning the device, flush the channel with 70% IPA to remove any dust and air bubbles. Follow this with at least three flushes of water. Make sure not to introduce any air

bubbles. Then flush with Grace's media to remove any water and any remaining IPA. Again, make sure not to introduce any air bubbles.

30. It's helpful to add some Grace's media to the 20 μL micropipette (not the full volume) so that you can pipette up and down, pushing and pulling the tissue into the channel. Make sure not to remove all the solution from the other side; otherwise, you will suck air in, and bubbles will impair imaging. It is not the end of the world if you get an air bubble, if it is not next to the tissue, it will not impair the imaging. Also, it is smart to place the tissue a little past the desired location because when you add the tubing, the insertion of it will push the tissue back just a little bit. This will take practice.
31. A new syringe is used for each experiment. Replace them at the beginning of experiments, but leave them attached to the tubing at the end. This is a good way not to lose any pieces.
32. Usually, a flow rate of ~2 μL/h. is sufficient but can and should be adjusted based on the design of the imaging chamber.

Acknowledgments

The work in this paper was supported by NIH Grant R35GM124935, and EMBRIO Institute, contract #2120200, a National Science Foundation (NSF) Biology Integration Institute.

References

1. Carafoli E, Krebs J (2016) Why calcium? How calcium became the best communicator. J Biol Chem 291(40):20849–20857. https://doi.org/10.1074/jbc.R116.735894
2. Brodskiy PA, Wu Q, Soundarrajan DK, Huizar FJ, Chen J, Liang P, Narciso C, Levis MK, Arredondo-Walsh N, Chen DZ, Zartman JJ (2019) Decoding calcium signaling dynamics during *Drosophila* wing disc development. Biophys J 116(4):725–740. https://doi.org/10.1016/j.bpj.2019.01.007
3. Brodskiy PA, Zartman JJ (2018) Calcium as a signal integrator in developing epithelial tissues. Phys Biol 15(5):051001. https://doi.org/10.1088/1478-3975/aabb18
4. Berridge MJ (2016) The inositol trisphosphate/calcium signaling pathway in health and disease. Physiol Rev 96(4):1261–1296. https://doi.org/10.1152/physrev.00006.2016
5. Smedler E, Uhlén P (2014) Frequency decoding of calcium oscillations. Biochim Biophys Acta 1840(3):964–969. https://doi.org/10.1016/j.bbagen.2013.11.015
6. Berridge MJ, Lipp P, Bootman MD (2000) The versatility and universality of calcium signalling. Nat Rev Mol Cell Biol 1(1):11–21. https://doi.org/10.1038/35036035
7. Zartman J, Restrepo S, Basler K (2013) A high-throughput template for optimizing *Drosophila* organ culture with response-surface methods. Development 140(3):667–674. https://doi.org/10.1242/dev.088872
8. Dye NA, Popović M, Spannl S, Etournay R, Kainmüller D, Ghosh S, Myers EW, Jülicher F, Eaton S (**2017**) Cell dynamics underlying oriented growth of the *Drosophila* wing imaginal disc. Development 144(23):4406–4421. https://doi.org/10.1242/dev.155069
9. Lerit DA, Plevock KM, Rusan NM (2014) Live imaging of *Drosophila* larval neuroblasts. J Vis Exp 89:e51756. https://doi.org/10.3791/51756

10. Tsao C-K, Ku H-Y, Lee Y-M, Huang Y-F, Sun YH (2016) Long term ex vivo culture and live imaging of *Drosophila* larval imaginal discs. PLoS One 11(9):e0163744. https://doi.org/10.1371/journal.pone.0163744
11. Handke B, Szabad J, Lidsky PV, Hafen E, Lehner CF (2014) Towards long term cultivation of *Drosophila* wing imaginal discs in vitro. PLoS One 9(9):e107333. https://doi.org/10.1371/journal.pone.0107333
12. Hales KG, Korey CA, Larracuente AM, Roberts DM (2015) Genetics on the fly: a primer on the *Drosophila* model system. Genetics 201(3):815–842. https://doi.org/10.1534/genetics.115.183392
13. Chorna T, Hasan G (2012) The genetics of calcium signaling in *Drosophila* melanogaster. Biochim Biophys Acta 1820(8):1269–1282. https://doi.org/10.1016/j.bbagen.2011.11.002
14. Levis M, Ontiveros F, Juan J, Kavanagh A, Zartman JJ (2019) Rapid fabrication of custom microfluidic devices for research and educational applications. J Vis Exp 153:60307. https://doi.org/10.3791/60307
15. Zartman JJ (2017) Organ culture methods for the *Drosophila* wing imaginal disc. In: Echalier G, Perrimon N, Mohr SE (eds) *Drosophila* cells in culture. Academic, pp 145–164
16. Restrepo S, Zartman JJ, Basler K (2016) Cultivation and live imaging of *Drosophila* imaginal discs. In: Dahmann C (ed) *Drosophila*: methods and protocols. Springer, New York, pp 203–213. https://doi.org/10.1007/978-1-4939-6371-3_11
17. Levis M, Sacco F, Velagala V, Ontiveros F, Zartman JJ (2024) Mechanical compression of *Drosophila* embryos using 2 rapid fabrication microfluidic devices. In: Tissue morphogenesis: methods and protocols; methods in molecular biology, vol 2805. Springer
18. Giovannucci A, Friedrich J, Gunn P, Kalfon J, Brown BL, Koay SA, Taxidis J, Najafi F, Gauthier JL, Zhou P, Khakh BS, Tank DW, Chklovskii DB, Pnevmatikakis EA (2019) CaImAn an open source tool for scalable calcium imaging data analysis. eLife 8:e38173. https://doi.org/10.7554/eLife.38173
19. Cantu DA, Wang B, Gongwer MW, He CX, Goel A, Suresh A, Kourdougli N, Arroyo ED, Zeiger W, Portera-Cailliau C (2020) EZcalcium: open-source toolbox for analysis of calcium imaging data. Front Neural Circuits 14:25. https://doi.org/10.3389/fncir.2020.00025
20. Wang Y, DelRosso NV, Vaidyanathan TV, Cahill MK, Reitman ME, Pittolo S, Mi X, Yu G, Poskanzer KE (2019) Accurate quantification of astrocyte and neurotransmitter fluorescence dynamics for single-cell and population-level physiology. Nat Neurosci 22(11):1936–1944. https://doi.org/10.1038/s41593-019-0492-2
21. Mahadevan AS, Long BL, Hu CW, Ryan DT, Grandel NE, Britton GL, Bustos M, Porras MAG, Stojkova K, Ligeralde A, Son H, Shannonhouse J, Robinson JT, Warmflash A, Brey EM, Kim YS, Qutub AA (2022) cytoNet: spatiotemporal network analysis of cell communities. PLoS Comput Biol 18(6): e1009846. https://doi.org/10.1371/journal.pcbi.1009846
22. Kaifosh P, Zaremba JD, Danielson NB, Losonczy A (2014) SIMA: python software for analysis of dynamic fluorescence imaging data. Front Neuroinform 8:80
23. Rueckl M, Lenzi SC, Moreno-Velasquez L, Parthier D, Schmitz D, Ruediger S, Johenning FW (**2017**) SamuROI, a python-based software tool for visualization and analysis of dynamic time series imaging at multiple spatial scales. Front Neuroinform 11:44. https://doi.org/10.3389/fninf.2017.00044
24. Zhang Y, Rózsa M, Liang Y, Bushey D, Wei Z, Zheng J, Reep D, Broussard GJ, Tsang A, Tsegaye G, Narayan S, Obara CJ, Lim J-X, Patel R, Zhang R, Ahrens MB, Turner GC, Wang SS-H, Korff WL, Schreiter ER, Svoboda K, Hasseman JP, Kolb I, Looger LL (2023) Fast and sensitive GCaMP calcium indicators for imaging neural populations. Nature 615(7954):884–891. https://doi.org/10.1038/s41586-023-05828-9
25. Weiß I, Bohrmann J (2019). Electrochemical Patterns during Drosophila Oogenesis: Ion-Transport Mechanisms Generate Stage-Specific Gradients of pH and Membrane Potential in the Follicle-Cell Epithelium | BMC Developmental Biology | Full Text. https://bmcdevbiol.biomedcentral.com/articles/10.1186/s12861-019-0192-x
26. Folk DG, Bradley TJ (2003). Evolved Patterns and Rates of Water Loss and Ion Regulation in Laboratory-Selected Populations of Drosophila Melanogaster. J Exp Biol 206:(16), 2779–2786. https://doi.org/10.1242/jeb.00498

Part V

Analysis of Calcium Data

Chapter 20

Computational Analysis of Calcium Flux Data Using R

Macarena Pozo-Morales and Sumeet Pal Singh

Abstract

Calcium imaging has emerged as a powerful tool for studying cellular dynamics, with applications spanning neuroscience, cell biology, and beyond. In this chapter, we present a comprehensive guide to the computational analysis of calcium flux data using the R programming language. Using an example of in vivo live imaging of GCaMP signal in zebrafish hepatocytes, we demonstrate techniques for segmentation, normalization, and quantification of calcium transients. We provide a step-by-step code example showcasing extraction of meaningful information from calcium imaging datasets. The code allows insights into the number of oscillating cells, number of oscillations per cell within a time frame, and generation of publication-ready plots for showcasing calcium dynamics. This chapter serves as a valuable resource for researchers seeking to leverage freely available computational tools for analyzing calcium flux data at cellular resolution and uncovering novel insights into cellular physiology.

Key words Calcium flux, caTools, Heatmap, Liver, Peak detection, R, Zebrafish

1 Introduction

Calcium ions (Ca^{2+}) play pivotal roles in regulating diverse cellular processes such as muscle contraction, neurotransmitter release, and gene expression. The ability to visualize and quantify calcium dynamics within living cells has revolutionized our understanding of intracellular signaling. Among the various techniques available, calcium imaging stands out for its noninvasive nature and high spatiotemporal resolution. In particular, genetically encoded calcium indicators (GECIs) [1] like GCaMP [2] have become indispensable tools for monitoring calcium fluctuations in real time. In this chapter, we focus on the computational analysis of calcium imaging data obtained from in vivo studies of zebrafish hepatocytes expressing GCaMP6s [3]. We showcase a Fiji-based workflow for segmentation and extraction of calcium flux values from confocal images of GECI-expressing samples, followed by R-based workflows for normalization and analysis of calcium imaging data. The workflow provides the necessary tools to extract quantitative

Caroline M. Gorvin (ed.), *Calcium Signaling: Methods and Protocols*, Methods in Molecular Biology, vol. 2861,
https://doi.org/10.1007/978-1-0716-4164-4_20,

insights from complex calcium flux datasets and unravel the intricacies of cellular signaling pathways in physiological and pathological contexts.

2 Materials

A brief introduction is provided for imaging of calcium flux in zebrafish hepatocytes using confocal microscopy, followed by the workflow to quantify calcium dynamics. However, the workflow can be applied for any GECI time-lapse images, and potentially can be expanded to any fluorescent indicator.

Prepare and store all reagents at room temperature (unless indicated otherwise). The reagents mentioned in the protocol are for reference; they can be interchanged with similar reagents from a different provider. Diligently follow all waste disposal regulations when disposing waste materials. Adult zebrafish of AB stain should be raised under standard conditions 14-h light/10-h dark cycle at 28 °C. Follow all guidelines and regulations for handing zebrafish as outlined by institutional and national ethical committees.

2.1 Mounting of Zebrafish Larvae for Imaging

1. Zebrafish transgenic line: *Tg(fabp10a:GCaMP6s; cryaa:mCherry)*^{ulb15} [3]. The line contains a transgene in which the fabp10a (fatty acid-binding protein 10a, liver basic) promoter drives GCaMP6s. In addition, the cryaa:mCherry serves as a marker to screen the transgenic animals using fluorescent red eyes at 3 days postfertilization (dpf) (*see* **Note 1**).
2. E3 media: 5 mM NaCl, 0.17 mM KCl, 0.33 mM CaCl2, 0.33 mM MgSO4, 1 mM HEPES. Prepare 1 L of 60× E3 stock by dissolving 17.53 g NaCl, 0.76 g KCl, 2.19 g CaCl2, 4.15 g MgSO4.xH2O, 11.91 g HEPES in 1 L of MilliQ water. Adjust pH to 7.2 and autoclave. Store at 4 °C. To make 10 L of 1× E3 medium, dilute 166 mL of 60× stock solution in 10 L of MilliQ water. Check pH and adjust to 7.2 if needed (*see* **Note 2**). Add 2.5 ml of 0.2% methylene blue.
3. Immobilizing agent: 4 mM tubocurarine chloride pentahydrate (abbreviated as tubocurarine) (93750; Sigma-Aldrich) in E3 media. Prepare fresh. (*see* **Note 3**).
4. Mounting Media: 1% low-melt agarose (50080; Lonza, Basel, Switzerland). Dissolve 1 g low-melt agarose in 100 mL of E3 medium (without methylene blue) by gently heating in a microwave. Make aliquots in 15 mL falcon tubes or 1.5 mL tubes. Store at 4 °C. On the day of the experiment, liquify one aliquot by gently heating it in a water bath at 60 °C. Keep the solution in a 37 °C incubator (or in a heat block).

5. Imaging apparatus: Glass-bottomed FluoroDish (FD3510-100; World Precision Instruments (WPI)).
6. Fine forceps: Dumont #5 (11295-10; Fine Science Tools).
7. Confocal microscope: A laser scanning confocal microscope equipped with 488 nm laser for excitation of GCaMP6s and a 40×/1.1 N.A and water correction lens for cellular resolution imaging (Zeiss LSM 780).

2.2 Extraction of Intensity Values

1. Image analysis software: Fiji (https://imagej.net/software/fiji/) [4].
2. Spreadsheet software: LibreOffice Calc (https://www.libreoffice.org/discover/calc/) or Microsoft Excel.

2.3 Analysis of Intensity Values

1. Statistical Software: R (https://cran.rstudio.com/) [5] and Rstudio (https://posit.co/download/rstudio-desktop/).
2. Packages: caTools (https://www.rdocumentation.org/packages/caTools/versions/1.17.1) [6], scorepeak (https://www.rdocumentation.org/packages/scorepeak/versions/0.1.2) [7], pheatmap (https://rdocumentation.org/packages/pheatmap/versions/1.0.12), beeswarm (https://rdocumentation.org/packages/beeswarm/versions/0.4.0), and RcolorBrewer (https://www.rdocumentation.org/packages/RColorBrewer/versions/1.1-3). Install the packages in R using the commands outlined on the RDocumentation (https://www.rdocumentation.org/) pages for the respective packages.

3 Methods

3.1 Confocal Imaging of GCaMP Signal in Zebrafish Hepatocytes

1. Immobilize 4 dpf *Tg(fabp10a:GCaMP6s)* larvae by incubation in 4 mM tubocurarine for 10 min in the incubator at 28 °C.
2. Transfer the larvae to glass bottom dish and remove excess E3 medium.
3. Cover the larvae with 100 μL of 1% low-melt agarose (sufficient to cover the animal).
4. Using fine forceps, orient the animal with the left side facing the bottom of the dish (the imaging side).
5. Leave the dish for 5 min at room temperature to allow the low-melt agarose to solidify.
6. Set up confocal imaging at 28 °C, using a 40×/1.1 N.A, water correction lens, excitation with 488 nm laser for GCaMP6s, and collection of 499–579 nm.

7. Set the imaging frame at 1024 × 1024 pixels, and the distance between confocal planes at 6 μm for a Z-stack coverage of, on average, 90 μm.
8. Acquire a time-lapse of 20 stacks with an interval of 45 s between each stack (*see* **Note 4**).

3.2 Segmentation and Data Extraction

1. Open the image file in Fiji using "File" - > "Open" (*see* **Note 5**).
2. Generate a maximum-intensity projection of the Z-stack using "Image" - > "Stacks" - > "Z Project..." tool. Select "Max Intensity" for "Projection type" in the Z project toolbox.
3. Manually select the cells for analysis. For this, use the region-of-interest (ROI) tool. Go to the "Analyze" menu, then select "Tools" and open the "ROI Manager." Utilize the "polygon selection tool" in the toolbar to manually draw the ROI.
4. Draw the ROI within the cytoplasm of individual hepatocyte. Add the selected region to the "ROI manager" by clicking the "Add [t]" button.
5. Under the "Analyze" menu, choose "Set measurements" and select "Mean grey value" to extract fluorescence intensity within the area.
6. In the "ROI Manager," click on "Multi Measure" to obtain the intensity measurements for cells throughout the time series. Ensure to select "Measure all nn slices" and "One row per slice" in the "Multi Measure" toolbox. Do not select "Append results" option. By running this command, a "Results" window will open that contains the intensity value per region for the time series.
7. Select the "Results" window and save the measurements using "File" - > "Save As..." option. Name the file with a .csv ending (e.g., "Liver1.csv"). The results will be saved in a comma-separated values (csv) file format. The file can be opened using LibreOffice calc (or Microsoft Office Excel) for viewing.

3.3 Quantification of Calcium Dynamics in R

1. Draft an R file in RStudio to perform the analysis. Initiate a new R file using "File" -> "New File" -> "R Script".
2. Initiate the packages using the following commands:

```
library(caTools)
library(scorepeak)
library(beeswarm)
library(pheatmap)
library(RColorBrewer)
```

3. Define a function to detect peak:

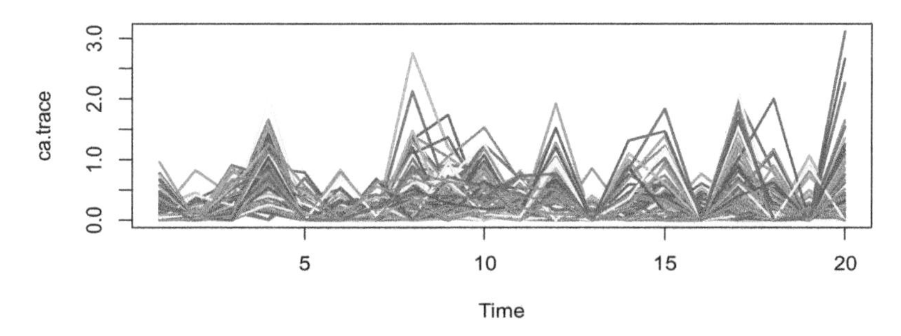

Fig. 1 Line trace of fluorescence intensities. Each line represents one region. X-axis represents the image number in the time-lapse

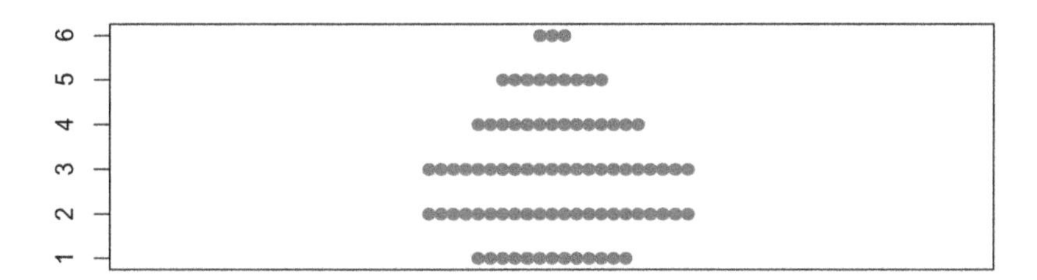

Fig. 2 Distribution of peaks detected in cells during the time-lapse

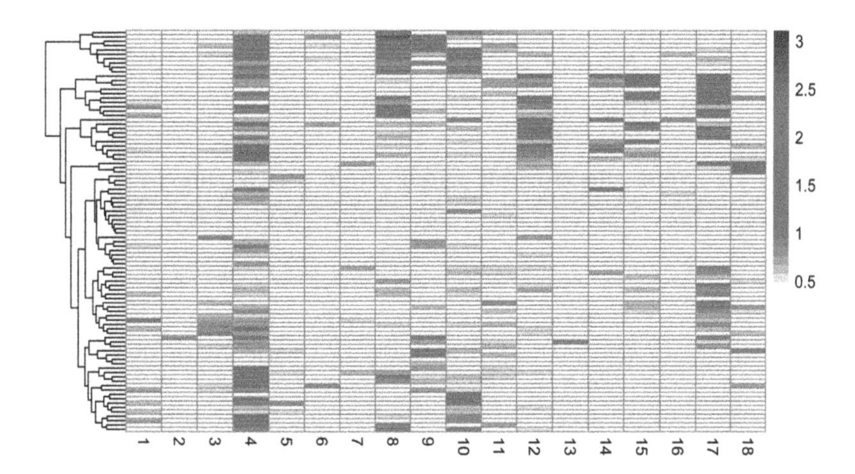

Fig. 3 Heatmap for the fluorescence intensities for all regions

```
## Function to detect peak
detect.peaks <- function(ca.trace, width1 = 3, width2 =
5, score_threshold = 0.03, threshold = 0.5){
 lp <- detect_localmaxima(ca.trace, w = width1)
 lp[1:2] <- FALSE
 lp[(length(lp)-2):length(lp)] <- FALSE
 score <- score_type1(ca.trace, w = width2)
  detected <- (score > score_threshold) & lp & (ca.trace >
threshold)
 return(detected)
}
```

4. Define a function to plot peak:

```
## Function to plot peak
plot.ts.peak <- function(ca.trace, col.trace = "black", ca.
peak, col.peak = "red"){
 plot.ts(ca.trace, plot.type = "single", col=col.trace, lwd =
2)
points(which(ca.peak), ca.trace[ca.peak], pch = 19, col = col.
peak)
}
```

5. Open the csv file generated by Fiji:

```
## Step 1: Choose a csv file generated by Fiji
ca.file <- file.choose()
print(paste0("file being processed: ", ca.file))
```

6. Read the data from the file:

```
## Step 2: Read the data from the file
ca.data <- read.csv(file = ca.file, header = T, stringsAsFac-
tors = F, row.names = 1)
## Only keep the data related to fluorescence intensities.
## These columns start with the pattern 'Mean'
ca.trace <- ca.data[,grep("^Mean", colnames(ca.data))]
```

7. Calculate fold changes for the calcium intensity values. The intensity values are normalized using the formula:

$$\left(F_t - F_{\min(t)}\right)/F_{\min(t)}$$

where F_t represents intensity at a given time and $F_{\min(t)}$ represents the local minima. Local minima is defined as the minimum intensity value between the current frame and two adjacent frames on either side of the time lapse. For this, the following code is used:

```
## Step 3: Calculate Fold Change for the Traces
min.trace <- as.data.frame(runmin(x = as.matrix(ca.trace), k
= 3))
ca.trace <- ca.trace - min.trace
ca.trace <- ca.trace / min.trace
```

8. Plot the calcium traces for different cells (Fig. 1):

```
## Plot the traces
suppressWarnings(my.colors <- brewer.pal(ncol(ca.trace),
"Set1"))
plot.ts(ca.trace, plot.type = "single", col=my.colors, lwd =
2)
abline(h = 0.5, lty = 2)
```

9. Save the plot using "Export" -> "Save as PDF..." tool (in the "Plots" panel of RStudio).
10. Detect peaks in the fluorescence traces from individual regions:

```
## Step 4: Detect Peaks
ca.peaks <- as.data.frame(apply(X = ca.trace, MARGIN = 2, FUN
= detect.peaks))
ca.peaks.num <- colSums(ca.peaks)
## Number of cells with at-least one peak
sum(ca.peaks.num>0)
ca.peaks.num.detect <- ca.peaks.num[ca.peaks.num > 0]
```

11. Quantify the distribution of peaks per cell:

```
## Distribution of peaks
peaks.cells <- as.data.frame(table(ca.peaks.num.detect))
colnames(peaks.cells) <- c("NumberPeaks", "NumberCells")
peaks.cells
```

12. Plot the distribution of peaks (Fig. 2):

```
beeswarm(x = ca.peaks.num.detect, pch = 19, col = "red")
```

13. Save the plot using "Export" -> "Save as PDF..." tool.
14. Generate a heatmap of the fluorescence intensity values for the entire organ (Fig. 3):

```
pheatmap(t(ca.trace[1:(nrow(ca.trace)-2),]),
 cluster_cols = F, cluster_rows = T,
 color = colorRampPalette(c("white", "red","red1", "red2",
"red4"))(30),
 breaks = seq(from = 0.4, to = max(ca.trace), length.out =
31),
 clustering_distance_rows = "minkowski", show_rownames = F)
```

15. Save the plot using "Export" -> "Save as PDF..." tool.

16. Save the characteristics (total, mean, and standard deviation of peaks) of the distribution in a csv file:

```
out.data <- data.frame(TotalCells = sum(ca.peaks.num>0),
 TotalPeaks = sum(ca.peaks.num.detect),
 Mean = mean(ca.peaks.num.detect),
 SD = sd(ca.peaks.num.detect))
## Save the characteristics
write.csv(x = out.data,
 file = paste0("Liver1","_Distribution.csv"))
```

17. The characteristics of the distribution can be used to compare samples (*see* **Note 6**).

4 Notes

1. The imaging protocol and analysis workflow should work with any zebrafish transgenic line expressing a GECI. Mounting characteristics will be specific to the organ being imaged.
2. HEPES is supplemented in the E3 medium to maintain its pH. Typically, the adjustment of pH for stock is sufficient, and pH adjustment is not needed after dilution. However, if pH is drastically different from 7.2 after dilution, consider discarding the stock solution and making a new one.
3. For the imaging of calcium flux in the zebrafish liver, immobilization using tubocurarine is critical. Anesthesia using tricaine (MS-222), the commonly used anesthesia for zebrafish, suppresses calcium oscillations in the zebrafish liver [3] and possibly in certain neurons [8] and, thus, should be avoided for these organs. However, tricaine-based anesthesia has been utilized for calcium imaging of pancreatic beta-cell calcium flux [9, 10]. Thus, a preliminary test to choose between tubocurarine and tricaine should be made.
4. For determining the time of the intervals between each stack, a preliminary test to check the speed of the calcium oscillations in the organ should be performed.
5. Confocal images typically open automatically in Fiji. However, "LSM Toolbox" can be used to open them as well. To do this, navigate to "Plugin | LSM Toolbox | Show LSM Toolbox." Within the toolbox, click on "Open LSM" and select the image file. If your file format is not supported, convert it to TIFF before proceeding with analysis.
6. To compare the number of oscillations between samples, it might be important to normalize the volume of the imaging plane. This can be done by analyzing the same volume in all the

samples, or by dividing the total number of oscillations by the volume of the organ. Volume can be measured by extracting the area of the liver in each imaging plane. For this, under the "Analyze" menu, choose "Set measurements" and select "Area." Using the ROI toolbox, manually outline the region of the liver in each plane and obtain the corresponding area with "Multi Measure." Volume, in μm^3, is calculated as (sum of the area) * (thickness of a single Z-stack = 6 μm). Divide the total number of oscillations with the volume to obtain the normalized number of oscillations per sample.

Acknowledgments

This work was supported by funding from FNRS (MISU 34772792, MISU-PROL 40005588 and CDR 40013427) to S. P.S.

References

1. Miyawaki A, Llopis J, Heim R et al (1997) Fluorescent indicators for Ca2+ based on green fluorescent proteins and calmodulin. Nature 388:882–887. https://doi.org/10.1038/42264
2. Zhang Y, Rózsa M, Liang Y et al (2023) Fast and sensitive GCaMP calcium indicators for imaging neural populations. Nature 615:884–891. https://doi.org/10.1038/s41586-023-05828-9
3. Pozo-Morales M, Garteizgogeascoa I, Perazzolo C et al (2023) In vivo imaging of calcium dynamics in zebrafish hepatocytes. Hepatology 77:789. https://doi.org/10.1002/hep.32663
4. Schindelin J, Arganda-Carreras I, Frise E et al (2012) Fiji: an open-source platform for biological-image analysis. Nat Methods 9: 676. https://doi.org/10.1038/nmeth.2019
5. R Core Team (2020) R: a language and environment for statistical computing. R Foundation for Statistical Computing, Vienna, Austria
6. Tuszynski J (2020) caTools: tools: moving window statistics, GIF, Base64, ROC AUC, etc
7. Ochi S (2019) Scorepeak: peak functions for peak detection in univariate time series
8. Turrini L, Fornetto C, Marchetto G et al (2017) Optical mapping of neuronal activity during seizures in zebrafish. Sci Rep 7:1–12. https://doi.org/10.1038/s41598-017-03087-z
9. Salem V, Silva LD, Suba K et al (2019) Leader β-cells coordinate Ca2+ dynamics across pancreatic islets in vivo. Nat Metab 1:615–629. https://doi.org/10.1038/s42255-019-0075-2
10. Singh SP, Chawla P, Hnatiuk A et al (2022) A single-cell atlas of de novo β-cell regeneration reveals the contribution of hybrid β/δ-cells to diabetes recovery in zebrafish. Development 149:dev199853. https://doi.org/10.1242/dev.199853

Index

Caroline M. Gorvin (ed.), *Calcium Signaling: Methods and Protocols*, Methods in Molecular Biology, vol. 2861, https://doi.org/10.1007/978-1-0716-4164-4,

Milton Keynes UK
Ingram Content Group UK Ltd.
UKHW052355061124
450475UK00006B/66

9 781071 641637